S0-BCG-954

Cell Electrophoresis

Edited by
Johann Bauer

CRC Press
Boca Raton   Ann Arbor   London   Tokyo

Library of Congress Cataloging-in-Publication Data

Cell electrophoresis/edited by Johann Bauer.
    p. cm.
   Includes bibliographical references and index.
   ISBN 0-8493-8918-6
   1. Electrophoresis. 2. Cell separation. I. Bauer, Johann, 1941-
QH585.5.E46C44   1944
574.87′6—dc20   94-15304
                                                                 CIP

This book contains information obtained from authentic and highly regarded sources. Reprinted material is quoted with permission, and sources are indicated. A wide variety of references are listed. Reasonable efforts have been made to publish reliable data and information, but the author and the publisher cannot assume responsibility for the validity of all materials or for the consequences of their use.

Neither this book nor any part may be reproduced or transmitted in any form or by any means, electronic or mechanical, including photocopying, microfilming, and recording, or by any information storage or retrieval system, without prior permission in writing from the publisher.

All rights reserved. Authorization to photocopy items for internal or personal use, or the personal or internal use of specific clients, may be granted by CRC Press, Inc., provided that $.50 per page photocopied is paid directly to Copyright Clearance Center, 27 Congress Street, Salem, MA 01970 USA. The fee code for users of the Transactional Reporting Service is ISBN 0-8493-8918-6/94/$0.00+$.50. The fee is subject to change without notice. For organizations that have been granted a photocopy license by the CCC, a separate system of payment has been arranged.

CRC Press, Inc.'s consent does not extend to copying for general distribution, for promotion, for creating new works, or for resale. Specific permission must be obtained in writing from CRC Press for such copying.

Direct all inquiries to CRC Press, Inc., 2000 Corporate Blvd., N.W., Boca Raton, Florida 33431.

© 1994 by CRC Press, Inc.

No claim to original U.S. Government works
International Standard Book Number 0-8493-8918-6
Library of Congress Card Number 94-15304
Printed in the United States of America 1 2 3 4 5 6 7 8 9 0
Printed on acid-free paper

PREFACE

In 1965, I wrote an introduction for a book also entitled *Cell Electrophoresis* which was edited by E. J. Ambrose, one of the grandfathers of this method. At that time, I wrote: "perhaps I can begin by observing that cell electrophoresis is in the 'teenage' stage of its development. Like the teenager who belligerently struggles with the problems of puberty under the doting eyes of his parents, the development of cell electrophoresis has not always been happy and carefree. But, just as ever-optimistic parents eternally ignore the disappointments of the past and dream bright futures for their children, so too the 'fathers' of cell electrophoresis are extremely optimistic about their offspring's future importance...and with just cause, I might add. The exact time when the teenager becomes an adult is impossible to define; he simply slides gradually into maturity."

If one compares the Ambrose book with this one, there is no question that it confirms my prediction: cell electrophoresis is now a mature creature.

Progress in establishing the theoretical foundation of the method has been especially pronounced. These advances have become so extensive that it seemed desirable to have an overview — not only for beginners, but also for experts. Such an overview should help to reduce the misinterpretation of results — as is always the case with theoretical foundations of sophisticated methods.

In addition, the clinical and/or biological applications of the method have also been expanded considerably. It is true that as one of the 'grandfathers of cell electrophoresis', one of our hopes has not yet been realized — namely, that it would be possible to reach a certain clinical diagnosis with only or at least at best with the aid of cell electrophoresis measurement. Thus, the common clinical application of the method is still limited.

This book, however, may serve the investigators of further biological/clinical cell problems in connection with their membrane charges and, so, perhaps fulfill this old hope.

G. Ruhenstroth-Bauer
Martinsried, Germany

THE EDITOR

Johann Bauer, Ph.D., obtained his training at the University of Munich and the Max Planck Institute for Biochemistry in Martinsried, receiving his diploma in 1978 and his Ph.D. in 1981. Following postdoctoral work at Duke University in Durham, North Carolina, he was appointed an Assistant at the Max Planck Institute in Martinsried.

Dr. Bauer is a member of the International Electrophoresis Society and the American Association of Immunologists and he has published more than 50 papers. His current major research interests relate to development and application of cell electrophoretical methods.

CONTRIBUTORS

Atul M. Athalye, Ph.D.
The BOC Group Technical Center
Murray Hill, New Jersey

Johann Bauer, Ph.D.
Max-Planck-Institut für Biochemie
Martinsried, Germany

Ingolf Bernhardt, D.Sc.
Department of Biology
Humboldt University
Berlin, Germany

Fritz G. Boese, Ph.D.
Max-Planck-Institut für
* Extraterrestrische Physik*
Garching, Germany

Hari H.P. Cohly, Ph.D.
Department of Surgery
University of Mississippi Medical Center
Jackson, Mississippi

Suman K. Das, M.D.
Department of Surgery
University of Mississippi Medical Center
Jackson, Mississippi

Ulrike Friedrich, Ph.D.
German Space Agency
Bonn, Germany

Michael V. Golovanov, Ph.D.
Cancer Research Center
Russian Academy of Medical Sciences
Moscow, Russia

Terese M. Grateful, Ph.D.
Merck & Co., Inc.
Merck Manufacturing Division
West Point, Pennsylvania

Nobuya Hashimoto, M.D., Ph.D.
Department of Internal Medicine
Jikei University School of Medicine
Tokyo, Japan

Jörn Heinrich, Ph.D.
ABIMED Analysen-Technik GmbH
Langenfeld, Germany

Keith A. Knisley, Ph.D.
Department of Cell Biology and Anatomy
Texas Tech University Health Sciences
* Center*
Lubbock, Texas

Edwin N. Lightfoot, Ph.D.
Department of Chemical Engineering
University of Wisconsin-Madison
Madison, Wisconsin

Dennis R. Morrison, Ph.D.
NASA/Johnson Space Center
Houston, Texas

Percy H. Rhodes, M.S.M.E.
NASA/Marshall Space Flight Center
Huntsville, Alabama

L. Scott Rodkey, Ph.D.
Department of Pathology
* and Laboratory Medicine*
University of Texas-Houston
Houston, Texas

Günter Ruyters, Ph.D.
German Space Agency
Bonn, Germany

Wolfgang Schütt, Ph.D., D.Sc.
Department of Physiology
Cornell University Medical College
New York

Motomu Shimizu, Ph.D.
Department of Cancer Therapeutics
Tokyo Metropolitan Institute
* of Medical Science*
Tokyo, Japan

George G. Slivinsky, Ph.D.
Institute of Zoology
National Academy of Sciences
Alma-Ata, Kazakhstan

Robert S. Snyder, D.Sc.
NASA/Marshall Space Flight Center
Huntsville, Alabama

Paul Todd, Ph.D.
Department of Chemical Engineering
University of Colorado
Boulder, Colorado

Fernando F. Vargas, D.D.S., Ph.D.
Center for Drug Research and Evaluation
FDA and Laboratory of Cell Biology
*   and Genetics*
National Institute of Health
Bethesda, Maryland

Horst Wagner, Ph.D.
Anorganische und Analytische Chemie
*   und Radiochemie*
Univesität des Saarlandes
Saarbrüchen, Germany

Carel J. van Oss, Ph.D.
Departments of Microbiology and
*   Chemical Engineering*
State University of New York
Buffalo, New York

CONTENTS

I. THEORETICAL CONSIDERATIONS

II. IMPROVEMENTS OF THE METHOD

III. BUFFERS AND THEIR EFFECTS ON CELLS

IV. THE IMPORTANCE OF DETERMINATION OF THE NEGATIVE SURFACE CHARGE DENSITY OF CELLS

V. CELL ELECTROPHORESIS IN SPACE

I
Theoretical Considerations

Chapter 1

Contributions to a Mathematical Theory of Free Flow Electrophoresis

Fritz G. Boese

CONTENTS

I. INTRODUCTION

The literature on the topic of the title is extensive, (see the articles by Arcus et al.,[1] Giannovario et al.,[2] Ivory,[3] Naumann and Rhodes,[4] and Saville[5] and the references cited therein). Two major themes are to be studied: (1) the determination of the set of stable working conditions in parameter space spanned by the various parameters and (2) determination of performance parameters such as resolution and throughput and their optimization over the set of stable states. The paper of Giannovario et al.[2] is devoted to the second group of problems. The mathematical analysis of the electrophoretic separation process in free-flow devices is as difficult as it is desirable. The problem lies where the thermodynamics, hydrodynamics, electrodynamics, and physical chemistry meet.

The cited paper[2] is of interest from several points of view. A first reading of the manuscript[2] gives the impression that it is possible to model the problem so that the mathematical questions that arise range from easy to handle to trivial and that the problem in its entirety can be treated successfully. Since the results given were not compared with results of real measurements, the only methodological way to evaluate the contents of the paper consists of a critical analysis of the argumentation presented, to lay open hidden assumptions, if any, exist to check all calculations, and to make comparisons.

One aim of this chapter is, therefore, to comment on some details given by Giannovario et al.[2] Some comments support their arguments, while others do not. Moreover, it shall be shown that, on the basis of the adopted model, useful partial results can be obtained. It will become evident that some of the equations that occur are still so simple that solutions in closed form can be obtained. A closed expression has the advantage of exhibiting parameter dependencies explicitly, which can hardly be overestimated in the context of optimization.

This chapter was previously published in *Journal of Chromatography,* 438, 145–170, 1988 and is reprinted here with permission of Elsevier Science Publishers.

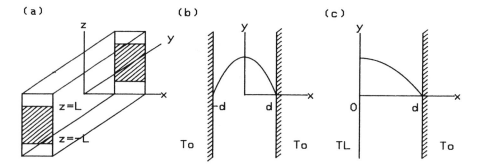

Figure 1-1 **(a)** A (finite) separation chamber and the orientation of the Cartesian coordinate system used; the qualitative temperature distribution **(b)** for double-sided cooling and **(c)** for cooling from back face. T_0 = Cooling temperature, TL = temperature of the laboratory. See Nomenclature for further symbol definitions.

Special emphasis has been placed on the behavior of the temperature and velocity distribution in the separation chamber while the electrical conductivity, thermal conductivity, and viscosity of the buffer electrolyte vary with temperature.

The expression "$p = a(b)c$" means that p runs from a to c in steps of b. The set of all integer numbers is denoted by **Z**. An interval with initial point a and terminal point b is denoted by "$[a,b]$". The expression "A: = B" or "B: = A" redefines A by B.

II. DISTORTION OF THE SAMPLE THREAD

The movement of the buffer and the sample material is a two-phase flow. The second phase, the sample material, moves relative to the carrier phase, the buffer electrolyte. During the passage of the sample from the inlet region to the collection ports the cross section of the sample thread changes its shape. A circular cross section deforms into a cross section of a less symmetrical form. Figure 1-1 shows the separation chamber together with the position and orientation of the Cartesian coordinate system used here. For the purpose of this section it is more convenient to measure z downstream such that $z = 0$ describes the plane near the sample inlet in which the sample flow is fully developed. $\mathbf{Q_0}$ denotes the cross section of the sample thread in the plane $z = 0$, while $\mathbf{Q_1}$ denotes the deformed cross section in the plane $z = l > 0$. The different streamlines which form the sample stream tube $\mathbf{Q_1}$ at $z = l$ may be parameterized by their position vector $\mathbf{u}$ in the plane $z = 0$. Let $x(\mathbf{u},l)$ be the position of the sample streamline at $z = l$ which has in $z = 0$ the position $\mathbf{u}$, i.e., $x(\mathbf{u},0) = \mathbf{u}$. Thus, $x(\mathbf{u},l) - \mathbf{u}$ measures the displacement of the sample streamline under consideration in planes orthogonal to the axial direction. Cross sections of the sample stream tube are important in two respects. First, the deformed cross section taken in planes $z = l$ determines the resolving capability of the separation process. Second, the optical detection system recognizes the sample concentration in the form of projections in planes $\mathbf{x} = const.$ and integrates over the depth coordinate x. It is reasonable to measure the deformation with the help of a single nonnegative quantity ΔS. The quantity

$$\Delta S(l) := \max_{\mathbf{u}, \, v \varepsilon \mathbf{Q_0}} \left\{ \left| x(\mathbf{u},l) - \mathbf{u} - \left[x(v,l) - v \right] \right| \right\} \tag{1-1}$$

measures the maximal change of the distance between pairs of points during the passage from $z = 0$ to $z = l$. For a rigid body motion $x(\mathbf{u},l)$ the deformation $\Delta S(l)$ vanishes. With the help of the mean buffer velocity, the length l can be expressed in terms of traveling time t of the sample. Let L be the vertical length of the separation path and let t_r be the corresponding residence time. In this context $\Delta S^* := \Delta S(t_r)$ is of interest, where $\Delta S^*(t)$ is defined as $\Delta S^*(t) := \Delta S(L/w)$, w being the mean axial buffer velocity.

The definition in Equation 1-1 differs from that given in Reference 2 insofar as in the latter only a special pair of streamlines — u,v — was considered. The choice made is influenced by prior knowledge or a prior expectation about the particle stream to be calculated later. It can be shown that this special pair of streamlines chosen by Giannovario et al.[2] does not enjoy the extremal property required in Equation 1-1. It is at least conceptually more satisfactory to start with Equation 1-1, although practicability may later dictate a special choice. In simple cases, $\Delta S(L)$ can be computed explicitly. Consider the principal contribution to the sample distortion, which is due to the velocity profile of the buffer electrolyte. It is assumed that no diffusion of particles of the sample thread into the surrounding buffer electrolyte takes places. Furthermore, the electroosmotic motion of the buffer electrolyte is neglected. At least a reduction of the osmotic motion of the buffer can be achieved by an appropriate coating of the walls of the separation chamber. Under these assumptions, charged particles follow the electric field. The electrophoretic mobility $\mu = \mu(T,pH)$ depends on the temperature $T = T(x)$ and the pH field, pH = pH(x,y). In order to discuss the main effects, μ is assumed to be constant. The particle under consideration has a velocity component $\mathbf{v} := \mu\mathbf{E}$ parallel to the electric field $\mathbf{E}$ with uniform field strength $|\mathbf{E}|$. Let $w(x)$ $(-d \leq x \leq d)$, be the axial velocity component of a particle at the depth position x. A particle which starts at time $t = 0$ from plane $z = 0$ will arrive at the plane $z = L > 0$ downstream at $t = L/w(x)$ with a displacement

$$d = d(x) = \frac{\mu|\mathbf{E}|L}{w(x)} \tag{1-2}$$

A mean electrophoretic displacement D is defined by $D := \mu|\mathbf{E}|L/\overline{w}$. The temperature-dependent distribution $w(x)$ is approximated by the isothermal velocity distribution

$$w(x) = \frac{3}{2} \cdot \overline{w} \cdot \left(1 - \frac{x^2}{d^2}\right) \tag{1-3}$$

where $\overline{w}$ is the mean velocity and $2d$ is the thickness of the chamber. Let $r < d$ and consider two different points (x_1,y_1) and (x_2,y_2) in the circular cross section $\mathbf{Q_0} := \{(x,y): x^2 + y^2 \leq r^2\}$. The distance of the image point pair under the mapping defined by Equation 1-2 is maximized there where the expression

$$D^2 := \left[\frac{1}{1 - \dfrac{x_1^2}{d^2}} - \frac{1}{1 - \dfrac{x_2^2}{d^2}}\right]^2 \tag{1-4}$$

attains its maximum over $|x_1|,|x_2| \leq r$. The maximal value is

$$D_{max} := \frac{r^2}{d^2 - r^2} \tag{1-5}$$

Using Equation 1-5 in Equation 1-1 gives

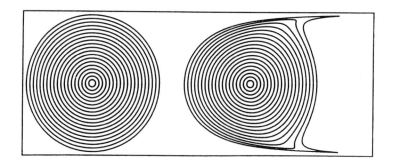

Figure 1-2 Distortion of a sample thread with circular cross section r/d = 0.05(0.05)0.95; sample mean displacement D = 1.33 mm. See Nomenclature for symbol definitions; see text for details.

$$\Delta S(L) = \frac{\mu |\mathbf{E}| Lr^2}{d^2 - r^2} \tag{1-6}$$

The left-hand part of Figure 1-2 shows the sequence of boundaries Q_0 for which r/d = 0.05(0.05)0.95. The right-hand part shows the corresponding distorted cross sections obtained at a distance L = 20 cm downstream. The other parameters used were

$$|\mathbf{E}| = 100 \frac{V}{cm}, \quad \mu = 10^{-4} \frac{cm^2}{Vs}, \quad \overline{w} = 1 \frac{cm}{s} \tag{1-7}$$

The effect of the diffusion is qualitatively easy to assess. The curves of the right-hand part of Figure 1-2 represent jumps in the concentration of the sample material. In the presence of diffusive motion into the buffer, concentration differences are diminished. The analytical form of the right-hand side of Equation 1-5 prompts criticism with respect to the common usage of the term "crescent phenomenon" (see Ivory[3,6] and many other authors). From Equation 1-4 and from Figure 1-2 one sees that the crescent formation is a boundary layer phenomenon. The usual qualitative illustrations of the sample streamlines found in many publications do not point out the boundary layer character of the distortion. When the electroosmotic motion of the buffer is modeled in the usual way (see Reference 3 and Section VI below), the form of the distortion remains untouched. If r/d is kept in reasonable limits, say $r/d < \frac{1}{2}$, then the distortion is small. Given the velocity profile (Equation 1-3) one can readily calculate the fraction r/d. One sees that maximal sample throughput and maximal resolution are conflicting objectives.

III. TEMPERATURE DEPENDENCE OF THE ZETA POTENTIALS

Figure 4 in Reference 2 is a graph representing the causal relationships between the different quantities which come into play. The placement of the knot "Power" is incorrect. Let $\sigma(T)$ denote the electrical conductivity and $T = T(x)$ the temperature distribution along the depth direction of the chamber considered. With the assumption of a constant electrical field strength, the power P dissipated in the volume V of the chamber is

$$P = |\mathbf{E}|^2 \int_V \sigma[T(x)] dV \tag{1-8}$$

Hence, the calculation of P requires the knowledge of the temperature distribution $T(x)$. Therewith the power knot is to be placed at the level of the viscosity and both zeta potentials. The placement of the latter is also misleading. Conceptually, the zeta potentials are temperature dependent. Therefore, one should find out if it is justified to assume temperature independence. The electrodiffusive double layer is the result of the competition of the diffusive thermal forces and the electrostatic forces acting on the ions. The simplest model predicts, for a symmetrical electrolyte with an ion valency z and N_0 ions per unit volume near the wall in a distance x from the wall for a temperature T, an electrostatic potential given by the equation

$$\psi(x,T):=\frac{4kT}{ze}\cdot\text{arctanh}\left[\tanh\left\{\frac{ze\psi_0}{4kT}\right\}e^{-\kappa x}\right],$$

$$\kappa:=ze\cdot\sqrt{\frac{2N_0}{k\varepsilon T}} \tag{1-9}$$

where κ is the Debye-Hückel parameter for the electrolyte under consideration, k the Boltzmann constant, e the elementary charge, ε the permittivity of the electrolyte, and ψ_0 the potential at the wall. Formula 1-9 may be found in all standard textbooks on the subject, e.g., that by Hunter.[7] Let δ_w be the distance of the slipping plane from the wall. Then, by definition, the wall zeta potential, $\zeta_w(T)$, is

$$\zeta_w(T):=\psi(\delta_w,T) \tag{1-10}$$

The zeta potential is a common notion, but disquiet has remained over the years concerning its details and fundamentals. Going a step further, we confine ourselves to a simpler approximation valid for small $|\psi_0|$ and set

$$\zeta_w(T):=\psi_0 e^{-\delta_w\kappa} \tag{1-11}$$

which is related to Figure 1 of Reference 2. From Equation 1-11 the temperature follows immediately:

$$\frac{\zeta_w(T)}{\zeta_w(T_0)}=e^{-D_w\cdot\left[\frac{1}{\sqrt{T\varepsilon(T)}}-\frac{1}{\sqrt{T_0\varepsilon(T_0)}}\right]},$$

$$D_w:=\delta_w ze\cdot\sqrt{\frac{2N_0}{k}} \tag{1-12}$$

Now the chamber considered is cooled from the front and back face. Since δ_w is small compared with the width, $2d$, of the chamber, it is justifiable to assume that the shear plane has the same temperature as the walls. According to the argument just given, the wall zeta potential may be considered constant when treating zone distortion, which is what was actually done by Giannovario et al.[2] Next we come to the zeta particle potential, $\zeta_p(T)$, at temperature T. With a degree of accuracy similar to that maintained in Equation 11, one obtains for sperical particles a radius of

$$\psi(r,T) := \frac{\Psi_0 ae^{-\kappa(r-a)}}{r}$$

$$\zeta_p(T) := \psi(\alpha + \delta_p, T)$$

(1-13)

This time, r is the distance from the center of the particle and δ_p describes the position of the shear surface of the particle. From Equation 1-13 the temperature dependence of the zeta particle potential follows immediately:

$$Z(T) := \frac{\zeta_p(T)}{\zeta_p(T_0)} = e^{-D_p \cdot \left[\frac{1}{\sqrt{T\varepsilon(T)}} - \frac{1}{\sqrt{T_0\varepsilon(T_0)}} \right]},$$

$$D_p := \delta_p ze \cdot \sqrt{\frac{2N_0}{k}}$$

(1-14)

We resort again to the simplest model. The von Smoluchowski equation calculates the electrophoretic mobility $\mu(T)$ at temperature T by

$$\mu(T) := \frac{\varepsilon(T)\zeta_p(T)}{\eta(T)}$$

(1-15)

where $\eta(T)$ is the dynamic viscosity of the buffer electrolyte at temperature T. In Figure 1-3 the ratio $C := C(T) := \eta(T)/\varepsilon(T)$ is called the zeta constant because it occurs in the form

$$\zeta_p = C\mu$$

(1-16)

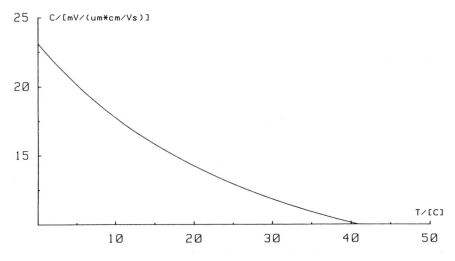

Figure 1-3 The temperature dependence of the zeta constant $C(T)$ defined in Formula 1-16 in the temperature interval [0°C, 50°C]. See Nomenclature for symbol definitions.

when calculating ζ_p given measured values of μ. It is evident that C decreases considerably with increase in temperature T in the interval [0°C, 50°C]. The monotonic behavior of $Z(T)$ from Equation 1-14 depends on that of $T\varepsilon(T)$. In the above temperature interval, the latter product decreases, in the case of water, about 3%. It is important to take T in the expression $T\varepsilon(T)$ as the thermodynamic temperature measured in Kelvin. The decrease in $\varepsilon(T)$ and the increase in $\zeta_p(T)$ compensate each other considerably in the temperature range of interest so that one speaks of a temperature-independent viscosity-mobility product (see Reference 8). The degree of compensation depends somewhat on the electrolyte concentration which enters $Z(T)$ via the Debye-Hückel parameter κ. From a practical point of view one may say that the temperature dependence of $\mu(T)$ is caused by $1/\eta(T)$. It becomes clear near the end of Reference 2 that $\zeta_p(T)$ was assumed to be constant, in contrast to the placement of the knot "zeta particle potential" in Figure 4 of that reference.

The above arguments support the suggestion that zone distortion due to a temperature-dependent electrophoretic mobility $\mu(T)$ should not be considered as a second-order effect. We note in passing that Formula 4 in Reference 2 implies the use of the CGS system of units whereas Equation 1-15 above involves the SI system.

IV. TEMPERATURE DISTRIBUTION IN THE SEPARATION CHAMBER

We now turn to a cornerstone of the paper by Giannovario et al.,[2] i.e., the calculation of the temperature distribution. To reduce the three-dimensional heat transfer problem to a one-dimensional problem, they assumed (implicitly) a chamber of infinite width and length. Moreover, to circumvent the treatment of a system of differential equations, the buffer fluid was assumed to be at rest. Therefore, no convective terms occur and the temperature field is decoupled from the velocity field. A flowing buffer is an essential for free-flow electrophoresis. This may lead to the impression that Giannovario et al.[2] have modeled rather roughly. Consideration of the amount of the heat transported by the fluid in relation to the total amount dissipated in the chamber shows that the assumption of a resting fluid is not unrealistic.

Consider a separation chamber of infinite height with infinitely long electrodes and of infinite width and thickness $2d$. Attach a Cartesian coordinate system to the chamber as in Figure 1-1. Thus, the depth from the central plane is measured such that x, the signed distance from the central plane, ranges over the interval $[-d, d]$. Clearly, the temperature distribution in a chamber of finite width will differ considerably near the electrode regions. In practice, the separation takes place in the middle of the chamber and the assumption made seems reasonable. In contrast to other workers, the more realistic case of temperature-dependent electrical conductivity, $\sigma(T)$, and temperature-dependent thermal conductivity, $k(T)$, was dealt with by Giannovario et al.[2]

Figures 1-4 to 1-6 show the temperature dependence of three buffers (names and compositions of the buffers do not matter here) in the temperature interval 0 to 20°C used in the present author's laboratory.

Figure 1-7 shows the temperature dependence of the thermal conductivity of the water in the temperature interval [0°C, 50°C]. The data for water are taken from Reference 9, while for Figures 1-4 to 1-7 are from References 10 and 11.

The error committed in equating $k(T)$ for water with that of the buffer used can be neglected unless highest precision is required. From Figures 1-4 to 1-7 it is evident that a linear approximation of $\sigma(T)$ and $k(T)$, as used in Reference 2, is reasonable. Since the application does not allow for very large temperature differences in the chamber, a linear

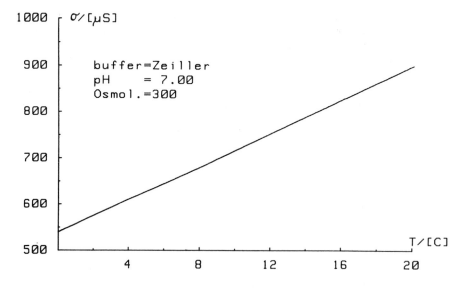

Figure 1-4 Electrical conductivity σ for Zeiller buffer in the temperature interval [0°C, 20°C]. See Nomenclature for symbol definitions.

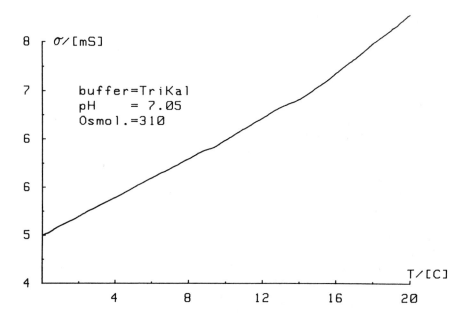

Figure 1-5 Electrical conductivity σ for Trikal buffer in the temperature interval [0°C, 20°C]. See Nomenclature for symbol definitions.

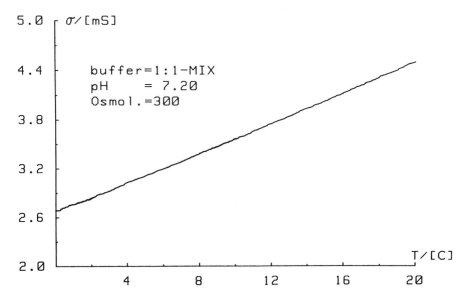

Figure 1-6 Electrical conductivity σ for the 1:1 mixture of the buffers from Figures 1-4 and 1-5 in the temperature interval [0°C, 20°C]. See Nomenclature for symbol definitions.

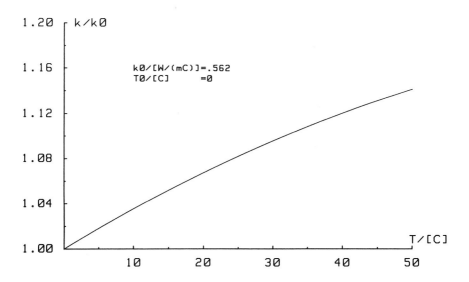

Figure 1-7 Thermal conductivity *k(T)* for water in the temperature interval [0°C, 50°C]. See Nomenclature for symbol definitions.

approximation can be expected to be sufficiently precise. Giannovario et al.[2] developed $k(T)$ and the volumetric heat production rate $Q(T)$ around the wall temperature T_0. Writing $y := T(x) - T_0$, they set

$$k(T) =: \alpha + \beta y =: \alpha(1 + \beta' y), \quad Q(T) =: Q(1 + \varepsilon y), \quad T_{tc} := \frac{1}{\beta'}, \quad T_{ec} := \frac{1}{\varepsilon} \quad \text{(1-17)}$$

The last two definitions in Equation 1-17 and β' have been added by the present author and will be used later. T_{tc} and T_{ec} are the doubling temperatures for the thermal and electrical conductivity, respectively, i.e., the temperature increments from the cooling temperature T_0 for which the conductivities would have double the value according to a linear law.

Two kinds of cooling are of interest: cooling from both the rear and front faces of the chamber and that from the back face only. The cooling jackets are normally made of a material with a high thermal conductivity, often copper. Slits in the copper blocks provide the possibility of observing the optical density (integrated along the depth direction) of the flowing material across the width of the chamber permanently. Cooling from the back face gives the possibility of inspecting the whole separation chamber and alleviates the maintenance of the chamber. It is assumed that the front face of the chamber is in contact with the resting air of the laboratory. Since air at rest is a good thermal isolator, perfect isolation is assumed. This means that heat flux through the front face vanishes pointwise. Hence, $T'(x) = 0$ at the x coordinate of the front face. In the case of double-sided cooling, the central plane plays the same role as the front face does in the case of cooling from the rear face. In this way it suffices to treat only one of the types of cooling.

Figure 1-8 shows the different laminae of a real chamber along the chamber width together with the temperature profile. In the context of this chapter, it is only important to note that the flowing electrolyte cannot be brought into direct contact with the temperature of the cooling block. Between both, a certain increase due to a nonvanishing

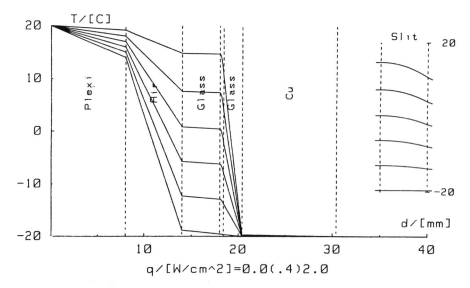

Figure 1-8 Construction of a free-flow separation chamber and a series of temperature profiles for different heat production rates q = 0(0.4)2. The front and rear faces are kept at fixed temperatures. See Nomenclature for symbol definitions.

thermal resistance always has to be taken into account. Let T_c be the cooling temperature; then T_0 is to be calculated. It is assumed here that this is already done.

It has been shown in Reference 11 that the fit

$$k(T) := k(T_0)\left[1 + a(T - T_0) + c(T - T_0)^2\right],$$

$$k(T_0) := 0.562 \frac{\text{W}}{\text{m}°\text{C}}, \quad a := 0.003732 \frac{1}{°\text{C}}, \quad c := 17.79 * 10^{-6} \frac{1}{(°\text{C})^2}, \quad T_0 := 0°\text{C} \quad \textbf{(1-18)}$$

yields a close approximation to the experimental data for water (see Reference 9). Typical values for ε lie in the interval

$$\frac{1}{60°\text{C}} < \varepsilon < \frac{1}{25°\text{C}} \qquad \textbf{(1-19)}$$

or, in terms of typical doubling temperatures,

$$T_{tc} = 250°\text{C}, \quad T_{ec} = 50°\text{C} \qquad \textbf{(1-20)}$$

It is evident from Figures 1-4 to 1-7 that both β and ε are positive: because $Q(T)$ is linear in $\sigma(T)$, linearization of $\sigma(T)$ also means a linearization of $Q(T)$. With the notation of Equation 1-17, the determination of the temperature distribution requires the solution of the equation

$$(\alpha + \beta y)y''(x) + \beta y'^2(x) + Q \cdot [1 + \varepsilon y(x)] = 0, \quad y'(0) = 0, \quad y(d) = 0 \qquad \textbf{(1-21)}$$

Looking for a symmetrical solution $y(-x) = y(x)$ shrinks the range of y to $[0,d]$. The prime always stands for the differentiation with respect to the position x. Now Equation 1-21 is a two-point boundary-value problem for a nonlinear second-order ordinary differential equation. This qualification of the problem at hand does not prevent us from trying to resolve Equation 1-21 and not resorting immediately to a numerical procedure to solve it, as was done in Reference 2. It will become evident that some insight, especially concerning parameter dependence, existence, and uniqueness of the solution, can be gained. To begin with, Equation 1-21 is written as

$$\left[\alpha y + \frac{1}{2}\beta y^2\right]'' + Q \cdot [1 + \varepsilon y] = 0, \quad y'(0) = 0, \quad y(d) = 0 \qquad \textbf{(1-22)}$$

Let $y(x;\beta,\varepsilon)$ be the positive solution of Equation 1-22 so far existing. When not stressing the parameter dependence, the term $y(x)$ will be used. The temperature difference

$$\Delta T := y(0;\beta,\varepsilon) \qquad \textbf{(1-23)}$$

between the wall and the central plane and the temperature gradient $y'(x)$ are of primary interest in addition to y itself. ΔT is expected to be highest temperature difference between any two points within the chamber volume.

The trivial case of constant conductivities, $\beta = \varepsilon = 0$, will serve henceforth as a reference case with which a dimensionless scaling will be performed. We have

$$y(x;0,0) = \frac{Q \cdot (d^2 - x^2)}{2\alpha}, \quad \Delta T = \frac{Qd^2}{2\alpha} \tag{1-24}$$

Let $\zeta := x/d$ be the dimensionless depth position and $\theta := y/\Delta T$ the dimensionless relative temperature difference. Then Equation 1-24 reads in dimensionless coordinates

$$\theta := 1 - \xi^2 \tag{1-25}$$

The first case of real interest is that of constant heat production, $\varepsilon = 0$, combined with increasing thermal conductivity, $\beta > 0$. Now Equation 1-21 may be written as

$$\left[(\beta y + \alpha)^2 \right]'' + 2\beta Q = 0, \quad y'(0) = 0, \quad y(d) = 0 \tag{1-26}$$

After twofold integration we arrive at

$$(\alpha + \beta y)^2 + \beta Q x^2 = \beta Q d^2 + \alpha^2 \tag{1-27}$$

From Formula 1-27 it can be seen immediately that the temperature distribution $y = y(x), 0 \leq x \leq d$, is an arc of an ellipse in the (x,y) plane with center at $(0, -\alpha/\beta)$, axes parallel to the coordinate axes, and semiaxes in the x and y directions with lengths $\sqrt{d^2 + \alpha^2 / \beta Q}$ and $\sqrt{a^2 + \beta Q d^2 / \beta}$, respectively. In the explicit form one finds

$$y(x;\beta,0) := \frac{\alpha}{\beta} \cdot \left[\sqrt{1 + \frac{\beta Q (d^2 - x^2)}{\alpha^2}} - 1 \right],$$

$$\Delta T := \frac{\alpha}{\beta} \cdot \left[\sqrt{1 + \frac{\beta Q d^2}{\alpha^2}} - 1 \right],$$

$$y'(x) = -\frac{Qx}{\alpha \cdot \sqrt{1 + \frac{\beta}{\alpha^2} \cdot Q(d^2 - x^2)}} \tag{1-28}$$

The dimensionless form of Formula 1-28 is Formula 1-29, where $\theta_0 := y(0;\beta,\varepsilon)/\Delta T$:

$$\theta = \frac{\sqrt{1 + 2\tau(1 - \xi^2)} - 1}{\tau}, \quad \tau := \frac{\beta \Delta T}{\alpha},$$

$$\theta_0 := \frac{\sqrt{1 + 2\tau} - 1}{\tau},$$

$$\theta'(\xi) = -\frac{2\tau\xi}{\sqrt{1 + 2\tau(1 - \xi^2)}} \tag{1-29}$$

The difference ΔT, viewed as a function of the chamber thickness $2d$, forms an arc of a hyperbola. Obviously, for $\beta \to 0$ we have $y(x;\beta,0) \to \mathbf{y}(x;0,0)$. A simple calculation shows that the function $y(x;\beta,0)$ decreases in β for each fixed x in $[0,d]$. This means that the effect of increasing thermal conductivity paired with a constant electrical conductivity lowers temperature differences in the chamber and is thus a stabilizing factor for the electrophoretic free-flow separation. This is just what one expects qualitatively by simple intuitive reasoning. The last statement was actually not shown in this extent. This can, however, be justified by a piecewise linear approximation of an increasing, more general conductivity $k(T)$. Note that boundary conditions at $x = \pm d$ imply that the cooling is assumed to be independent of the amount of heat generated in the chamber. In practice, the cooling temperature is (approximately) maintained up to a prescribed maximal heat production. The next thing to be considered is the case of constant thermal conductivity together with linearly increasing volumetric heat production caused by an increasing electrical conductivity. This time Equation 1-21 appears as

$$(1+\varepsilon y)'' + \frac{\varepsilon Q}{\alpha} \cdot (1+\varepsilon y) = 0, \quad y'(0) = 0, \quad y(d) = 0 \tag{1-30}$$

A formal solution of Equation 1-30 can be obtained for all parameter values for which

$$2d \cdot \sqrt{\frac{\varepsilon Q}{\alpha \pi^2}} \neq 1 + 2k, \quad k \varepsilon Z \tag{1-31}$$

holds and is given by

$$y(x;0,\varepsilon) := \frac{1}{\varepsilon} \cdot \left[\frac{\cos\left(x \cdot \sqrt{\frac{\varepsilon Q}{\alpha}}\right)}{\cos\left(d \cdot \sqrt{\frac{\varepsilon Q}{\alpha}}\right)} - 1 \right],$$

$$\Delta T := \frac{1}{\varepsilon} \cdot \left[\frac{1}{\cos\left(d \cdot \sqrt{\frac{\varepsilon Q}{\alpha}}\right)} - 1 \right] \tag{1-32}$$

In dimensionless form Formula 1-32 can be written as

$$\theta = \frac{2\left[\dfrac{\cos(\zeta\, \tau)}{\cos(\tau)} - 1\right]}{\tau^2}, \quad \tau := \sqrt{2\varepsilon\Delta T},$$

$$\theta_0 = \frac{2\left[\dfrac{1}{\cos(\tau)} - 1\right]}{\tau^2} \tag{1-33}$$

It is not difficult to show that the expression

$$\left(1-\xi^2\right)\cdot\frac{\tan(\tau)}{\tau} < \tau < \frac{1-\xi^2}{\cos(\tau)} \tag{1-34}$$

holds. Both bounds of Formula 1-32 increase with increasing ε. One sees immediately from Equation 1-26 that $y(x;0,\varepsilon) \to y(x;0,0)$ for $\varepsilon \to 0$. By elementary but cumbersome calculations one can show that $y(x;0,\varepsilon)$ increases with increasing ε for each fixed y with $0 \le x < d$ (see Reference 12 for details). This means, in other words, that an increasing heat production combined with a constant thermal conductivity increases the temperature differences in the chamber provided the operation conditions allow a steady temperature distribution. Clearly, nothing else is expected and Equation 1-26 contains the quantification in the form of a cosine profile. A positive solution throughout the whole range of definition of Equation 1-25 is only possible in the parameter set defined by

$$4d^2\varepsilon Q < \alpha\pi^2 \tag{1-35}$$

A qualitative understanding of Equation 1-35 is easily obtained. For a given thermal conductivity α inequality Equation 1-35 indicates that neither the thickness $2d$ of the chamber nor the rise coefficient ε of the heat production rate nor the heat production rate Q itself is allowed to be arbitrarily large. Suppose that one of the named quantities is large enough. In this case the heat produced cannot be entirely transported away. Consequently, the chamber becomes warmer, which causes augmented heat production. In this way a kind of thermal explosion is taking place. Under these circumstances a steady temperature distribution is not possible, in contrast to the underlying assumption. Another way of reflecting on this is to argue that the heat flux through the front and rear faces of the chamber increases linearly as does the heat production rate. If the rise coefficient of the latter increases, one reaches a state at which more heat is generated than flows away. Given a fixed separation chamber and a fixed buffer electrolyte, only the heat production rate Q can be chosen freely via the choice of the applied voltage. In this situation, Equation 1-35 says that if

$$Q < Q_{max} := \frac{\alpha\pi^2}{4d^2\varepsilon} \tag{1-36}$$

is violated, no steady-state operation is possible. If the latter is the case, then the temperature in the separation chamber will increase up to the boiling point of the buffer and then remains constant as long as buffer fluid evaporates if no damage or emergency halt interrupts the operation of the system earlier. The maximal heat production rate Q_{max} was calculated under the assumption of a constant thermal conductivity, $\beta = 0$. With $\beta > 0$, as applies with a real buffer electrolyte, Q_{max} is expected to be slightly larger. Hence, Q_{max} in Equation 1-36 can be regarded as a lower bound for Q_{max} calculated for $\beta > 0$.

An explicit knowledge of Q_{max} in terms of the parameters involved provides the opportunity to derive a lower bound for resolution of a free-flow chamber. The local resolution of a given chamber at depth position x is defined as the minimal difference $\Delta\mu(x)$ of electrophoretic mobilities of charged particles which is necessary to achieve the situation where the positions of the respective concentration density modes are at the height of the outlet at a distance of s, where s is the spacing of the axes of two neighboring collection vessels.

Let h be the height usable for separation, b the width usable for separation, $w(x)$ the velocity at position x in the chamber, and $\sigma_0 := \sigma(T_0)$ the electrical conductivity at the cooling temperature T_0. A simple kinematic argument shows that the two subpopulations of particles with mobilities $0 < \mu_1 < \mu_2$ can only be separated if

$$\mu_2 - \mu_1 \geq \Delta\mu_0 := \frac{4w(x)s\sigma_0^2 d^2\varepsilon}{\alpha h\pi^2},$$

$$\mu_2 < \mu_{2\max} := \frac{4w(x)b\sigma_0^2 d^2\varepsilon}{\alpha h\pi^2} \tag{1-37}$$

The local resolution $\Delta\mu_0(x)$ given in Equation 1-37 takes into account only that the sample material remains in the usable area and that Equation 1-36 is fulfilled. Since no other band-broadening effects are taken into account, $\Delta\mu_0$ must be considered as a lower bound for the local resolution $\Delta\mu(x)$ in a real device. Because our definition of resolution is local with respect to the depth position x of the chamber, it allows the possibility of calculating a global resolution $\Delta\mu$ for any velocity profile $w(x)$. This is viewed as an alternative approach in the concept of resolution.

When $\beta\varepsilon > 0$, both conductivities increase with increasing temperature. This case promises to prove more difficult. The greater temperature differences due to the augmented heat production are compensated to a certain extent by an increase in thermal conductivity. Therefore, the temperature profile is expected to depend in a certain way on the ratio of both temperature coefficients.

The monotonic behavior of $y(x;\beta,\varepsilon)$ with respect to each of the three arguments is easily seen for $\beta\varepsilon > 0$. Equation 1-21 is integrated twice and the obtained integral equation for y in fixed point form is written as

$$y(x) = \frac{Q \cdot \left[d^2 - x^2 + 2\varepsilon\left\{ \int_0^x (d-t)y(t)dt - \int_0^x (x-t)y(t)dt \right\} \right]}{2\alpha + \beta y(x)} \tag{1-38}$$

It is obvious that the function on the right-hand side of Equation 1-38 decreases in x and β and increases in ε for $y \geq 0$. From this fact it can be inferred that the nonnegative solutions of Equation 1-38 existing so far decrease in x for all x in $(0,d)$ and all $\beta,\varepsilon > 0$. Moreover, the solutions decrease in ε and decrease in β for the same range of arguments.

This means that our expectation for ΔT from Equation 1-23 is now confirmed for all cases under consideration. The monotonicity of y in x allows us to derive a useful inclusion for ΔT. One can take $x = 0$ in Equation 1-38 and obtain

$$\Delta T \cdot \left(\frac{2\alpha}{\beta} + \Delta T \right) = \frac{Qd^2}{\beta} + \frac{2\varepsilon}{\beta} \cdot \int_0^d (d-t)y(t)dt \tag{1-39}$$

Using the trivial inclusion

$$0 \leq y(x;\beta,\varepsilon) \leq \Delta T \tag{1-40}$$

in Equation 1-38 gives, with $q := Qd^2/\beta$,

$$T_{tc} \cdot \left[\sqrt{1 + \frac{\beta d^2 Q}{\alpha^2}} - 1 \right] < \Delta T < T_{tc} \cdot \left[\sqrt{(1-q)^2 + \frac{\beta d^2 Q}{\alpha^2}} - (1-q) \right] \tag{1-41}$$

For q > 1 the dissipated power must satisfy the inequality

$$Qd^2\beta \le \alpha^2(1-q) \tag{1-42}$$

which is a quadratic inequality for Q.

For our present purpose a new transformed dependent variable, $Y(x)$, is introduced having the meaning of a dimensionless thermal conductivity:

$$Y(x) := 1 + \beta' y(x) \tag{1-43}$$

With the new dependent variable we obtain

$$Y^{2''} + \frac{2Q}{\alpha} \cdot (\beta' - \varepsilon) + \frac{2\varepsilon Q}{\alpha^2} \cdot Y(x) = 0, \quad Y'(0) = 0, \quad Y(d) = 1 \tag{1-44}$$

After multiplication of Equation 1-44 with $2Y^{2'}$ we have, after integration, the first-order differential equation

$$Y^{2'^2} + A \cdot Y^2 + B \cdot Y^3 + C = 0, \quad Y'(0) = 0, \quad Y(d) = 1,$$

$$A := \frac{4Q(\beta' - \varepsilon)}{\alpha}, \quad B := \frac{8\varepsilon Q}{3\alpha^2} \tag{1-45}$$

where C is an integration constant. To select from the one-parameter family of differential equations those which satisfy the boundary condition at $x = 0$,

$$C := -A \cdot Y_0^2 - B \cdot Y_0^3 \tag{1-46}$$

is chosen, where $Y_0 := Y(0)$ is yet unknown. Now we have to deal with

$$Y^{2'^2} = A \cdot \left(Y_0^2 - Y^2\right) + B \cdot \left(Y_0^3 - Y^3\right), \quad Y(d) = 1,$$

$$= B\left[Y_0^3 - Y^3 - c \cdot \left(Y_0^2 - Y^2\right)\right], \quad c := \frac{3}{2} \cdot \left(1 - \frac{\beta'}{\varepsilon}\right) \tag{1-47}$$

The distribution of the zeros of the expression in brackets in Equation 1-47.

$$F(Y) := Y_0^3 - Y^3 - c \cdot \left(Y_0^2 - Y^2\right) \tag{1-48}$$

will be of importance. Obviously, the differential equation in Formula 1-47 is only for such real Y defined for which $F(Y) \geq 0$ holds. The point $Y = Y_0$ is for all values of the parameter c a zero of F. The function F possesses by virtue of the relationships

$$F'(Y) = Y \cdot (2c - 3Y), \quad F''(Y) = 2c - 6Y \tag{1-49}$$

a local minimum in $Y = 0$ with $F(0) = Y_0^2 \cdot (Y_0 - c)$ and a local positive maximum in $Y = 1 - \beta'/\varepsilon$. Therefore, F is nonnegative in the interval $[1, Y_0]$. For $Y_0 = c$, F has a double zero in the origin, which is possible only if $\varepsilon \geq \beta'$ holds. F has no further zeros for $Y_0 > c$. In the case $Y_0 < c$, which is possible only for $3\beta' < \varepsilon$, F shows three real zeros, one being Y_0 and the other two being

$$Y_j := \frac{1}{2} \cdot \left[c - Y_0 - (-1)^j \sqrt{D} \right], \quad j = 1, 2, \quad Y_2 < Y_1 < Y_0,$$

$$D := \left(c - Y_0 \right) \cdot \left(3Y_0 + c \right) \tag{1-50}$$

The cases $Y_0 < 0$, $Y_0 = c$, and $Y_0 > c$ are called the low temperature case, the critical case, and the high temperature case, respectively. These terms are based on the ΔT with which they are related. The critical case is $\Delta T = 50°C$ when the typical values of Equation 1-17 are used. Hence, it can be assumed that the low temperature case encompasses most applications of free-flow electrophoresis, if not all. We start with the critical case $Y_0 = c$. From Equation 1-48 it follows that

$$4Y'^2 = B \cdot (c - Y),$$

$$\left[\sqrt{c - Y} \right]'^2 = \frac{B}{16},$$

$$Y = c - \frac{\varepsilon q x^2}{16} \tag{1-51}$$

Equation 1-51 written in the original variable y gives for $\varepsilon > 3\beta'$ the parabolic temperature distribution

$$y(x; \beta, \varepsilon) = \frac{1}{2\beta'} - \frac{3}{2\varepsilon} - \frac{\varepsilon q x^2}{6\beta'},$$

$$\Delta T = \frac{1}{2\beta'} - \frac{3}{2\varepsilon},$$

$$d := \frac{\sqrt{3\varepsilon - \beta'}}{\sqrt{\beta' q \varepsilon^2}} \tag{1-52}$$

Note that the last equation is an additional relation between β', ε, q, and d in addition to $\varepsilon > 3\beta'$.

We now turn to the low temperature case. Solving Equation 1-43 for $Y^{2'}$ yields, in the case $Y_0 < c$, for the positive branch of the root

$$\frac{dY^2}{\sqrt{BF(Y)}} = \pm dx, \quad Y(d) = 1,$$

$$x = \frac{2}{\sqrt{B}} \int_1^Y \frac{t}{\sqrt{F(t)}} dt,$$

$$d = \frac{2}{\sqrt{B}} \int_1^{Y_0} \frac{t}{\sqrt{F(t)}} dt \qquad (1\text{-}53)$$

The integrals occurring in Equation 1-49 are elliptical ones and can be expressed by the Legendre normal integrals

$$F(x,m) := \int_0^x \frac{1}{\sqrt{(1-u^2)(1-mu^2)}} du, \quad E(x,m) := \int_0^x \sqrt{\frac{1-u^2}{1-mu^2}} du \qquad (1\text{-}54)$$

For definition and theory, see Reference 13 or 14; for numerical routines, see the IMSL manual[15] or consult other mathematical program libraries. In the case under consideration, the normal substitution

$$t = Y_0 - (Y_0 - Y_1)u^2 \qquad (1\text{-}55)$$

allows the conversion to the Legendre normal form. The temperature profile is obtained in parameterized form where the dimensionless thermal conductivity is the parameter in the form

$$x = \frac{4Y_1}{B(Y_0 - Y_2)} \cdot F(u,m) + \frac{1}{\sqrt{4(Y_0 - Y_2)}} \cdot E(u,m),$$

$$u^2 := \frac{Y_0 - Y}{Y_0 - Y_1}, \quad m := \frac{Y_0 - Y_1}{Y_0 - Y_2} < 1 \qquad (1\text{-}56)$$

The unknown value Y_0 is the solution of the first formula of Equation 1-56 with $x = d$. If $d < d_{max}$, where d_{max} is the maximum of the right-hand side of the first line of Equation 1-56 taken over $Y_0 \geq 1$, then this equation for Y_0 is solvable. The quantity d_{max} depends on the parameters $\alpha, \beta,$ and q. Having determined Y_0, we have the temperature profile in the form of $x = x(Y)$, $Y \in [1, Y_0]$. Further details would be inappropriate here.

Consider the closed first quadrant of the (β', ε) parameter plane. The elementary temperature profiles given above cover the whole margin of the first quadrant and Equation 1-52 the interior of the quadrant. The left-hand part of Figure 1-9 shows seven elementary profiles (four of them fall nearly together in the scaling used) for the parameter values $\varepsilon = 0$, $\beta' = 0(0.004)0.012$ and $\beta' = 0$, $\varepsilon = 0(0.02)0.08$. The temperature is made dimensionless with the help of $T_0 := \Delta T$. The effect of the increasing thermal conductivity is so small on the scale of the ordinate that the four profiles appear as a broadened line.

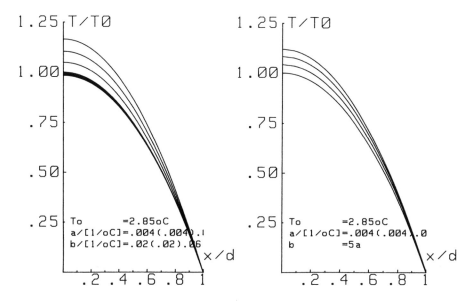

Figure 1-9 A family of elementary temperature profiles (left) and (right) elliptical profiles. See Nomenclature for symbol definitions.

The effect of an increasing electrical conductivity is more prominent. A unit on the ordinate scaling corresponds to $T_0 = 2.85°C$. The right-hand part of Figure 1-9 shows the elliptical profiles on the line $\varepsilon = 5\beta'$ in the (β',ε) plane, which correspond to the same β' value as the left-hand part. The temperature-lowering effect of the thermal conductivity is visible.

The argumentation in this section is largely based on an earlier paper.[12] Some of the recent authors refer to the increasing heat production with increasing temperature as the "autothermal effect" and attribute this term to Lynch and Saville.[16] This effect was already treated by Semjonow in 1923. The interested reader is also referred to the book of Frank-Kamenetzki.[17]

V. NONISOTHERMAL VELOCITY DISTRIBUTION OF THE BUFFER FLOW

Giannovario et al.[2] mentioned that the isothermal temperature profile of the buffer electrolyte is parabolic. Indeed, for a chamber of infinite width this profile is an exact solution of the Navier-Stokes equations. For a chamber of finite width this no longer holds true, however. In Formula 13 of Reference 2 (in which the quantity d is not explained) the authors speak of a chamber of finite width. Hence, it is not clear what the authors really had in mind. Figures 1-10 shows an isothermal velocity profile for a chamber of rectangular cross section in perspective representation. This profile is based on Formula 13 of Reference 2. In order to achieve an acceptable visual impression of the flow profile a very thick chamber has been used. In real applications the chambers are two orders of magnitude thinner. Figure 1-11 shows the curves of constant buffer velocity for a slimmer chamber. Figures 1-10 and 1-11 stem from earlier work.[16] It is justifiable to say that, apart from the lateral margins, the parabolic velocity is a good approximation of the exact isothermal velocity profile obtained from the Navier-Stokes equations. One can show that the Navier-Stokes equations (see also Section VI) for the Newtonian fluid with a constant mass density reduce in the case under consideration to Equation 1-11, which expresses

Figure 1-10 Velocity profile of the buffer flow in a chamber with rectangular cross section in a perspective representation.

Figure 1-11 Curves of constant velocity for a chamber with rectangular cross section, thickness:width = 1:10.

the equilibrium of the viscous stress and the pressure force for all x. Giannovario et al.[2] departed correctly from the equation

$$\left[\eta(T(x))w'(x)\right]' + p = 0, \quad w'(0) = 0, \quad w(d) = 0 \tag{1-57}$$

where $\eta(T)$ is the temperature-dependent dynamic viscosity of the buffer electrolyte, $w(x)$ is the component of the buffer flow in the length direction of the chamber, and p is a constant (modified) pressure gradient driving the buffer flow. The solution to Equation 1-57 is

$$w(x) = p \cdot \int_x^d \frac{t}{\eta[T(t)]} dt \tag{1-58}$$

which gives the profile $w(x)$ apart from an integration. For constant viscosity, $\eta = \eta_0$, one naturally reproduces the parabolic isothermal Poiseuille velocity profile

$$w(x) = \frac{p(d^2 - x^2)}{2\eta_0} = \frac{3\overline{w}}{2}\left(1 - \frac{x^2}{d^2}\right), \quad \overline{w} := \frac{pd^2}{3\eta_0} \tag{1-59}$$

The viscosity of a fluid varies significantly with temperature. Figure 1-12 shows the temperature dependence of the dynamic viscosity $\eta(T)$ in the temperature interval [0°C, 50°C] for water. Again, since no critical point of $\eta(T)$ lies in the interval in question, for deviations from wall temperature a linear approximation is exactly enough:

$$\eta(T) = \eta_0\left[1 - \gamma(T - T_0)\right], \quad \eta_0 := \eta(T_0) = \frac{a}{(T_0 + b)^2},$$

$$a = 506K, \quad b = -150K \tag{1-60}$$

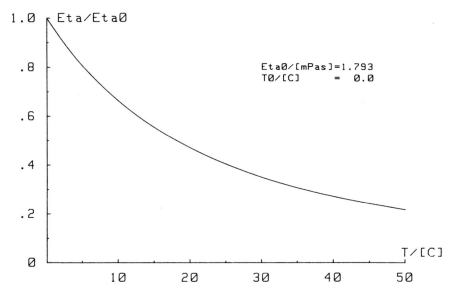

Figure 1-12 Dynamic viscosity $\eta(T)$ for water in the temperature interval [0°C, 50°C]. See Nomenclature for symbol definitions.

One has for water and $T_0 = 0°C$ the values $\eta_0 := 1.793$ mPa s and $\gamma = 0.033/°C$. The profiles 1-57 and 1-60 provide the opportunity to define a Reynolds number, Re, for the buffer flow by using the equation

$$Re := \frac{d\overline{w}\rho_0}{\eta_0} \tag{1-61}$$

Using the typical values $2d = 1$ mm, $\overline{w} = 1$ cm/s, $\rho_0 = 1$ g/cm^3, and $\eta_0 = 1.793$ mPa s yields $Re = 2.79$, which means creeping flow far below the critical Reynolds number of the separation chamber.

Equation 1-58 is integrated using Equation 1-60 and the parabolic temperature profile $y(x;0,0)$ from Equation 1-24, resulting in, for double-sided cooling, the velocity profile

$$w(x;\gamma) = \frac{\alpha p}{\eta_0 \gamma Q} \cdot \ln\left[\frac{1}{1 - \frac{\gamma Q(d^2 - x^2)}{2\alpha}}\right], \quad d^2 < \frac{2\alpha}{\gamma Q} \tag{1-62}$$

The dimensionless form of Equation 1-62 can be written as

$$\frac{w(x;\gamma)}{w(0;0)} = -\frac{\ln\left[1 - e \cdot \left(1 - \zeta^2\right)\right]}{e}, \quad e := \frac{\gamma Q d^2}{2\alpha} = \gamma \Delta T \tag{1-63}$$

As $\eta(T)$ decreases with T, it is plain that $-w(x;\gamma)$ of Equation 1-62 increases with γ. Naturally, for $\gamma \rightarrow 0$, $w(x;y)$ in Equation 1-45 tends to $w(x)$ of Equation 1-59. Since the function $\ln[1/(1-t)]$ has a singularity at $t = 1$, for larger values of γ the increase in the

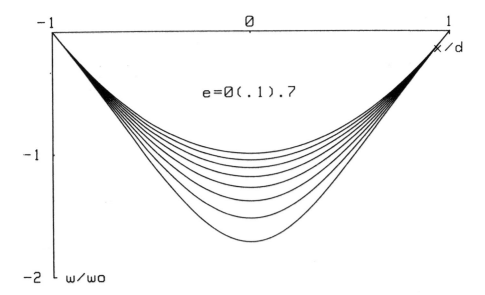

Figure 1-13 Velocity profiles for a parabolic temperature distribution combined with a dynamic viscosity decreasing linearly with temperature for $e = 0(0.1)0.17$ (see Equation 62). See Nomenclature for symbol definitions.

velocity in the central part of the chamber due to the variable viscosity may be substantional. Figure 1-13 shows the corresponding velocity profiles for $e = 0(0.1)0.7$. Perhaps the profiles $w(x,\gamma)$ are new. According to the above discussion, Profile 62 is an exact solution of the Navier-Stokes equations for a Newtonian fluid with a viscosity linear with temperature. The name of the dimensionless group e is presently not known to the author. Equation 1-57 does not account for buoyancy forces caused by a temperature-dependent mass density $\rho = \rho(T)$ of the buffer. For our purpose it is justifiable to equate the density of the buffer electrolyte with that of pure water. The density $\rho = \rho(T)$ of degasified water at atmospheric pressure is known with high precision and can be described, for example,[9] by the equation

$$\rho(T) = \rho_0^* \cdot \left[1 - \frac{a(T-b)^2(T+c)}{T+d} \right] \qquad \text{(1-64)}$$

where

$$\rho_0^* := 0.9999739 \frac{g}{cm^3}, \quad a := \frac{1}{508929.2} \frac{1}{°C^2}, \quad b := 3.9863°C,$$

$$c := 288.9414°C, \qquad d := 68.12963°C \qquad \text{(1-65)}$$

Figure 1-14 shows $\rho(T)$ from Equation 1-64 in the temperature interval [0°C, 50°C]. A wall temperature is chosen and $\rho(T)$ is developed from Equation 1-64 in a power series in the variable $(T - T_0)$ of the shape

$$\rho(T) = \rho_0 \left[1 - \beta(T-T_0) + \cdots \right], \quad \beta := 2\alpha \cdot (T_0 - b) \qquad \text{(1-66)}$$

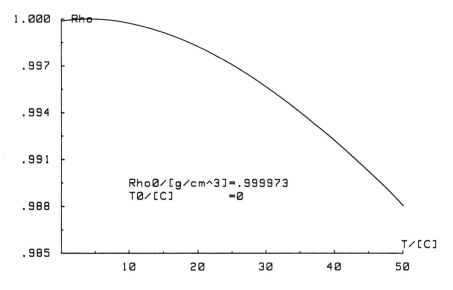

Figure 1-14 Density $\rho = \rho(T)$ in the temperature (T) interval [0°C, 50°C].

The expression for β in Equation 1-66 is a good approximation for β of Equation 1-64. In our case it is appropriate to adopt the usual Boussinesq approximation, which amounts to the retention of the linear term in Equation 1-66 only. The quantity β is known as the temperature coefficient of the mass density $\rho(T)$. Figure 1-15 shows its dependence on the reference temperature $T = T_0$ in the temperature interval [0°C, 50°C]. In the presence of thermal buoyancy, Equation 1-57 appears (see Section VI), as

$$\left[\eta(T(x))w'(x)\right]' + p - \rho_0 g\beta\left(T(x) - T_0\right) = 0 \tag{1-67}$$

We now take $T(x): = T_0 + y(x;0,0)$ and $\eta(T)$ from Equation 1-60 and arrive at

$$w(x;\gamma,\beta) := w(0;0,0)\int_{\xi^2}^1 \frac{1 - \theta \cdot \left(1 - \dfrac{u}{3}\right)}{1 - e \cdot (1 - u)}\,du \tag{1-68}$$

where e is the dimensionless group from Equation 1-63 and θ is a second dimensionless group,

$$\theta := \frac{\rho_0 g\beta\Delta T}{p}, \quad w(0;0,0) := \frac{pd^2}{2\eta_0} \tag{1-69}$$

Evaluation of the integral in Equation 1-68 yields the dimensionless form of the buffer flow profile in the axial direction:

$$\frac{w(x;\gamma,\beta)}{w(0;0,0)} = \frac{1}{3e} \cdot \left\{\theta(1 - \xi^2) - [3e - \theta(1 + 2e)] \cdot \frac{\ln\left[1 - e(1 - \xi^2)\right]}{e}\right\} \tag{1-70}$$

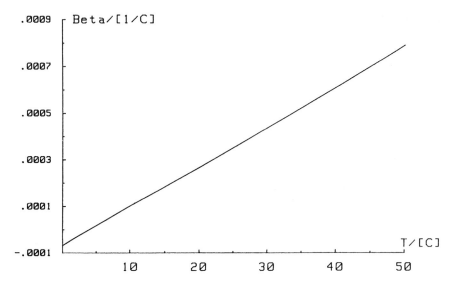

Figure 1-15 Temperature coefficient β = β(T) in the temperature interval [0°C, 50°C].

For β → 0 Equation 1-70 becomes Equation 1-63. Moreover, for a constant viscosity, γ → 0, we arrive at

$$\frac{w(x;0,\beta)}{w(0;0,0)} = \left(1 - \xi^2\right) \cdot \frac{6 - \theta\left(4 + 1 - \xi^2\right)}{6} \tag{1-71}$$

A positive value θ has a destabilizing effect on the axial flow profile. For $0 \leq \theta < 1$ in Equation 1-71 the initial parabolic profile becomes flatter with increasing θ. For $1 \leq \theta \leq$ $^3/_2$ a pair of local maxima occurs at $\xi^2 = 3(1 - 1/\theta)$. The pressure gradient dominates the flow for $0 \leq \theta < ^6/_5$. The pressure force and the buoyancy force are of similar magnitude for $^6/_5 \leq \theta \leq ^3/_2$, and a thermic counterflow occurs in the interval specified by $\xi^2 \leq 3(1 - 1/\theta)$. For $\theta \geq ^6/_5$ the buoyancy force dominates. Figure 1-16 shows a sequence of profiles from Equation 1-71 with θ = 0(0.2)1.6 and e = 0. Figure 1-17 shows the corresponding profiles for θ values, but with e = 0.2.

VI. ELECTROOSMOTIC MOTION OF THE BUFFER ELECTROLYTE

The electroosmotic motion is introduced in the Navier-Stokes equations by means of a prescribed tangential velocity v_{eo} near the front and rear faces of the chamber. Neglecting the distance δ_w of the slipping plane from the walls, one usually prescribes an electroosmotic velocity directly at the walls. Under this boundary condition the Navier-Stokes equations were to be solved. The prescribed tangential electroosmotic velocity v_{eo} causes the buffer motion. Giannovario et al.,[2] however, made a driving force *F* of uncertain origin responsible for this motion. In Equation 1-73 it is evident that F from Equation 15 in Reference 2 is a constant lateral pressure gradient. It is formed and maintained as a consequence of the electroosmotic motion of the buffer. Giannovario et al.[2] did not make clear whether or not the velocity field found is a solution of the Navier-Stokes equations. We shall make this point clear and specify the underlying assumptions.

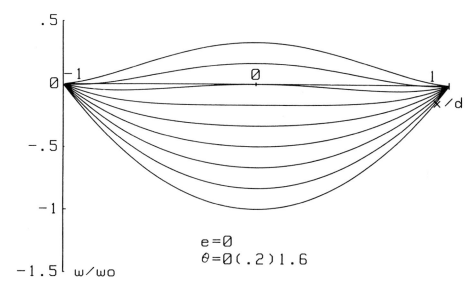

Figure 1-16 Distribution of the axial component of the buffer velocity according to Equation 1-70 with $e = 0$ and $\theta = 0(0.2)1.6$. See Nomenclature for symbol definitions.

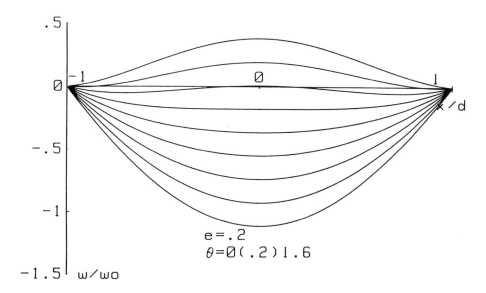

Figure 1-17 Distribution of the axial component of the buffer velocity according to Equation 1-70 with $e = 0.2$ and $\theta = 0(0.2)1.6$. See Nomenclature for symbol definitions.

Consider a separation chamber of infinite length, infinite width, and thickness $2d$. Attach a Cartesian coordinate system to the chamber as shown in Figure 1-1. Assume a constant pressure gradient maintained by a pump and a tangential electroosmotic velocity v_{eo} at the front and rear faces of the chamber as driving forces for the buffer electrolyte. The flow field of the buffer is assumed to have the form

$$\mathbf{v} := [0, v(x), w(x)] \tag{1-72}$$

When a constant mass density ρ is assumed for the buffer, the flow field (Equation 1-72) satisfies the continuity equation (conservation of mass) $\nabla \cdot \mathbf{v} = 0$. The Navier-Stokes equations for the flow field (Equation 72) reduce this to the system

$$\frac{\partial p}{\partial x} = 0,$$

$$[\eta(T(x))v'(x)]' = \frac{\partial p}{\partial y}, \quad v(-d) = v(d) = v_{eo},$$

$$[\eta(T(x))w'(x)]' = \frac{\partial p}{\partial z} - \rho[T(x)]g, \quad w(-d) = w(d) = 0 \tag{1-73}$$

where g is the component of the gravity vector $\mathbf{g}$ falling in the z direction. It should be noted that ρ is assumed to be constant in the continuity equation, but not in Formula 1-73. Both assumptions are only approximately compatible when the mass density differences in Formula 1-73 are small. From Formula 1-73 it follows that the hydrodynamic pressure p is of the form

$$p = p_0 + by + cz \tag{1-74}$$

where p_0, b, and c do not depend on the coordinates x, y, and z. It is inferred that under our assumptions the electroosmotic and the pressure-driven motion of the buffer electrolyte are decoupled. Therefore, the last differential equation in Formula 73 can be dropped.

First the isothermal electroosmotic flow profile $v(x)$ of the buffer electrolyte is considered. The second equation of Formula 1-73 yields a one-parameter family of profiles:

$$v(x) = v_{eo} \cdot \left[1 + s\left(1 - \frac{x^2}{d^2}\right)\right] \tag{1-75}$$

where s is the free parameter. In a chamber of finite width with a flow field independent of z, the net flow through a plane $y = $ const. vanishes. Under this requirement the parameter s turns out to be $s = -\frac{3}{2}$ or

$$v(x) = v_{eo} \cdot \left[1 - \frac{3}{2}\left(1 - \frac{x^2}{d^2}\right)\right] \tag{1-76}$$

Under the assumption of the von Smoluchowski equation (Equation 1-15),

$$v_{eo}(T) := \frac{\varepsilon(T)\zeta w(T)|\mathbf{E}|}{\eta(T)} \tag{1-77}$$

where ζ_w is the zeta wall potential. Assuming a wall temperature $T = 10°C$, one obtains, in the case of water, with parameters $\varepsilon(T) = 83.82 \times 8.854 \times 10^{-12}$ As/(Vm), $\eta(T) = 13.169 \times 10^{-4}$ Ns/m², as well as $|\mathbf{E}| = 100$ V/cm, and $\zeta_w = 5$ mV

$$v_{eo}(T) = 28.2 \frac{\mu m}{s} \tag{1-78}$$

which amounts to ca. 2 mm/min. The magnitude of v_{eo} may be adjusted by appropriately coating the inner surface of the separation chamber.

We now come to the temperature-dependent form of the profile for Equation 1-76. A viscosity linear with temperature is assumed according to Equation 1-60 and a parabolic temperature distribution according to Equation 1-24. These assumptions lead to the temperature-dependent electroosmotic velocity distribution

$$v(x) := v_{eo} \cdot \left(1 + s \ln\left[1 + e\left(1 - \frac{x^2}{d^2}\right)\right]\right), \quad s := -\frac{1}{\int_0^1 \ln\left[1 - e\left(1 - u^2\right)\right] du} \tag{1-79}$$

Figure 1-18 shows the profile $v(x)$ from Equation 1-79 in its dimensionless form, namely

$$\frac{v(x)}{v_{eo}} = 1 + s \ln\left[1 - e\left(1 - \xi^2\right)\right] \tag{1-80}$$

The temperature dependence is not so strong when e is kept well below unity.

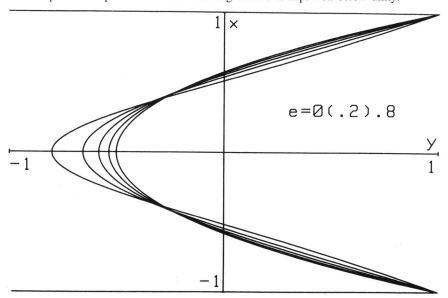

Figure 1-18 Distribution of the lateral (electroosmotic) velocity in dimensionless form according to Equation 1-80 with $e = 0(0.2)0.8$.

VII. CALCULATION OF THE RESIDENCE TIME

The effect of diffusion of the sample particles in the depth direction causes a residence time $t_r = t_r(x)$ dependent on the distance x from the central plane. To account for this effect, Giannovario et al.[2] make the *ad hoc* setting

$$L = t_r \cdot \left(V_0(x) - \frac{A(x)t_r}{2} \right) \qquad (1\text{-}81)$$

where $V_0(x)$ is the lateral particle velocity and $A(x) := |V_0'(x)|$. L is the length of the separation path. If and only if

$$2LA(x) \le V_0^2(x) \qquad (1\text{-}82)$$

is obeyed, two solution branches exist for the residence time t_r, the one of interest is

$$t_r = t_r(x) := \frac{V_0(x) \cdot \left[1 - \sqrt{1 - \dfrac{2LA(x)}{V_0^2(x)}} \right]}{A(x)} \qquad (1\text{-}83)$$

Since Equation 1-83 gives the explicit residence time t_r, there is absolutely no need to solve Equation 1-81 for t_r by means of Newtonian iteration. Moreover, the numbers given for initial values $t_r^0 := L/V_0$ in Table 1 of Reference 2 are false; the error decreases with increasing thickness of the chamber.

The quantity ΔX to be found in the last row of Table 1 is not formally explained. Figure 2 in Reference 2, in which ΔX occurs, gives the idea, however, that the quantities ΔX and $\Delta x'$ (the latter is defined as the increase in the sample diameter due to radial diffusion) are the same. Such inconsistency in notation does not alleviate the reader's task. The deceleration values $A(x)$ in Table 1 of Reference 2 have, in the sense of the definition above, $A(x) := |V_0'(x)|$, an incorrect sign. Note that Formula 1-81 only makes sense in our definition of $A(x)$, or the sign of $A(x)$ from Table 1 and that in Formula 31 of Reference 2 are incompatible. Furthermore, the moduli of the $A(x)$ are so large that the solvability condition 1-82 is eclatantly violated.

Formula 42 in Reference 2, viewed from the standpoint of dimensioned quantities, shows on the right-hand side quantities bearing the dimension of length and on the left-hand side an electrophoretic mobility which is, according to Table 2 in the article of Giannovario et al.,[2] not related to a quantity with the dimension of length. Naturally, the ultimate aim of all electrophoretic separation methods is to transform electrophoretic mobility differences to length differences, and the larger the latter the better. The whole of paper[2] was focused on the conversion factor. Hence, the reader cannot be expected to assume unities in which the conversion factor becomes unity numerically.

VIII. SUMMARY AND CONCLUSION

The mathematical model of free-flow-electrophoresis considered is that of Giannovario et al.[2] Thermal analysis in the separation chamber is considered, and it is shown that the temperature profile across the depth of the chamber can be calculated explicitly up to a quadrature. Thereby, the thermodynamic properties of the buffer electrolyte, i.e., the electrical conductivity, the thermal conductivity, and the viscosity, are assumed to vary linearly with temperature. The temperature dependence of these quantities and other quantities is presented in the form of plots. In several special cases, explicit expressions for the temperature-dependent quantities have been obtained. The influence of the temperature on the motion of the buffer electrolyte and the sample material was investigated.

In this way some steps have been made to put the arguments of Reference 2 on a sound base. However, they are still far from being complete. Thus, some further efforts seem to be necessary.

NOMENCLATURE

Q = cross section of separation chamber
ΔS = deformation measure of sample stream
L = length of separation path
D = electrophoretic displacement
E = electrical field strength vector
$2d$ = chamber thickness
μ = electrophoretic mobility
σ = electrical conductivity
T = temperature
P = dissipated power in the chamber
V = chamber volume
ψ = electrostatic potential
κ = Debye-Hückel parameter
e = elementary charge (in Section III)
ε = electrolyte permittivity
ζ_w = wall zeta potential
δ_w = distance from slipping plane to wall
ζ_p = particle zeta potential
w = velocity of the buffer flow
C = zeta constant
η = buffer viscosity
Q = volumetric heat production
k = thermal conductivity
θ = dimensionless temperature
ρ = mass density
α = temperature coefficient of thermal conductivity
c = second temperature coefficient of thermal conductivity
Re = Reynolds number
β = temperature coefficient of mass density
e = a dimensionless group (in Section V)
p = pressure
v = velocity vector of buffer flow
v_{eo} = electroosmotic velocity
t_r = residence time
a = radius of sperical particles (see Section III)

REFERENCES

1. **Arcus, A., McKinney, A.E., Livesy, J.H., Mecalf, W.S., Vaughan, S., and Keey, R.B.,** Continuous-flow, support-free, electrophoretic separation in thin layers: towards large scale operation, *J. Chromatogr.,* 202, 157, 1980.
2. **Giannovario, J.A., Griffin, R.N., and Gray, E.L.,** A mathematical theory of free-flow electrophoresis, *J. Chromatogr.,* 153, 329, 1978.
3. **Ivory, C.F.,** Continuous flow electrophoresis: the crescent phenomenon revisited. I. Isothermal effects, *J. Chromatogr.,* 195, 165, 1980.
4. **Naumann, R.J. and Rhodes, P.H.,** Thermal considerations in continuous flow electrophoresis, *Sep. Sci. Technol.,* 19, 51, 1984.
5. **Saville, D.A.,** *The Fluid Mechanics of Continuous Flow Electrophoresis in Perspective of Physico-Chemical Electrophoresis,* Vol. 1, Pergamon Press, London, 1980, 297.
6. **Ivory, C.F.,** Continuous flow electrophoresis: the crescent phenomena revisited. II. Nonisothermal effects, *Electrophoresis,* 2, 31, 1981.
7. **Hunter, R.J.,** *Zeta Potential in Colloid Science,* Academic Press, London, 1981.
8. **Mehrishi, J.N. and Seaman, G.V.F.,** Temperature dependence of the electrophoretic mobility of cells and quartz particles, *Biochim. Biophys. Acta,* 112, 154, 1966.
9. **Schäfer, K.,** *Landoldt-Börnstein, Transportphänomene II, Kinetik, homogene Glasgleichgewichte,* Springer-Verlag, Berlin, 1968.
10. **Boese, F.G.,** Automatische Leitfähigkeitsmessung in Abhängigkeit von der Temperatur für Elektrolyte der Free-Flow-Elektrophorese, Internal Report, Max-Planck-Institut für Biochemie, Martinsried, Germany, 1986.
11. **Boese, F.G.,** Temperaturabhängigkeit der thermodynamischen Stoffgrößen, die bei der Free-Flow-Elektrophorese von Bedeutung sind, Internal Report, Max-Planck-Institut für Biochemie, Martinsried, Germany, 1986.
12. **Boese, F.G.,** Zur thermischen Analyse der trägerfreien Elektrophorese, in *Elektrophorese Forum '87,* Radola, B.J., Ed., Technische Universität München, Munich, Germany, 1987, 175.
13. **Erdélyi, A.,** *Higher Transcendental Functions,* Vol. 2, Dover Publications, New York, 1970.
14. **Abramowitz, M. and Stegun, I.,** *Handbook of Mathematical Functions,* Dover Publications, New York, 1970.
15. **Anon.,** User's Manual SFun/Library: Fortran Subroutines for Evaluating Special Functions, Version 2.0, IMSL Inc., Houston, 1987.
16. **Lynch, E.D. and Saville, D.A.,** Heat transfer in the thermal entrance region of an internally heated flow, *Chem. Eng. Commun.,* 9, 201, 1981.
17. **Frank-Kamenetzki, D.,** *Diffusion and Heat Exchange in Chemical Kinetics,* University Press, Princeton, NJ, 1955.
18. **Boese, F.G.,** Isotherme viskose Poiseuille-Strömung im rechteckigen unendlich langen Rohr, angetrieben durch einen Druckgradienten, Internal Report of Max-Planck-Institut für Biochemie, Martinsried, Germany, 1986.

Chapter 2

Numerical Description of Zone Electrophoresis in the Continuous Flow Electrophoresis Device

Terese M. Grateful, Atul M. Athalye, and Edwin N. Lightfoot

CONTENTS

I. INTRODUCTION

The growth of biotechnology has led to an increasing need for separation methods on the analytical, preparative, and production scale. Electrophoresis has emerged as a powerful separative tool for analysis, but its potential as a production-scale separation method has not yet been realized. Limitations imposed by complex convective heat and mass transfer processes and the lack of reliable quantitative descriptions for these processes have hindered development in even the most promising equipment in current use.

Continuous flow electrophoresis (CFE)[1] is an attractive method for the scale-up of electrophoresis. Previous work on the description of this system has primarily focused on limiting ranges of operating conditions[2,3] (with the exceptions of a previous presentation of the finite difference scheme of Section III.A.[4] and a Monte Carlo simulation[5]). Although the solutions from these prior efforts do provide useful insight into the operation of the CFE device under limiting conditions, there is a need for a more general approach. Because of the complexity of the transport processes involved, we must resort to numerical solutions, rather than limiting analytical expressions, to predict the behavior

0-8493-8918-6/94/$0.00+$.50
© 1994 by CRC Press, Inc.

within the device. Such a numerical solution not only must agree with the established limiting results, but also must provide a description when these asymptotic solutions are inadequate.

The focus of this undertaking is the formulation and validation of a basic, three-dimensional, numerical description of mass transfer behavior within the CFE device. Such a description will permit the investigation of the effects of changes in operating conditions (e.g., buffer conditions, electric field strength, sample residence times) without the expense and effort of actual experiments. Once a reliable description of the mass transfer behavior has been developed, it can be extended later to account for secondary effects such as free convection, electrohydrodynamic flows, and electrokinetic effects.

II. THEORY

This chapter will focus on the most common configuration for the CFE device, typically referred to as the Hannig device, represented schematically in Figure 2-1. The CFE apparatus consists of a rectangular flow chamber with dimensions such that the length of the chamber (L) is much greater than both the width (w) and the thickness $(2d)$ of the chamber. Separation buffer flows in the axial (x) direction, and an electric field is imposed across the width of the chamber $(z$ direction), perpendicular to the electrolyte flow. The chamber is typically cooled through one or both faces $(y = \pm d)$.

The sample is introduced continuously into the chamber at the upstream end, and the trajectory of each sample component within the chamber is determined by the vector sum of the species-dependent motion —electrophoresis — and the nonselective fluid motion — axial buffer flow and electroosmotic flow. Spatial variations in the lateral dimension exist in both the axial buffer flow and the electroosmotic flow, and this position-dependent velocity greatly limits the resolution attainable in this system and also complicates its description.

A. GOVERNING EQUATIONS

For this discussion, the case of isothermal operation will be examined initially, followed by a brief discussion of strategies for adapting the numerical methods presented here to nonisothermal conditions. The current discussion also will be limited to conditions under which nonlinear concentration effects are minimized (i.e., electrokinetic effects[6] and electrohydrodynamic spreading[7] are of secondary importance).

With the assumption of constant fluid properties, the following steady-state form of the continuity equation can be used:

$$v_x \frac{\partial c}{\partial x} + v_z \frac{\partial c}{\partial z} = D\left(\frac{\partial^2 c}{\partial x^2} + \frac{\partial^2 c}{\partial y^2} + \frac{\partial^2 c}{\partial z^2} \right) \tag{2-1}$$

where D is the effective binary diffusivity of the component of interest in the buffer medium, and v_x and v_z represent the x and z components of the species velocity.

Comparing terms in Equation 2-1, it is clear that the relative importance of mass transport resulting from axial convection and axial diffusion can be determined by examination of the axial Péclet number Pe_x, where

$$Pe_x \equiv \frac{Lv_o}{D} \tag{2-2}$$

Under conditions of operation typical in the CFE device, Pe_x will be very large; that is,

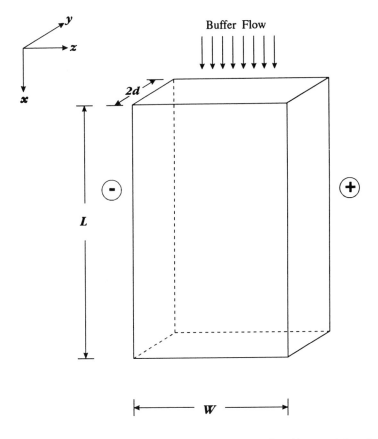

Figure 2-1 Continuous flow electrophoresis apparatus. See Nomenclature for symbol definitions.

$$Pe_x \gg 1 \tag{2-3}$$

so that it is reasonable to neglect mass transfer induced by axial diffusion as unimportant relative to that caused by axial convection, giving

$$v_x \frac{\partial c}{\partial x} + v_z \frac{\partial c}{\partial z} = D\left(\frac{\partial^2 c}{\partial y^2} + \frac{\partial^2 c}{\partial z^2}\right) \tag{2-4}$$

By neglecting axial diffusion, x becomes essentially "time-like" in behavior in the sense that the equation now can be treated as an initial-value problem with respect to the axial coordinate.

The boundary conditions for Equation 2-4 are

1. $c \rightarrow 0$ rapidly as $z \rightarrow \pm\infty$
2. $\partial c/\partial y = 0$ $y = \pm d$
3. $c(x,y,z) = c_{in}(y,z)$ $x = 0$

where $c_{in}(y,z)$ is a specified input condition. For this development an impulse input of the form

$$c_{in}(y,z) = c_o \delta(z) \qquad (2-5)$$

will be used, where c_o is the mass input per unit area, $\delta(z)$ is the unit impulse function defined by $\delta(z) = 0$ for $z \neq 0$, and $\int_{-\infty}^{\infty} \delta(z)dz = 1$. Of course, alternative input conditions exist, and the implementation of these alternatives in the following numerical methods is reasonably straightforward; results from one such case will be shown later.

The assumption of isothermal conditions allows the expression of the axial velocity component to be given as

$$v_x(y) = v_o \left(1 - (y/d)^2\right) \qquad (2-6)$$

where v_o is the maximum axial velocity.

The species velocity in the z direction will consist of both the electrophoretic velocity and the flow of the buffer induced by electroosmosis and can be written as

$$v_z(y) = v_{ef} + v_{eo}\left(1 - \frac{3}{2}\left(1 - (y/d)^2\right)\right) \qquad (2-7)$$

where v_{ef} is the electrophoretic velocity of the species of interest and v_{eo} is the electroosmotic velocity. The terms v_{ef} and v_{eo} also can be expressed in terms of the electric field strength E since $v_{ef} = \mu_{ef}E$ and $v_{eo} = \mu_{eo}E$. The terms μ_{ef} and μ_{eo} are the electrophoretic and electroosmotic mobilities, respectively, which are functions of both the buffer properties and the species properties (in the case of μ_{ef}) and the wall properties (for μ_{eo}). (Note: positive values of v_{ef} and v_{eo} indicate movement in the positive z direction.)

Nonisothermal conditions primarily affect the velocity profiles, even to the point of inducing free convection which destroys the potential for separations in the device. Then, in order to account for nonisothermal conditions, it is often only necessary to use alternative velocity profiles that can be obtained by solving the equations of change for energy and momentum simultaneously. In the numerical methods presented here the velocity profiles obtained then can be included in the computer code.

B. SCALING OF EQUATIONS

Equations 2-4 to 2-7 can be rewritten in dimensionless terms in an effort to reduce the number of parameters necessary for investigation. To this end the following new parameters can be defined:

$$x^* = \frac{x/d}{\text{Pe}}, \qquad y^* = y/d, \qquad z^* = z/d,$$

$$v_x^*(y^*) = \frac{v_x(y)}{v_o}, \qquad v_z^*(y^*) = \frac{v_z(y)}{v_o}\,\text{Pe}, \qquad c^* = dc/c_o$$

where $\text{Pe} = v_o d/D$ is a Péclet number. The governing equation now takes the form

$$v_x^* \frac{\partial c^*}{\partial x^*} + v_z^* \frac{\partial c^*}{\partial z^*} = \left(\frac{\partial^2 c^*}{\partial y^{*2}} + \frac{\partial^2 c^*}{\partial z^{*2}}\right) \qquad (2-8)$$

where

$$v_x^*(y^*) = 1 - y^{*2} \tag{2-9}$$

and

$$v_z^*(y^*) = \text{Pe}_{ef}\left\{1 + k\left(1 - \frac{3}{2}\left(1 - y^{*2}\right)\right)\right\} \tag{2-10}$$

Here $\text{Pe}_{ef} = v_{ef}d/D$ is the electrophoretic Péclet number and $k = v_{eo}/v_{ef}$.
   The boundary conditions become

1. $c^* \rightarrow 0$ \qquad rapidly as $z^* \rightarrow \pm\infty$

2. $\dfrac{\partial c^*}{\partial y^*} = 0$ \qquad $y^* = \pm 1$

3. $c^*(x^*,y^*,z^*) = \delta(z^*)$ \qquad $x^* = 0$

   At this point, two quantities of potential interest can be defined: the average concentration c_{avg}^*, given by

$$c_{avg}^*\left(x^*,z^*\right) = \frac{\displaystyle\int_{-1}^{1} c^*\left(x^*,y^*,z^*\right)dy^*}{\displaystyle\int_{-\infty}^{\infty}\int_{-1}^{1} c^*\left(x^*,y^*,z^*\right)dy^*\,dz^*} \tag{2-11}$$

and the bulk, or flow-averaged, concentration c_b^*, defined as

$$c_b^*\left(x^*,z^*\right) = \frac{\displaystyle\int_{-1}^{1} v_x^* c^*\left(x^*,y^*,z^*\right)dy^*}{\displaystyle\int_{-\infty}^{\infty}\int_{-1}^{1} v_x^* c^*\left(x^*,y^*,z^*\right)dy^*\,dz^*} \tag{2-12}$$

These values then can be calculated at a given dimensionless chamber length L^*, where

$$L^* = \frac{L/d}{\text{Pe}} \equiv \text{Gz}^{-1} \tag{2-13}$$

and Gz is the Graetz number. It is now evident that the influence of the experimental parameters can be summarized in terms of the following dimensionless groups: L^* (or Gz), Pe_{ef}, and k.

C. SPATIAL MOMENTS
Before proceeding to the numerical solution of the governing equations, it is worthwhile to look for additional insight into system behavior by analogy with other familiar transport models. The linearity of this problem suggests that spatial moments of the concentration profile can yield useful information about the expected location and overall character of the solute peak; it shall be shown that this is indeed the case.
   The nth normalized moment $M_n(x^*)$ of the bulk, or flow-averaged, concentration is defined as is done in elementary statistics:

$$M_n(x^*) = \int_{-\infty}^{\infty} z^{*n} c_b^*(x^*,z^*)dz^*, \qquad n = 1,2,\dots \tag{2-14}$$

The physical interpretation of these moments is similar to that of their counterparts in statistics. Thus, the first moment, $M_1(x^*)$, gives the *mean* migrational position of the bulk concentration profile at an axial location x^*. The second moment, $M_2(x^*)$, is related to the (dimensionless) variance σ^{*2}, or the spread of the profile about the mean, by the expression

$$\sigma^{*2}(x^*) = M_2(x^*) - \left[M_1(x^*)\right]^2 \qquad (2\text{-}15)$$

The higher moments provide information about the deviation of the concentration profile from a Gaussian shape in the migration coordinate z^*.

Before the results for the moments are presented, some qualitative aspects of their expected behavior shall be mentioned briefly. First note that the time-like nature of the axial coordinate makes this problem mathematically analogous to the dispersion of an initial solute distribution in flow through a channel or a tube. The generalized Taylor dispersion theory of Brenner[8,9] then suggests that, in the limit of long times (i.e., for large x^*), the profile should asymptotically approach a Gaussian shape in the separation coordinate z^*, with the mean position and variance both increasing linearly with time (or x^*) to leading order.

The calculation of the moments can be carried out following the classical method of Aris.[10] Here only the results of the somewhat lengthy calculation[11] for the impulse input condition given in Equation 2-5 are provided. The mean position of the flow-averaged concentration profile for all x^* is given by the equation

$$M_1(x^*) = \frac{3}{2}x^* \, Pe_{ef} \qquad (2\text{-}16)$$

while the asymptotic variance can be written to leading order as

$$\sigma^{*2}(x^*) \simeq 3x^* \left(1 + \frac{2}{105}(1+k)^2 Pe_{ef}^2\right) \quad \text{for large } x^*. \qquad (2\text{-}17)$$

It is evident that the mean and the asymptotic variance of the solute profile are indeed linear with respect to the axial position. The variance and higher moments also have been calculated exactly for smaller axial distances by a numerical procedure[11] similar to that used in Section III.B. It can be shown that the above expression for the variance is identical in form to that for transient dispersion in Poiseuille flow between parallel plates.

While the variance is explicitly available only for large x^*, the results given here do furnish a means for conservatively estimating the expected location and spread of the solute profile in the migration direction for all axial positions. It thus becomes possible to set reasonable bounds *a priori* (about 4 to 5 σ^* on either side of the mean) on the region chosen for spatial discretization in a numerical scheme. Moreover, as shown below, the form of Equation 2-16 suggests making a transformation into a moving coordinate frame, while the trends seen in Equation 2-17 give useful insight into the effects of process parameters on separation performance.

D. MOVING FRAME

Keeping in mind that the efficiency of numerical methods is often limited by computation time and memory, the results of the moments analysis can be used to our advantage. By using a moving reference frame, our frame of observation can be limited to the active region of the chamber. The first moment $M_1(x^*)$ presented in the previous section gives us the location of the mean of the solute peak, which serves as an ideal basis for our "moving reference frame".

This moving frame can be described in terms of a new variable ζ, where

$$\zeta \equiv z^* - \frac{3}{2}x^* \text{Pe}_{ef} \tag{2-18}$$

so that the reference frame moves at the mean speed of the solute peak.

Rewriting Equation 2-8 in terms of ζ, $c^*(x^*,y^*,z^*)$ is replaced with $c^*(x^*,y^*,\zeta)$ to obtain

$$v_x^* \frac{\partial c^*}{\partial x^*} + v_\zeta^* \frac{\partial c^*}{\partial \zeta} = \left(\frac{\partial^2 c^*}{\partial y^{*2}} + \frac{\partial^2 c^*}{\partial \zeta^2} \right) \tag{2-19}$$

where

$$v_x^* = \left(1 - y^{*2}\right) \tag{2-20}$$

and

$$v_\zeta^* = (k+1)\text{Pe}_{ef}\left(1 - \frac{3}{2}\left(1 - y^{*2}\right)\right) \tag{2-21}$$

with boundary conditions

1. $c^* \to 0$ rapidly as $\to \pm\infty$
2. $\partial c^*/\partial y^* = 0$ $y^* = \pm 1$
3. $c^*(x^*,y^*,\zeta) = \delta(\zeta)$ $x^* = 0$

The above formulation suggests that the system behavior can be expressed in terms of three new parameters: Gz, $(k+1)\text{Pe}_{ef}$, and $\text{Gz}^{-1}\text{Pe}_{ef}$, an alternative set to that obtained in Section II.B.

E. EFFECTS OF PROCESS PARAMETERS
Although this new set of parameters could be used to define our system, for physical reasons, which will soon become apparent, the original three parameters (Gz, Pe_{ef}, and k) will be used to explore the mass transfer behavior in the CFE device.

1. Graetz Number
In the determination of the mass transfer behavior, the effect of the value of the Graetz number Gz can be seen by looking at the definition of Gz:

$$\text{Gz} \equiv \text{Pe}\frac{d}{L} = \frac{d^2/D}{L/v_o} \approx \frac{\tau_{diff,y}}{\tau_{conv,x}} \tag{2-22}$$

where $\tau_{diff,y}$ is the time constant for diffusion in the lateral coordinate y, the dimension over which the velocities vary, and $\tau_{conv,x}$ is an estimate of the residence time in the chamber.

It is evident that, with respect to Gz, two limiting behaviors exist. At high values of Gz the residence time is sufficiently short that diffusion effects are minimal, and convective or "hydrodynamic" effects dominate. This condition gives a solute profile that has a fairly narrow peak but a long tail, at times referred to as the "crescent effect".[12] The tail is composed of material which entered the chamber near the wall, had a very long residence time, and, hence, spent more time under the influence of the electric field. At low values of Gz the residence time is sufficiently long that the effects of the nonuniform velocity profiles are mitigated by the effects of lateral diffusion. With a longer residence time the solute peak will be broader, with a significantly shorter tail.

2. Electrophoretic Péclet Number

The effect of electrophoretic velocity v_{ef} on the concentration profile can be seen through variations in the Péclet number Pe_{ef}, where

$$Pe_{ef} = \left(v_{ef} / v_o\right)Pe = v_{ef}d / D \tag{2-23}$$

From Equation 2-17 it is expected that, for long times, increases in v_{ef} will lead to increases in the spreading of the concentration band. At short times, when diffusion is less important, an increase in electrophoretic velocity also leads to an increase in band spreading and worsens the tailing seen in the profiles.

3. Relative Electroosmotic Velocity

Electroosmosis has a significant effect on the behavior seen in CZE in terms of relative electroosmotic velocity, k:

$$k = v_{eo} / v_{ef} \tag{2-24}$$

Strickler and Sacks[12] showed clearly that as $k \to -1$, electroosmosis balanced the effect of the nonuniform residence time in the chamber and, thus, minimized band spreading for systems with short residence times or high Gz. Looking at Equation 2-17, it is evident that a similar effect is expected for long residence times (or low values of Gz).

III. NUMERICAL SOLUTIONS

This section presents two numerical schemes capable of predicting the mass transfer behavior in the CFE device as described in Section II in terms of the moving reference frame derived in Section II.D. The first method is a finite difference scheme which was previously shown to be capable of describing mass transport over the parameter space of interest[4] by comparison to analytical solutions available for the limiting conditions of high[3] and low[2] Graetz number. The second method presented is based on the method of lines. It too is applicable over the required parameter space and has the advantage of greatly reducing the required computational time.

A. FINITE DIFFERENCE SCHEME

The approach used here is similar to one previously used by Biscans et al.[13] This method, however, is not limited by the restrictions of the Biscans solution (small sample input stream, high Gz),[4] and our numerical approach does not require an iterative numerical solution.

For this scheme it is recognized that the lack of an axial diffusion term allows us to treat the axial dimension x^* as "time-like", and Equation 2-19 is thus parabolic. This suggests the use of an alternating direction implicit (ADI) method. The ADI method consists of discretization in the time-like variable and manipulation of the resulting equations to decouple the two remaining dimensions. These operations reduce a two-dimensional problem to a series of one-dimensional problems.[14] Here a modified version of an ADI scheme proposed by Peaceman and Rachford[15] is used to find the numerical solution to Equation 2-19. The Thomas algorithm is then used to solve the resulting tridiagonal system of equations.[14]

This numerical scheme, although perhaps not optimal, has some definite advantages: it is unconditionally stable, has well-defined limitations, is easy to implement, and does not require computer software other than any standard programming language. It also permits the decoupling of the effects of the three variables, eliminating the need for an iterative method.

1. Alternating Direction Implicit Scheme

We begin by rewriting Equation 2-19 in the form[14]

$$\frac{\partial c *}{\partial x *} = A_1 c * + A_2 c *$$

(2-25)

where A_1 and A_2 are linear operators, defined as

$$A_1 c* = v_x^{*-1} \frac{\partial^2 c *}{\partial \zeta^2} - v_x^{*-1} v_\zeta^* \frac{\partial c *}{\partial \zeta^2}$$

(2-26)

and

$$A_2 c* = v_x^{*-1} \frac{\partial^2 c *}{\partial y *^2}$$

(2-27)

By using a Taylor series expansion in the axial dimension and rearranging the equation, an expression which is second-order accurate is obtained:[14]

$$\left(1 - \frac{\Delta x *}{2} A_1\right)\left(1 - \frac{\Delta x *}{2} A_2\right) c * \left(x_{j+1}^*, y*, \zeta\right)$$

$$= \left(1 + \frac{\Delta x *}{2} A_1\right)\left(1 + \frac{\Delta x *}{2} A_2\right) c * \left(x_j^*, y*, \zeta\right)$$

(2-28)

where x_j^* and x_{j+1}^* are the discretized values of $x*$ at steps j and $j + 1$, and $x_{j+1}^* = x_j^* + \Delta x *$. If A_{1h} and A_{2h} are then defined as finite difference approximations of A_1 and A_2, respectively, Equation 2-28 can be "decomposed" into the following two-step scheme proposed by Peaceman and Rachford:[15]

$$\left(1 - \frac{\Delta x *}{2} A_{1h}\right)\tilde{C}^{j+1/2}(m, n) = \left(1 + \frac{\Delta x *}{2} A_{2h}\right) C^j(m, n)$$

(2-29)

$$\left(1 - \frac{\Delta x *}{2} A_{2h}\right) C^{j+1}(m, n) = \left(1 + \frac{\Delta x *}{2} A_{1h}\right)\tilde{C}^{j+1/2}(m, n)$$

(2-30)

where $C^j(m,n)$ and $C^{j+1}(m,n)$ are approximations of $c*(x_j^*, y_m^*, \zeta_n)$ and $c*(x_{j+1}^*, y_m^*, \zeta_n)$, respectively, and $\tilde{C}^{j+1/2}(m,n)$ is an intermediate variable. The variables y_m^* and ζ_n are the discretized variables at steps m and n, respectively.

Once the concentration profile on a given x^* plane (initially, the $x^* = 0$ boundary condition) is known, the right side of Equation 2-29 is known and is used to calculate the values on the left side. Then Equation 2-30 is used in an analogous manner.

2. Implementation of Finite Difference Scheme

In order to implement this finite difference scheme, finite difference expressions are needed for A_{1h} and A_{2h}. For this purpose the second-order finite difference approximations are chosen, given by the equations

$$A_{1h}C^j(m,n) = v_x^{*-1}\left(\frac{C^j(m,n+1) - 2C^j(m,n) + C^j(m,n-1)}{(\Delta\zeta)^2}\right)$$

$$-v_x^{*-1}v_\zeta^*\left(\frac{C^j(m,n+1) - C^j(m,n-1)}{2\Delta\zeta}\right) \tag{2-31}$$

$$A_{2h}C^j(m,n) = v_x^{*-1}\left(\frac{C^j(m+1,n) - 2C^j(m,n) + C^j(m-1,n)}{(\Delta y^*)^2}\right) \tag{2-32}$$

where Δy^* and $\Delta\zeta$ are the step sizes in y^* and ζ, respectively, giving a tridiagonal system that can be solved easily using the Thomas algorithm or other similar algorithm.[14]

The y^* range for the finite difference scheme is defined by $y_0^* \le y_m^* \le y_M^*$ where $y_0^* = -1$ and $y_M^* = 1$. The range for ζ is $\zeta_0 \le \zeta_n \le \zeta_N$ where the boundaries at ζ_0 and ζ_N are set such that they do not significantly impact the solution.

The boundary conditions can then be written as

1. $C^j(m,0) = C^j(m,N) = 0$
2a. $-3C^j(0,n) + 4C^j(1,n) - C^j(2,n) = 0$
2b. $-3C^j(M,n) + 4C^j(M-1,n) - C^j(M-2,n) = 0$

with the "initial" condition

3. $C^j(m,n) = 1$, $x^* = 0$, $-1 \le y^* \le 1$, $\zeta = 0$

which approximates the impulse mass input shown in Equation 2-5. This initial condition can be easily adapted to account for alternative input conditions.

Boundary condition 1 approximates the "infinite boundary" condition given for Equation 2-19. Boundary conditions 2a and 2b are second-order, one-sided approximations for boundary condition 2.

In order for the solution of this numerical scheme to show the correct convective-dispersive character, the step size in ζ is limited to[14]

$$\Delta\zeta \le 2v_\zeta^{*-1} \tag{2-33}$$

This condition places a lower limit on the time it takes to determine the solution by imposing a rather large minimum size for the matrix of concentrations.

B. SOLUTION BY THE METHOD OF LINES

An alternative to the finite-difference approach described above is the method of lines (MOL), which is presented in this section. Presented first is an explanation of the basic method, followed by brief outlines of the schemes used for the axial integration and lateral discretization. Finally, the implementation of this approach using a numerical package called PDESAC is discussed.

1. Basic Approach

The method of lines is applicable to the solution of initial-boundary-value problems, which involve a time-like coordinate and one or more spatial coordinates. In essence, this method consists of first discretizing the governing equations and boundary conditions in the spatial coordinates alone. The only derivatives then remaining in the resulting set of coupled equations are with respect to the time-like coordinate. This system of ordinary differential equations (ODEs) now can be integrated forward in the time-like coordinate from a specified initial state. Note that upon discretization most spatial boundary conditions will yield algebraic equations that have to be solved simultaneously with the ODEs. In the present problem the axial position serves as the time-like independent variable, and it will be referred to as such in the discussion below.

The methods used for spatial discretization will be looked at first. When dealing with two or more position variables, an effective strategy is to employ a "local" method, such as finite differences, in one spatial coordinate and to adopt a more global approach, such as certain forms of the method of weighted residuals, in the other coordinate(s). For the latter, experience suggests that it often suffices to use a relatively small number of basis functions in a trial expansion for the unknown variable if no locally sharp spatial gradients are expected in the relevant coordinate.[16] This approach has the advantage of generating a system of equations that has both a relatively small bandwidth and a manageable size for practical calculations.

In this work the above scheme is followed by using centered finite differences in the migration direction (with a uniform mesh for simplicity) and orthogonal collocation in the lateral direction. This strategy was suggested by the work of Wang and Stewart[17] on Taylor dispersion in Poiseuille flow. The success of this approach on two other transport problems[18,19] led to its application to continuous-flow zone electrophoresis.[11]

Finally we turn to the integration of the ODEs generated following spatial discretization. It is generally recognized that accurate and stable integration in a time-like coordinate can be done quite efficiently with an implicit, variable-order, multistep method.[20] Several integrators of this kind are now available as canned packages for nonlinear initial-value problems involving a coupled set of algebraic equations and first-order ODEs.[21,22] Most of these select the number and sizes of the time steps automatically and adaptively in order to keep the local error in the integration below a user-specified tolerance.

Having explained the method of lines, a description of the use of orthogonal collocation to prepare the problem at hand for solution by the package PDESAC is presented.

2. Orthogonal Collocation

Collocation consists of fitting a trial solution to the governing equations at a selected set of points. The collocation nodes are chosen on the basis of an optimality criterion and are often the zeroes of a member of an orthogonal set of polynomials; the method is then called orthogonal collocation.[23,24]

It is usually convenient to build the expected symmetry of the unknown solution into its trial form. For the present problem, the lateral symmetry (about the plane $y^* = 0$) that is guaranteed by the governing equation and boundary conditions is exploited. Hence, the lateral coordinate y^* is replaced by a new symmetry-adapted variable, $\eta = y^{*2}$. (Note that this transformation will not be appropriate if the velocity profiles or the input condition are not themselves symmetric; for example, temperature gradients across the chamber can potentially disturb the velocity profiles from the symmetric form here.) For simplicity, the notation c^* will be retained for the concentration distribution in the coordinates (x^*, η, ζ).

N lateral collocation points, labeled $\eta_i (i = 1,2,...,N)$, are chosen between 0 and 1, and the node $\eta_{N+1} = 1$ is added in order to account for the boundary condition at the wall. These points are selected here to be zeroes of the Jacobi polynomial $P_N^{(\alpha,\beta)}(\cdot)$, with $\alpha = 1$ and $\beta = -0.5$; this choice is dictated by the form of the integral in the flow-averaged concentration that is finally sought.[25]

The specific form of orthogonal collocation employed here is based on Lagrange interpolation[26] and leads to the so-called method of ordinates.[23] First the concentration distribution is expressed in terms of its lateral ordinates $c_j^*(x^*,\zeta) \equiv c^*(x^*,\eta_j,\zeta)$ ($j = 1,2,...,N + 1$) in the trial form

$$c^*\left(x^*,\eta,\zeta\right) \doteq \sum_{j=1}^{N+1} \ell_j(\eta)c_j^*\left(x^*,\zeta\right) \quad \text{on } 0 < \eta \le 1 \tag{2-34}$$

where the quantities $\ell_j(\cdot), j = 1,2,...,N + 1$, are "global" basis functions known as Lagrange interpolation polynomials. It is thus seen that derivatives of the trial solution with respect to η evaluated at the collocation nodes turn into weighted sums over the ordinates.

Upon forcing this trial solution to satisfy the governing equation (Equation 2-19) at the N interior nodes $\{\eta_i; i = 1,2,...,N\}$, the following set of coupled partial differential equations (PDEs) for the N interior ordinates is obtained:

$$v_x^* \frac{\partial c_i^*}{\partial x^*} + v_\zeta^* \frac{\partial c_i^*}{\partial \zeta} = \frac{\partial^2 c_i^*}{\partial \zeta^2} + \sum_{j=1}^{N+1} F_{ij} c_j^* \quad \text{for } i = 1,2,...,N \tag{2-35}$$

The no-flux boundary condition at the wall takes the form

$$\sum_{j=1}^{N+1} a_j c_j^* = 0 \tag{2-36}$$

while the flow-averaged concentration can be expressed as a weighted sum (quadrature) over the ordinates:

$$c_b^*\left(x^*,\zeta\right) = \sum_{i=1}^{N} w_i c_i^*\left(x^*,\zeta\right) \tag{2-37}$$

Note that the matrix $\mathbf{F}$, the vectors $\mathbf{a}$ and $\mathbf{w}$, and the nodes $\{\eta_1, \eta_2,...,\eta_N\}$ can be calculated quite efficiently by means of standard routines available, for example in the book by Villadsen and Michelsen.[25]

The rapid-decay boundary conditions at $\zeta = \pm\infty$ (condition 1 for Equation 2-19) now apply to every ordinate. The initial state with respect to the axial coordinate can be set up by evaluating both sides of a general input condition, as in condition 3 below Equation 2-4, at each of the N interior collocation nodes. We shall see in the next section that three to five interior nodes are usually sufficient for a uniform sample input in the lateral direction. When the input sample occupies only a part of the chamber thickness or is otherwise not uniform in the lateral coordinate, more collocation points may be necessary in order to capture the shape of the initial profile accurately. As a general rule, when the number of required collocation points exceeds about ten, it may be advantageous to discretize in the lateral dimension via collocation on finite elements (rather than on "global" basis functions like the Lagrange polynomials used here) or even via finite differences. It should be noted, however, that the shape of the input condition affects the final profile only for moderate to high values of Gz.

3. Implementation Using PDESAC

Finally, the actual implementation of the method of lines using the software package PDESAC (Parabolic Differential Equation Sensitivity Analysis Code) is outlined. This routine solves initial-boundary-value problems written as a set of coupled (parabolic) PDEs in one time-like variable and one position.[20] PDESAC automatically sets up a centered finite-difference scheme in the position variable (on a user-specified grid) for all the PDEs in the set and then does the subsequent time-like integration by means of an extended version of the stiff, implicit integrator DASSL.[21] Details of the underlying procedure can be found from the authors of PDESAC.[20] Other competitive numerical packages, such as the parabolic PDE system solver by the Numerical Algorithms Group (NAG),* are also available.

The procedure for combining the orthogonal collocation scheme described above with the routine PDESAC is as follows. First, the collocated boundary condition (Equation 2-36) is used to eliminate the ordinate at the wall, $c_{N+1}^*(x^*,\zeta)$, from the N interior equations (Equation 2-35) in favor of the N interior ordinates $\{c_i^*(x^*,\zeta), i = 1,2,\ldots,N\}$. These N PDEs are now simultaneously integrated forward in the time-like variable x^*, starting from the "initial" state, by means of PDESAC to determine the N interior ordinates on a uniform grid in the separation coordinate ζ. The bulk concentration $c_b^*(x^*,\zeta)$ then can be calculated by means of Equation 2-37 at each successive desired value of axial position.

IV. RESULTS

The capability of the finite difference scheme to describe the mass transfer behavior presented in Section II has been described previously, so this section focuses on two points: the number of collocation points N necessary in the method of lines to achieve the same result as predicted by the finite difference scheme, and, more importantly, the behavior to expect in the system as a function of the critical parameters Gz, Pe$_{ef}$, and k. Table 2-1 lists estimates of practical ranges for these parameters; the ranges were determined using parameter values and concentration profiles given in references, as well as estimates of solution properties.[4] These ranges were used as guidelines.

First the effect of Gz on the mass transfer behavior will be explored. The results of a series of computer simulations for a single component, injected at $z^* = 0$, are presented in Figures 2-2 to 2-6 in the stationary frame. In these figures the value of Gz is decreased from 100 down to 1. This variation can be accomplished experimentally by gradually increasing the residence time, either by using chambers of increasing length or by decreasing the flow rate through a chamber of a given length. The remaining parameters are held constant at values of Pe$_{ef}$ = 50 and $k = 0$ (no electroosmosis).

Table 2-1 **Experimental parameter ranges**

Parameter	Approximate Range
Gz	1–100
$|Pe_{ef}|$	10–1000
$|k|$	0.1–1

---

* Numerical Algorithms Group (NAG), 1101 31st St., Suite 100, Downers Grove, IL 60615-1263.

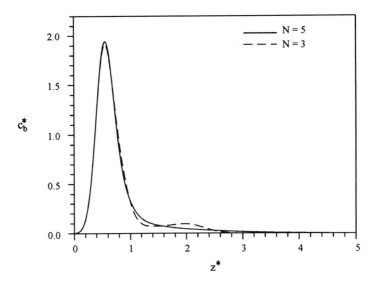

Figure 2-2 Comparison of the bulk averaged concentration c_b^*, calculated using the method of lines for $N = 3, 5$. $Gz = 100$, $Pe_{ef} = 50$, $k = 0.0$. See Nomenclature for symbol definitions.

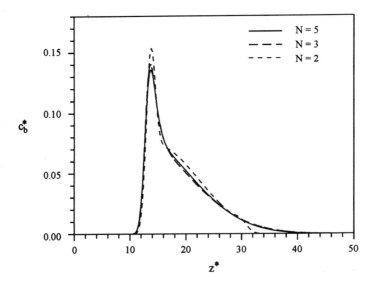

Figure 2-3 Comparison of the bulk averaged concentration c_b^*, calculated using the method of lines for $N = 2, 3, 5$. $Gz = 4$, $Pe_{ef} = 50$, $k = 0.0$. See Nomenclature for symbol definitions.

Figures 2-2 to 2-5 describe the bulk concentrations c_b^*, defined in Equation 2-12, as a function of chamber position in the stationary frame (z^*) for each value of Gz. These results are shown for the method of lines given in Section III.B, using various numbers of collocation points (N). Note that the concentration profile predicted for $N = 5$ is indistinguishable from the result obtained using the finite difference scheme presented in

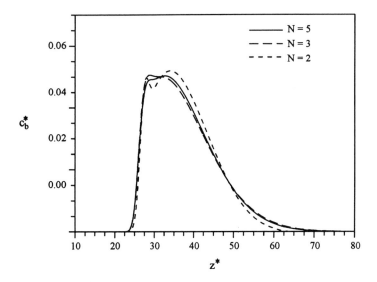

Figure 2-4 Comparison of the bulk averaged concentration c_b^*, calculated using the method of lines for $N = 2, 3, 5$. Gz = 2, $Pe_{ef} = 50$, $k = 0.0$. See Nomenclature for symbol definitions.

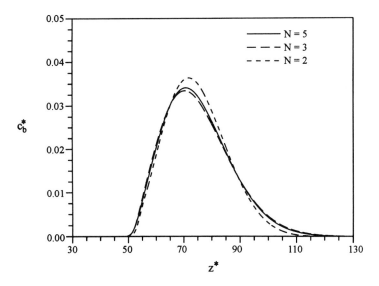

Figure 2-5 Comparison of the bulk averaged concentration c_b^*, calculated using the method of lines for $N = 2, 3, 5$. Gz = 1, $Pe_{ef} = 50$, $k = 0.0$. See Nomenclature for symbol definitions.

Section III.A. For $N = 2$ and $N = 3$ the profiles are not as accurate, but even these results are quite good.

In Figure 2-2, Gz = 100; at this high value of Gz the concentration profile seen is primarily a function of hydrodynamics, giving a purely "convective peak". The peak is sharp-fronted and has a long tail, as discussed in Section II.E.1.

48

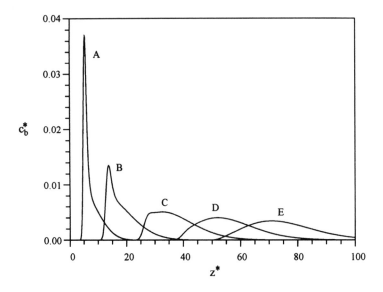

Figure 2-6 Effect of the dimensionless length of the chamber Gz on the bulk averaged concentration c_b^*. Curves are for Gz = (A) 10, (B) 4, (C) 2, (D) 1.3, (E) 1. Pe_{ef} = 50, k = 0.0. See Nomenclature for symbol definitions.

In Figure 2-3, Gz = 4 and the solute peak shape is no longer defined solely by hydrodynamic forces. The hint of a double peak is now apparent, with a broadening effect at the higher z^* values (i.e., farther from the injection location, $\zeta^* = 0$). In Figure 2-4, with Gz = 2, the situation is reversed, with dispersive effects dominating and the convective peak now appearing as a shoulder on the predominant dispersive peak. In Figure 2-5, Gz = 1 and the concentration profile closely approaches a purely dispersive shape; the convective character is no longer evident.

Figure 2-6 summarizes the progression of the concentration profile from the convection-dominated regime to the dispersion-dominated regime. The experimental parameters are the same as in Figures 2-2 to 2-5, with Gz now ranging from 10 to 1. It is interesting to note that the "double peak" seen at intermediate values of Gz is similar to that sometimes seen in chromatography[27] and in other mathematically similar systems.[28]

The concentration profiles described above are all for a single species undergoing electrophoresis. Figure 2-7 shows the concentration profiles for species with different electrophoretic mobilities — Pe_{ef} = 23, 50, or 53 — with Gz = 1.3 and k = 0.0. It is apparent that only one species can be purified to any significant extent under these conditions. Also, the effect of convective dispersion on the spreading of the peak is apparent, with the components with higher mobility having the broader peaks, as expected when there is no electroosmosis.

Although the preceding profiles were calculated using conditions of no electroosmosis (k = 0), electroosmosis does occur under normal experimental conditions, and it is thus important to be able to include its effects in any description of the system. Figure 2-8 shows the results obtained for k = 0.0, –0.5, and –1.0, with Gz = 1.3 and Pe_{ef} = 23. As was predicted using Equation 2-17 for low Gz, dispersion is minimized as the value of the electrophoretic velocity approaches a value equal and opposite to the electroosmotic velocity ($k \rightarrow -1$).

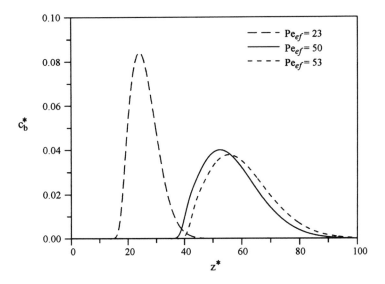

Figure 2-7   Comparison of the effect of electrophoretic velocity on the bulk averaged concentration profile c_b^*, calculated using the finite difference scheme, for $Pe_{ef} = 23$, 50, and 53. $Gz = 1.3$, $k = 0$. See Nomenclature for symbol definitions.

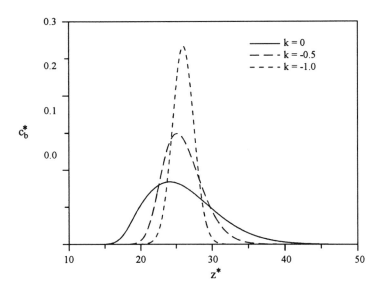

Figure 2-8   Comparison of the effect of electroosmotic flow on the bulk averaged concentration c_b^*, calculated using the finite difference scheme for $k = 0.0, -0.5, -1.0$. $Gz = 1.3$, $Pe_{ef} = 23$. See Nomenclature for symbol definitions.

Thus far we have shown the mass transfer behavior of the system over a range of the key governing parameters cited in Section II.E: Gz, Pe_{ef}, and k. In Figures 2-9 and 2-10 the initial condition imposed on the system is relaxed. Specifically, the condition

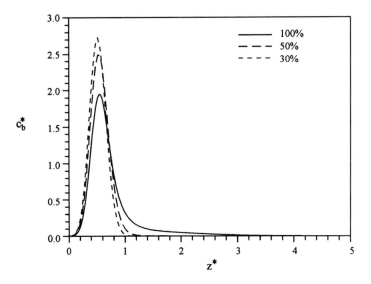

Figure 2-9 Comparison of the effect of sample application on the bulk averaged concentration profile c_b^*, calculated using the finite difference scheme, for $\alpha = 100\%$, 50%, 30%. $\mathrm{Pe}_{ef} = 50$, $\mathrm{Gz} = 100$, and $k = 0$. See Nomenclature for symbol definitions.

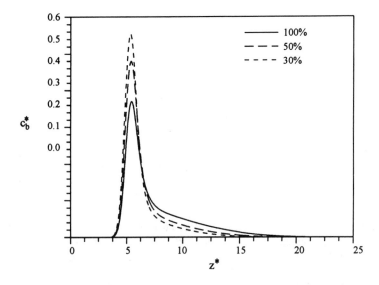

Figure 2-10 Comparison of the effect of sample application on the bulk averaged concentration profile c_b^*, calculated using the finite difference scheme, for $\alpha = 100\%$, 50%, 30%. $\mathrm{Pe}_{ef} = 50$, $\mathrm{Gz} = 10$, and $k = 0$. See Nomenclature for symbol definitions.

$$c*\left(x*, y*, z*\right) = \delta(z*), \quad x* = 0, \quad -1 \le y* \le 1 \tag{2-38}$$

is replaced by the more general condition

$$c*\left(x*, y*, z*\right) = \delta(z*), \quad x* = 0, \quad -\alpha \le y* \le \alpha \tag{2-39}$$

where $\alpha \le 1$.

In Figures 2-9 and 2-10, Gz = 100 and 10, respectively, with $Pe_{ef} = 50$ and $k = 0$. Results obtained using the finite difference calculations are shown for $\alpha = 100\%$, 50%, and 30%. It is apparent that by restricting the sample to a narrower region the effects of tailing lessen. This effect is diminished as the value of Gz decreases because diffusion will "force" material toward the walls and therefore reduce the improvement seen in the solute profile. It is important to note that the improvement realized by narrowing the sample band has a price — the total throughput of material is greatly decreased.

V. DISCUSSION

This chapter has demonstrated the application of numerical methods for the quantitative description of mass transfer behavior in the CFE device. Analytical solutions, such as those of Ivory[3] and Reis et al.,[2] have the general advantage of ease of use; however, these solutions are applicable only under limiting conditions.[4] The use of the inappropriate model will give an error not only in the peak location, but in the peak shape as well. The numerical methods presented here are more generally applicable and, thus, can be used to describe the mass transfer behavior over the useful range of operating parameters.

The finite difference scheme of Section III.A is simple to program, unconditionally stable with respect to parameter values, and has no numerical limitation other than that stated in Equation 2-33. This scheme also has the advantage of not requiring additional computer software. It can be easily modified to incorporate nonuniform, nonrectangular sample inputs and to describe behavior under nonisothermal conditions, which are primarily manifested as changes in the velocity profiles. However, the inclusion of nonlinear concentration effects will require an iterative method for solution.

The method of lines of Section III.B has a significant advantage in terms of its speed of computation. Nonlinear concentration effects and other velocity profiles do not require any special handling. However, for sample inputs occupying only a fraction of the thickness of the chamber, this method may require an increased number of collocation points. While the implementation in this work used a specific software package (PDESAC) for solving the equations resulting from orthogonal collocation in the lateral dimension, other widely available routines also could be used. The use of collocation does involve some trial and error in order to find the minimum number of collocation points required to obtain a desired accuracy.

When using numerical methods the computation time required is typically an area of concern. In applying the finite difference scheme no real optimization of the numerical method or computer code was performed. It is useful to note, however, that for a sample run, such as can be seen in Figure 2-7 or 2-8, the code took 9.5 cpu hours to execute on a VAX station 3100 (Digital Equipment Corporation). This run was composed of 1000 steps in $x*$ and 40 and 1750 points in $y*$ and $z*$, respectively. In these particular runs, we did not take advantage of the moving frame formulation, which would have reduced the required computation time by about a factor of three. For a similar run using the method of lines, the computation time on the same computer system was approximately 6 cpu minutes for $N = 3$ and 15 cpu minutes for $N = 5$, without significant loss in accuracy for the faster run.

NOMENCLATURE

c	=	concentration of species of interest
c^*	=	dimensionless concentration, dc/c_o
c_{avg}^*	=	average dimensionless concentration
c_b^*	=	bulk, or flow-averaged, dimensionless concentration
c_0	=	mass input per unit area
d	=	half-thickness of the chamber
D	=	effective binary diffusivity of species of interest
Gz	=	Graetz number, Ped/L
k	=	relative electroosmotic velocity, v_{eo}/v_{ef}
L	=	length of the electrode region
$M_n(x^*)$ =		nth spatial moment of flow-averaged concentration c_b^*
N	=	number of collocation nodes in the lateral direction
Pe	=	Péclet number, $v_o d/D$
Pe_{ef}	=	electrophoretic Péclet number, $v_{ef}d/D$
Pe_x	=	axial Péclet number, $v_0 L/D$
v_o	=	maximum axial velocity in the chamber
v_{ef}	=	electrophoretic velocity of species of interest
v_{eo}	=	electroosmotic velocity
v_x	=	axial or x-component of the buffer velocity
v_x^*, v_z^*	=	dimensionless velocities
v_z	=	z-component of species velocity (stationary frame)
v_ζ^*	=	ζ-component of species velocity (moving frame)
w	=	width of the chamber
x,y,z	=	rectangular coordinates
x^*	=	$(x/d)Pe^{-1}$
y^*,z^*	=	$y/d, z/d$
α	=	percent chamber thickness used for sample application
$\delta(z)$	=	unit impulse function
η	=	symmetry-adapted lateral coordinate, $\eta = y^{*2}$
σ^{*2}	=	scaled spatial variance

REFERENCES

1. **Hannig, K.,** Free-flow electrophoresis, in *Methods in Microbiology,* Norris, J. R. and Ribbons, D. W., Eds., Academic Press, New York, 1971, chap. 8.
2. **Reis, J., Lightfoot, E. N., and Lee, H.-L.,** Concentration profiles in free-flow electrophoresis, *AIChE J.,* 20, 362, 1974.
3. **Ivory, C. F.,** Continuous flow electrophoresis. The crescent phenomena revisited. I. Isothermal effects, *J. Chromatogr.,* 195, 165, 1980.
4. **Grateful, T. M. and Lightfoot, E. N.,** Finite difference modelling of continuous-flow electrophoresis, *J. Chromatogr.,* 594, 341, 1992.
5. **Clifton, M. J., Jouve, N., de Balmann, H., and Sanchez, V.,** Conditions for purification of proteins by free-flow zone electrophoresis, *Electrophoresis,* 11, 913, 1990.
6. **Bier, M., Palusinski, O. A., Mosher, R. A., and Saville, D. A.,** Electrophoresis: mathematical modeling and computer simulation, *Science,* 219, 1281, 1983.
7. **Rhodes, P. H., Snyder, R. S., and Roberts, G. O.,** Electrohydrodynamic distortion of sample streams in continuous flow electrophoresis, *J. Colloid Interface Sci.,* 129(1), 78, 1989.

8. **Frankel, I. H. and Brenner, H.,** On the foundations of generalized Taylor dispersion theory, *J. Fluid Mech.,* 204, 97, 1989.

9. **Brenner, H.,** Macrotransport processes, *Langmuir,* 6(12), 1715, 1990.

10. **Aris, R.,** On the dispersion of a solute in a fluid flowing through a tube, *Proc. R. Soc. London,* A235, 67, 1956.

11. **Athalye, A. M.,** University of Wisconsin, unpublished data, July 1992.

12. **Strickler, A. and Sacks, T.,** Focusing in continuous-flow electrophoresis systems by electrical control of effective cell wall zeta potentials, *Ann. N.Y. Acad. Sci.,* 209, 497, 1973.

13. **Biscans, B., Alinat, P., Bertrand, J., and Sanchez, V.,** Influence of flow and diffusion on protein separation in a continuous flow electrophoresis cell: computation procedure, *Electrophoresis,* 9, 84, 1988.

14. **Strikwerda, J. C.,** *Finite Difference Schemes and Partial Differential Equations,* Wadsworth and Brooks/Cole Advanced Books and Software, Pacific Grove, CA, 1989.

15. **Peaceman, D. W. and Rachford, H. H., Jr.,** The numerical solution of parabolic and elliptic differential equations, *J. Soc. Ind. Appl. Math.,* 3, 28, 1955.

16. **Finlayson, B. A.,** *The Method of Weighted Residuals and Variational Principles,* Academic Press, New York, 1971.

17. **Wang, J. C. and Stewart, W. E.,** New descriptions of dispersion in flow through tubes: convolution and collocation methods, *AIChE J.,* 29(3), 493, 1983.

18. **Athalye, A. M.,** Effects of High Sample Loading on Nonadsorptive Protein Chromatography, Ph.D. thesis, University of Wisconsin, Madison, 1993.

19. **Raths, K. R., Athalye, A. M., and Lightfoot, E. N.,** Diffusion and flow in stacked adsorptive membrane devices, manuscript in preparation.

20. **Stewart, W. E., Caracotsios, M., and Sørensen, J. P.,** *Computational Modelling of Reactive Systems,* Butterworths, Reading, MA, manuscript in preparation.

21. **Petzold, L. R.,** A Description of DASSL: A Differential/Algebraic System Solver, Technical Rep. 82-8637, Sandia, 1982.

22. **Press, W. H., Teukolsky, S. A., Vetterling, W. T., and Flannery, B. P.,** *Numerical Recipes: The Art of Scientific Computing,* Cambridge University Press, New York, 1992.

23. **Villadsen, J. V. and Stewart, W. E.,** Solution of boundary-value problems by orthogonal collocation, *Chem. Eng. Sci.,* 22, 1483, 1967.

24. **Stewart, W. E.,** Simulation and estimation by orthogonal collocation, *Chem. Eng. Educ.,* 18(4), 204, 1984.

25. **Villadsen, J. V. and Michelsen, M. L.,** *Solution of Differential Equation Models by Polynomial Approximation,* Prentice-Hall, Englewood Cliffs, NJ, 1978.

26. **Abramowitz, M. and Stegun, I., Eds.,** *Handbook of Mathematical Functions,* Dover Publications, Mineola, NY, 1972.

27. **Weber, S. G. and Carr, P. W.,** The theory of the dynamics of liquid chromatography, in *High Performance Liquid Chromatography,* Brown, P. R. and Hartwick, R. A., Eds., John Wiley & Sons, New York, 1984, 1.

28. **Lenhoff, A. M. and Lightfoot, E. N.,** Convective dispersion and interphase mass transfer, *Chem. Eng. Sci.,* 41(11), 2795, 1986.

II
Improvements of the Method

Chapter 3

Theoretical and Experimental Studies on the Stabilization of Hydrodynamic Flow in Free Fluid Electrophoresis

Percy H. Rhodes and Robert S. Snyder

CONTENTS

I. INTRODUCTION

Continuous flow electrophoresis (CFE) has significant potential for high-resolution separation of particles, cells, and macromolecules on a preparative scale. It is a bulk zone electrophoresis technique which takes place within a flowing curtain of aqueous electrolyte buffer in a rectangular chamber. Figure 3-1 schematically shows a typical system.

The buffer fluid flows in the x (axial) direction, while a uniform electric field in the z (lateral) direction is maintained by electrode systems at the sides. A narrow gap in the y (transverse) direction exists between the broad faces of the chamber. Sample (with proteins, cells, or particles dispersed or dissolved in buffer) is continuously injected through an inflow port (nozzle) at the center plane of the curtain flow, as a narrow circular filament, and is fractionated by the electric field. The fractionated sample components are subsequently collected in a uniform array of collection ports at the fluid exit of the chamber.

For many years it was thought that CFE performance was limited mainly by buoyancy-driven thermal convection. In accord with this premise, microgravity space electrophoresis experiments were performed on Apollo,[1] in the Apollo-Soyuz Test Project (ASTP),[2] and on the space shuttle.[3] The series of CFE experiments performed on the space shuttle by the McDonnell-Douglas Aeronautics Company showed that there was significant hydrodynamic separation degradation even in the microgravity environment of space. The CFE experiments on STS-6 and STS-7[4,5] showed clear evidence of sample spreading.

Strickler and Sacks,[6] in an experimental and theoretical study, showed how a circular sample was distorted into a crescent shape (illustrated in Figure 3-1) under the combined

0-8493-8918-6/94/$0.00+$.50
© 1994 by CRC Press, Inc.

57

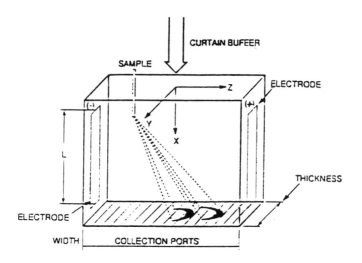

Figure 3-1 Scheme of a continuous flow electrophoresis chamber. (From Rhodes, P.H., Snyder, R.S., Roberts, G.O., and Baygents, J.C., *Appl. Theor. Electrophoresis*, 213, 87, 1991. With permission.)

effects of wall electroosmosis and the Poiseuille flow profile down the chamber. They proposed a rather elaborate method to control or "focus" these disturbances. While it was known that additional sample stream distortions occurred under the action of an electric field, the cause of these disturbances was not known.

Rhodes et al.[7] first demonstrated the importance of electrohydrodynamic (EHD) flows in CFE. Until this time it was universally accepted that the reason for the discrepancy in actual and predicted separation performance was buoyancy-driven thermal convection, which would be eliminated by operation in a microgravity environment. In addition, it was argued that thick chambers could be used in space to reduce sample band spreading attributed to "wall effects" characteristic of the thin chambers on Earth. This rationale was supported by the fact that electrophoresis in porous gels had become the standard technique to separate and identify the various protein constituents in a complex mixture. The porous gel effectively eliminated flow due to thermal convection, and sedimentation was not a problem as long as the proteins were soluble in the electrophoresis buffer. Porous gels could not be used for biological cells because the limiting pore size was an order of magnitude too small.

This chapter will first look at the theoretical and experimental studies as pertain to the role of buoyancy-induced thermal convection on CFE performance. Then the nongravitational factors which affect CFE operations — mainly EHD — will be discussed.

II. ROLE OF CONVECTION IN CONTINUOUS FLOW ELECTROPHORESIS PERFORMANCE

Saville[8] extensively studied the role of buoyancy-induced thermal convection in the operation of both narrow- and wide-gap CFE flow chambers. He found that buoyancy effects appeared if the structure of the steady rectilinear flow in the CFE chamber was altered. The relative roles of the buoyancy and viscous forces are given by the dimensionless group

$$N_3 = \frac{g\beta\sigma_0 E_0^2 d^4}{2k_0\nu_0 u_0} \tag{3-1}$$

where β is the coefficient of thermal expansion, v_0 is the kinematic viscosity, u_0 is the average axial velocity, k_0 is the thermal conductivity of the buffer, E_0 is the electric field strength, σ_0 is the buffer electrical conductivity, and d is the buffer flow chamber half-thickness (y-direction dimension). Changing from a narrow ($d = 0.08$ cm) to a wide gap ($d = 0.25$ cm) changes N_3 from 0.24 to 30, showing a vastly increased role for bouyancy.

Bouyancy-induced instabilities can arise from either lateral or vertical temperature gradients. The onset of the instability that leads to this flow is governed in part by the size of the Grashof number,

$$G_r = \frac{g\beta\sigma_0 E_0^2 d^5}{2k_0 v_0^2}$$

(3-2)

which, although it is less than unity for a narrow-gap device, increases dramatically due to the d^5 factor.

Axial temperature gradients arise from uneven heating or cooling and from entrance effects associated with the development of the temperature field. A vertically oriented chamber with downward flow (as in Figure 3-1) sometimes leads to a configuration where cold fluid lies above warmer fluid, and the instability in this case arises when the Rayleigh number

$$Ra = \frac{g\beta\dfrac{dT}{dx} d^4}{v_0 \alpha_0}$$

(3-3)

exceeds a critical value. Here α_0 is the thermal diffusivity of the buffer fluid.

The Rayleigh number characterizing the entrance effect can be given in terms of the operating conditions N_3 as

$$Ra_e = \frac{3}{2} N_3$$

(3-4)

Thus, the chamber is very sensitive to axial temperature gradients, whatever their source.

Therefore, three types of temperature gradients — i.e., axial, lateral, and transverse — can affect the flow stability of a CFE chamber. Transverse gradients result from joule heating in the chamber and are characterized by the temperature difference ΔT between the chamber walls and the chamber center plane

$$DT = \frac{1}{2}\frac{\sigma_0 E_0^2 d^2}{k_0}$$

(3-5)

This expression is required to balance the rate of heat generation with conduction to the walls, where essentially all heat rejection takes place. This temperature difference gives rise to dual circulations in the narrow plane at the chamber. Due to the fact that these disturbances are essentially uniform and are attenuated by the proximity of the vertical walls (wall effects), they tend to be second order disturbances. Lateral temperature gradients also occur in the chamber due to uneven heating and/or conduction at the chamber walls. Lateral temperature gradients give rise to fluid circulations in the broad plane of the chamber and, hence, influence flow as a first order effect, as the next section will show. Axial temperature gradients cause flow disruption when an unstable temperature gradient results. This can occur most readily when the chamber is operated in downflow. For this reason many CFE chambers are operated in upflow with little or no cooling required.

III. ROLE OF TEMPERATURE GRADIENTS IN CONTINUOUS FLOW ELECTROPHORESIS FLOW

Fully developed laminar flow must be maintained in the electrophoresis chamber if true separations based on electrophoretic mobility are to be realized. While this base flow is parabolic in the narrow plane (thickness) of the chamber, it is ideally uniform across the width of the plane of separation, except at the lateral edges, where it is zero due to the no-slip condition. Bouyancy-induced convection due to fluid temperature gradients will distort this uniform flow. Moreover, very small lateral gradients cause a significant deformation of the uniform base flow and mask effects due to axial gradients. This section will show the separate effects of lateral and axial gradients on the base flow.

IV. EXPERIMENTAL APPARATUS AND PROCEDURES

Figure 3-2 shows the Plexiglas® flow chamber with cooling chambers on the front and rear and the arrangement of the thermocouple probes. A total of 18 thermocouples, mounted in sets of three on movable bars which were adjusted into or out of the flow curtain, were used to define the temperature field. An additional thermocouple was mounted in the flow inlet to monitor the temperature there. The buffer entered at the top of the chamber through a single entry port and exited at the other end through a similar opening. Coolant entered the cooling chambers at the bottom and was channeled back and forth in a serpentine fashion by means of horizontal baffles in the chamber. Figure 3-3 shows the experimental apparatus, and Figure 3-1 defines the chamber orientation and reference axes.

The velocity profiles at the center plane were determined by a fluorescence method. Three small porous fibers (0.2 mm diameter) were stretched across the width of the flow chamber at the center plane, as shown in Figure 3-2. Fluorescence of the fluid medium immediately surrounding the fiber was achieved by passing an activator fluid through the fiber. Then subsequent movement of the fluorescent thread of fluid traced the velocity field in the plane of the fiber.

The data-taking sequence started by establishing the flow field in the chamber. The flow was judged when temperature changes of <0.10 °C were observed over a period equal to the average fluid residence time of the chamber. Then temperature and flow data were recorded and a sequence of photographs of the flow field was made. The photographs and printed data formed a simultaneous record of the flow events.

Saville and Ostrach[9] examined the stability of fully developed flow with an adverse axial gradient (cooler fluid above warmer fluid) using techniques in which the problem was separated into the sum of steady and disturbance flows. Symmetrical and antisymmetrical disturbance flows were found to be possible at Rayleigh numbers of about six. The results obtained here generally confirm their results. The following sequence of figures shows the significant features. On the left in each figure is a photograph of the chamber with the flow delineated by three fluorescing flow fronts. On the right is a schematic representation of the flow fronts along with the temperature as measured by the 18 thermocouples. The Rayleigh numbers are based on the temperatures shown along the center plane.

Figure 3-4 shows a flow with very small lateral temperature gradients. Note that the distortion to the base flow occurring at the top of the chamber reflects the lateral temperature distribution; flow is retarded in regions where the local temperature is above

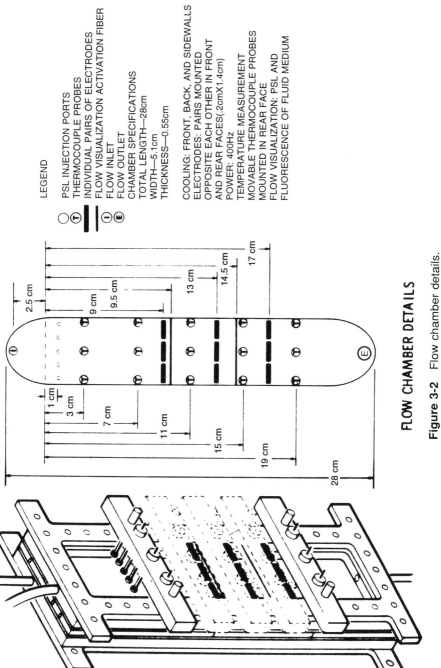

FLOW CHAMBER DETAILS

Figure 3-2 Flow chamber details.

LEGEND

PSL INJECTION PORTS
THERMOCOUPLE PROBES
INDIVIDUAL PAIRS OF ELECTRODES
FLOW VISUALIZATION ACTIVATION FIBER
FLOW INLET
FLOW OUTLET
CHAMBER SPECIFICATIONS
TOTAL LENGTH—28cm
WIDTH—5.1cm
THICKNESS—0.55cm

COOLING: FRONT, BACK, AND SIDEWALLS
ELECTRODES: PAIRS MOUNTED
OPPOSITE EACH OTHER IN FRONT
AND REAR FACES(.2cmX1.4cm)
POWER: 400Hz
TEMPERATURE MEASUREMENT
MOVABLE THERMOCOUPLE PROBES
MOUNTED IN REAR FACE
FLOW VISUALIZATION: PSL AND
FLUORESCENCE OF FLUID MEDIUM

Figure 3-3 Experimental apparatus.

the average for that particular cross section, and vice versa. Even though uniform base flow is reestablished at the middle of the chamber, the very small lateral gradient present there (0.05 °C/cm) provides a signature on the base flow profile. The flow profile at the bottom is retarded in the center and in fact may be at a point of instability due to the adverse axial temperature gradient.

Figure 3-5 shows another situation with larger lateral and axial temperature gradients. Note the significant symmetrical flow distortions of the base flow at the top and middle of the chamber; at the bottom the flow is reversed, possibly as the result of instability, as the local Rayleigh number suggests.

The very strong reversal at the top of the chamber in Figure 3-6 dominates flow throughout the chamber, as shown by the middle and lower profiles. This dominance is readily apparent once it is noted that the flows in the chamber are opposite from what one would expect based on the lateral gradients shown. In this case, therefore, axial gradients

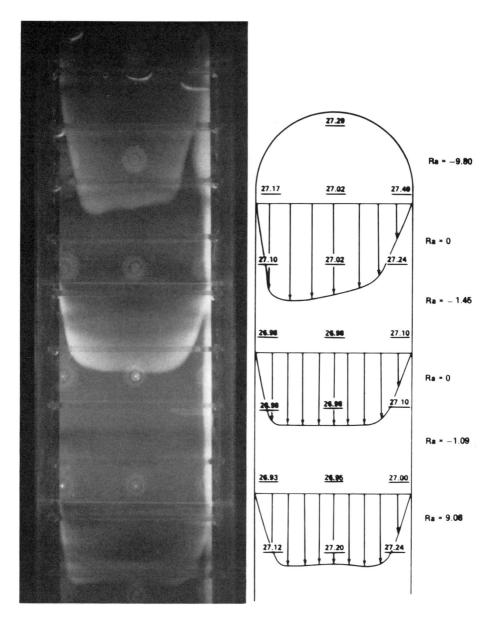

Figure 3-4 Effects of small lateral temperature gradients on the uniform base flow.

predominate and control the flow in the chamber much as the lateral gradients controlled the flow configuration shown in Figure 3-4.

In Figure 3-7 a very strong antisymmetrical flow instability dominates flow in the chamber. Note also the instability in the lower part of the chamber where the Rayleigh number is 8.33.

The results indicate that very small lateral and axial temperature gradients will deform the uniform base flow. Adverse axial gradients which give rise to Rayleigh numbers greater than critical can cause flow reversal and essentially destroy the base flow.

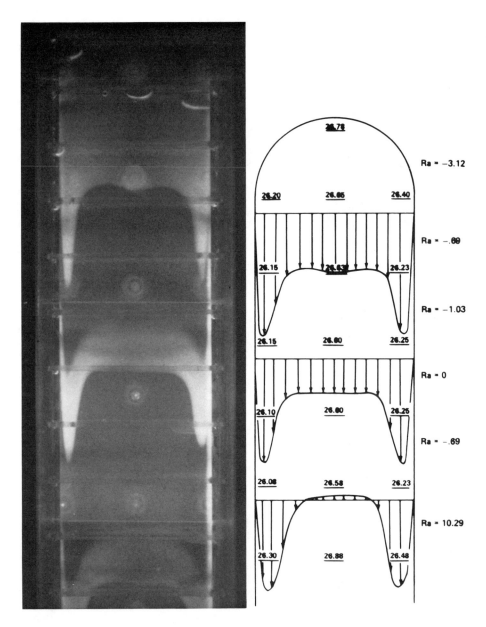

Figure 3-5 Effects of axial and lateral temperature gradients on uniform base flow.

Lateral temperature gradients influence the development of the disturbance flows, i.e., determine whether a symmetrical or antisymmetrical mode will be initiated. Furthermore, the disturbance flows themselves influence the lateral gradients.

V. ELECTROHYDRODYNAMIC EFFECTS IN CONTINUOUS FLOW ELECTROPHORESIS

It is well known that when a sample stream deviates from the chamber center plane the sample stream distribution and, hence, resolution degrade. However, it was not known

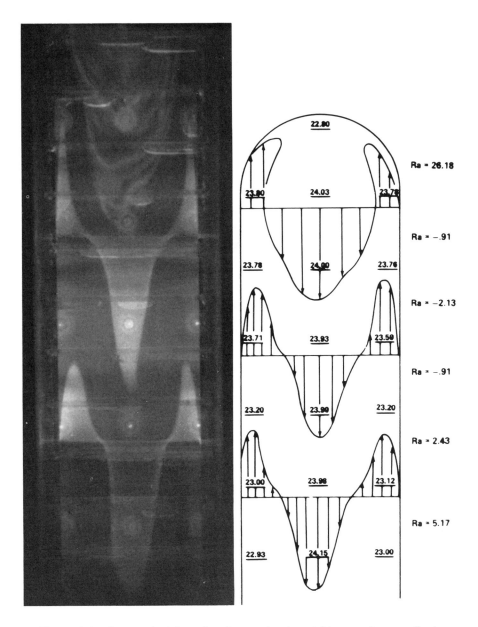

Figure 3-6 Symmetrical flow disturbance due to axial temperature gradients.

why the sample stream would deviate from the center plane once placed there at injection in stable flow. This section will show that this phenomenon is a result of EHD flows.

Experimentally, an alternating voltage was used, thus effectively eliminating electroosmosis and electrokinetics, which are linear in the applied field (see below). All effects seen were therefore electrohydrodynamic.

EHD flows are quadratic in the applied field and are associated with the charge and dipole distributions induced by the interaction of the applied field with variations of conductivity and dielectric constant. The theory predicted, and the experiments confirmed, that a low-conductivity circular sample is flattened into a ribbon perpendicular to

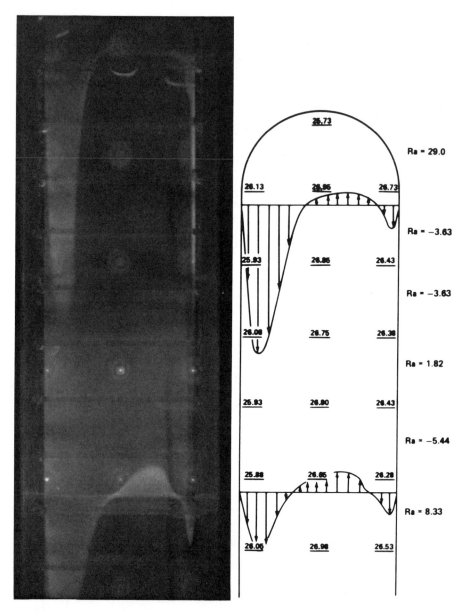

Figure 3-7 Antisymmetrical flow disturbance due to large axial gradients.

the applied field, while a high-conductivity circular sample is flattened into a ribbon aligned with the field. With direct current, the EHD flows (and the more complicated flows associated with more complex conductivity distributions) are superimposed on the electroosmosis and electrokinetic effects.

For a circular sample stream with interior conductivity σ_i and dielectric constant K_i, in a buffer with exterior conductivity σ_e and a dielectric constant K_e, the radial EHD velocity at the interface is derived in the form

$$u = FD \cos 2\theta \qquad (3\text{-}6)$$

where the amplitude function is

$$F = \frac{aE^2 K_e}{12(\mu_i + \mu_e)(R+1)^2} \qquad (3\text{-}7)$$

and the discriminant function is

$$D = R^2 + R + 1 - 3S \qquad (3\text{-}8)$$

In these expressions electrostatic units have been used, and θ is the polar coordinate angle measured from the electric field direction, a is the radius of the circular sample, E is the applied electric field, μ_i and μ_e are the interior and exterior viscosities, respectively, and the conductivity and dielectric constant ratios are, respectively,

$$R = \frac{\sigma_i}{\sigma_e} \qquad (3\text{-}9)$$

and

$$S = \frac{K_i}{K_e} \qquad (3\text{-}10)$$

The angular dependence, $\cos 2\theta$, implies distortion of the circular sample cylinder first to an ellipse and then to a ribbon. The orientation of the ribbons depends on the sign of D. The amplitude function F includes the square of the applied electric field E; thus, the effect is quadratic in the applied field. The electric forces are balanced by viscosity; thus, on CFE scales the fluid inertia is negligible.

The discriminant function D is zero unless at least one of the ratios R and S differs from unity. It is positive for large interior conductivity and for small interior dielectric constants. For the special case $S = 1$ the discriminant function reduces to

$$D = (R-1)(R+2) \qquad (11)$$

so that the sign of $(R - 1)$ determines the mode of filament distortion.

VI. EXPERIMENTS ON SAMPLE STREAM DISTORTION

Extensive experiments have been performed using the vertically mounted CFE chamber shown in Figure 3-8. The chamber is 40 cm long (axial), 5 cm wide (lateral), and 0.32 cm thick (transverse). An array of parallel entry ports serve to smooth the inlet flow, while the effluent is currently collected in a single centrally located exit port. Sample is injected at the chamber center plane through a 200-μm (I.D.) capillary. The chamber is equipped with a cross-section illuminator[6] located near the fluid exit end. The cross-section illuminator is an optical arrangement which forms a thin inclined blade of light that sections the flow curtain. This plane of light is scattered from the flowing white polystyrene latex (PSL) sample particles (diameter 0.4 μm) and reveals the sample cross section

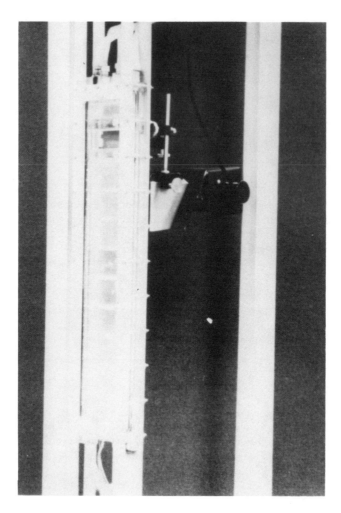

Figure 3-8 Continuous flow electrophoresis-type chamber with cross-section illuminator. See text for details. (From Rhodes, P.H., Snyder, R.S., Roberts, G.O., and Baygents, J.C., *Appl. Theor. Electrophoresis,* 213, 87, 1991. With permission.)

and its distortion. For increased flow stability and reduced cooling requirements, the CFE chamber was operated in upflow. Latex percentages of up to 0.5 by volume were used. Since the buffer flow was up, larger percentages led to gravitational settling of the bulk sample in the chamber due to its excessive mean density.

Figure 3-9 shows typical sample stream distortions seen by using the cross-section illuminator. In these cases, the dielectric constant ratio $S = 1$ and an alternating electric field is used. Figure 3-9a shows a small sample conductivity case, with the conductivity ratio $R < 1$ and the resulting discriminant function D negative. The sample stream is elongated in the transverse direction, perpendicular to the electric field. In Figure 3-9b the sample and buffer conductivities are matched so that $R = 1$ and the discriminant $D = 0$. In Figure 3-9c the sample conductivity is high, $R > 1$, D is positive, and the circular sample column has been deformed into a ribbon aligned with the electric field. The ribbon thickness is slightly exaggerated in this case by the optics. In reality, the sample cross-section area is conserved since diffusion of the latex particles is negligible.

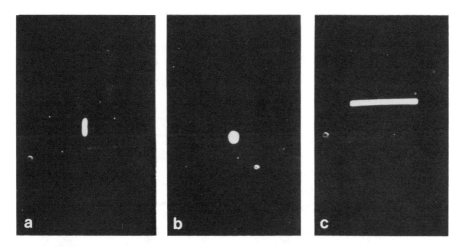

Figure 3-9 Sample stream distortions as seen with the cross-section illuminator. **(a)** $R = 0.41$; **(b)** $R = 1$; **(c)** $R = 1.85$. Sample material, polystyrene latex; sample flow, 0.048 ml/min; sample concentration, 0.05%; phosphate buffer, pH 7.29, buffer flow = 50 ml/min; voltage gradient, (a) 10.4 V/cm and (b,c) 21.4 V/cm; voltage frequency, 2 Hz. (From Rhodes, P.H., Snyder, R.S., Roberts, G.O., and Baygents, J.C., *Appl. Theor. Electrophoresis*, 213, 87, 1991. With permission.)

The E^2 factor in the amplitude function F given by Equation 3-7 is four times as large in Figures 3-9b and c as it is in Figure 3-9a because of the doubled electric field. Note that a smaller electric field was used for the experiment shown in Figure 3-9a to stop the sample from getting too close to the wall, where the residence time in the chamber becomes excessive. The R values for Figures 3-9a, b, and c are 0.41, 1.0, and 1.85, respectively. The corresponding values of the product FD (the amplitude in Equation 3-1) are -0.715, 0, 1.61, where the discriminant function D has been divided by the factor $(R + 1)^2$ in the definition of F and multiplied by 4 for the E^2 factor for Figure 3-9c. This partially explains why the ribbon is wider for Figure 3-9c than it is for Figure 3-9a.

Figure 3-10 shows three sample streams in a horizontally mounted chamber with a thickness of 0.6 cm. The R values from left are again <1, 1, and >1. As before, the left and right sample streams are distorted into ribbons perpendicular and parallel, respectively, to the electric field, while the center sample stream remains circular. This confirms the experiments of Figure 3-9, and the analysis above.

A. COUPLING OF ELECTROHYDRODYNAMICS WITH OTHER HYDRODYNAMIC EFFECTS

The distortion of a low-conductivity sample into a ribbon aligned perpendicular to the electric field may appear at first sight to allow improved separation and collection. This is not the case, however, because of other hydrodynamic and electrokinetic effects on the sample.

For example, consider a low-conductivity sample stream which has been elongated in the transverse direction, perpendicular to the electric field. The parabolic flow profile (Poiseuille flow) in the axial direction results in longer residence times near the walls. In addition, the electric field interacts with the charge double layer on the walls to produce a flow in the field (lateral) direction, with a return pressure-drive flow in the center (the electroosmosis flow profile). These effects combine with the electrophoresis to produce the same crescent shapes as discussed by Strickler and Stack.[6] The crescent problem

Figure 3-10 Sample stream distortion as a function of conductivity ratio R. See text for details. (From Rhodes, P.H., Snyder, R.S., Roberts, G.O., and Baygents, J.C., *Appl. Theor. Electrophoresis,* 213, 87, 1991. With permission.)

becomes much worse as the sample spreads toward the walls. Note that there is no crescent in Figure 3-9a or 3-10 because an alternating voltage was used; for electrophoretic separation a direct voltage is essential.

B. IMPORTANCE OF ELECTROHYDRODYNAMICS

The EHD flow speed in CFE is proportional to the relative variation in the conductivity and the square of the electric field. In fact, balancing the viscous force with the EHD force yields the magnitude estimate

$$\frac{\mu u}{L^2} = \left(\frac{KE^2}{4\pi L}\right)\left(\frac{\Delta\sigma}{\sigma} + \frac{\Delta K}{K}\right) \tag{3-12}$$

where μ is the viscosity (0.01 cgs units for water), u is the flow speed estimate (cm/s), L is the length scale (in CFE, the sample radius), E is the electric field in electrostatic units ($^1/_{300}$ times the field in volts per centimeter), K is the dielectric constant (about 80 for water), $\Delta\sigma/\sigma$ is the relative variation in the conductivity, and $\Delta K/K$ is the relative variation in the dielectric constant. This is in precise agreement with the exact result given by Equations 3-6 through 3-10 for the particular case of a circular distribution in an infinite fluid. The chamber width plays a relatively minor role. If it is not much more than $2L$, it will limit the flow more than if it is large compared with $2L$. Relative variations in the dielectric constant may be important for particle suspensions, while conductivity variations are always dominant for samples not containing particles.

For comparison, the separation speed estimate is $E\Delta U$, where ΔU is the difference between the effective mobilities (normally mU) of the species of interest at the buffer pH. Here m is the mean ionization and U is the intrinsic mobility. The separation time estimate is the distance L divided by this speed. The estimated rate at which sample material is separated, in millimoles per second per centimeter of chamber length, is $LcE\Delta U$, where c is the sample concentration (molarity, or millimoles per cubic centimeter).

The ratio of the EHD flow speed estimate to the separation speed is

$$r = \left(\frac{KEL}{4\pi\mu\Delta U} \right) \left(\frac{\Delta\sigma}{\sigma} + \frac{\Delta K}{K} \right) \tag{3-13}$$

For good separation this must be less than unity; otherwise the sample shape is distorted and the separated regions tend either to overlap or to be advected into regions close to the walls. Plainly, increasing the effective mobility difference ΔU is beneficial; this can be optimized by the choice of the pH (and perhaps by additives), but this optimum is fixed. In addition, varying the pH may be the primary method for matching the wall and sample mobilities to minimize crescent formation. Increasing u by adding sucrose is not helpful because ΔU will decrease in proportion. Decreasing the product EL affects the throughput by the same factor.

Thus, the key to decreasing the EHD flow r is the term in the second set of parentheses, the sum of the relative variations in the conductivity and the dielectric constant. It is obviously sensible to modify the sample so that its conductivity and dielectric constant match those of the buffer. Relative variations in the dielectric constant are negligible, except for samples consisting of suspensions of particles or cells. Even in this case the separation does not lead to significant changes in the relative variation of the dielectric constant. On the other hand, while it is easy to modify the sample so that its conductivity is equal to that of the buffer, the separation always modifies the conductivity distribution and increases $\Delta\sigma$. In certain cases the changes in the conductivity distribution are very significant.

C. CONDUCTIVITY CHANGES IN CONTINUOUS FLOW ELECTROPHORESIS WITH TWO-COMPONENT BUFFER

CFE is normally performed in a two-component buffer consisting of either a strong monovalent base with an excess of a weak monovalent acid (e.g., sodium acetate plus acetic acid), or a strong acid with an excess of a weak base (e.g., ammonium chloride plus ammonia), or a mixture of a weak acid and a weak base in arbitrary proportions. The buffer pH is determined by charge neutrality, since the mean degree of ionization m of the weak acid or base is a function of the pH and the positive and negative charge densities must cancel. The sample consists of the buffer plus relatively small concentrations of the materials to be separated.

It can be shown that as the result of the charge cancellation the two buffer components carry fixed proportions of the current, determined only by their relative intrinsic mobilities U. Hence, with arbitrary initial distributions of the two components (subject only to the limitation that the concentration of the weak acid or base must everywhere exceed that of any strong base or acid), the two concentrations of a moving fluid element are not changed by the direct action of the electric field. They stay constant at a point which moves with the fluid.

When sample components are mixed with the two buffer components, the sample components have a mean ionization at the buffer pH which results in their separate mobilities and in the effective mobility difference ΔU. This also modifies both the charge balance and the proportions of the current carried by the buffer components (whether or not the sample components carry a significant proportion of the current). The proportion carried by the buffer component moving in the same direction as the sample is decreased, while the proportion carried by the other component is increased. As a result, the concentration of each component is decreased at the sample boundary toward which the sample components are moving. The conductivity also decreases. Similarly, the concentration of each component is increased at the opposite sample boundary, and the conductivity also increases.

As the sample region moves, it leaves the high-conductivity region behind and moves into the low-conductivity region. As it continues to move relative to the fluid, fluid elements have their conductivity progressively drop as they enter the sample region and then increase as they pass out the back so that the sample region propagates as a region of decreased conductivity. Assuming the sample does not contribute significantly to the conductivity, the concentration deficits at the center of the moving sample region can be estimated as

$$\frac{c_s U_1 U_2}{U_s \left(U_1 + U_2 \right)}$$

This is usually more than the sample concentration because the sample mobility in the denominator is small. If the sample and the buffer conductivities are initially matched, then

$$\frac{\Delta \sigma}{\sigma} = \frac{c_s U_1 U_2}{C U_s \left(U_1 + U_2 \right)} \tag{3-14}$$

where C is the buffer concentration. Thus, r can be kept small while increasing the throughput product by increasing C in proportion.

D. NUMERICAL RESULTS WITH TWO-COMPONENT BUFFER

Numerical results from simulations of steady-state, three-dimensional, continuous-flow electrophoresis in a rectangular chamber[10,11] demonstrate the importance of EHD flows even at relatively low sample loadings and their quadratic dependence on the applied electric field. They go a long way toward explaining laboratory and microgravity observations which had not been understood previously. In these computations a millimolar sodium barbiturate buffer and a micromolar hemoglobin sample were used. The wall and sample mobilities were matched to eliminate crescent formation. The matched conductivity of the sample ensured that the initial EHD flow upon entry to the chamber was negligible. Fields of 1, 10, and 100 V/cm were used to compare the EHD flows in the three cases.

As predicted by the above theory, the electrokinetics modified the conductivity distribution and, thus, the electric field distribution. Regions of high and low conductivity were initially produced behind and ahead of the electrokinetically moving protein sample. Due to this electrokinetic motion of the protein through the fluid, the high-conductivity region was left behind and the protein was left at the center of a region of reduced conductivity (the peak reduction was about 10% for all three strengths, consistent with the above estimate for the concentration decrease). EHD flows were largest while the sample was still near the high-conductivity region; they continued even after this region was left behind. With only a single sample component and, thus, no separation of ΔU, the EHD flow ratio r could not be defined. However, over respective times of 10,000, 1,000, and 100 s, the three cases moved the sample the same nominal distance of several diameters, with the EHD flows being negligible, noticeable, and dominant, respectively.

E. BUFFER CHOICES TO REDUCE ELECTROHYDRODYNAMICS

As shown above, there are problems with two-component buffers where the same species (e.g., sodium acetate) provides both the buffering and the conductivity. The problems are minimized if the buffer is chosen so that the mean charge of the sample components at that pH is close to zero. In that case, there is still a ΔU and the sample components separate by moving in opposite directions. Conductivity differences induced by the electrokinetics are minimized. However, this method may be incompatible with the requirement to match the sample mobility with the wall mobility to eliminate crescent formation. If the sample is uncharged, its mean mobility is zero. It is, therefore, necessary to find a wall or coating giving a similar zero mobility or to use a moving-wall CFE apparatus,[12] which eliminates pressure-gradient-driven flow profiles in the chamber.

Lauer and McManigill[13] reported using a buffer consisting of potassium chloride (to provide the conductivity and make it independent of the pH) and an ampholyte (zwitterion) of low mobility (to control the pH without affecting the conductivity). The application involved capillary electrophoresis, but the method seems attractive for CFE as well if a suitable ampholyte is available. The sample would not change the proportion of the current carried by potassium and chloride ions, so their concentrations would not be modified and the conductivity, if initially constant, would remain constant.

The same result can be achieved using a two-component buffer, provided the two intrinsic mobilities are very different (e.g., sodium and a very large-molecular-weight weak organic acid). There is a practical difficulty with both of these methods in that the subsequent separation of the desired species from the large-molecule, low-mobility buffer component may be difficult.

VII. CONCLUSIONS

Electrohydrodynamic distortion of the originally circular sample filament causes resolution degradation in CFE both directly and indirectly by working in combination with the accompanying pressure-driven and electroosmotic flows in the chamber. Matching the sample constituent mobility with that of the wall can significantly reduce the crescent distortion if R is kept less than 1 (sample conductivity less than buffer conductivity). A new method of direct control of electroosmosis by using an external electric field[14] conceivably could be used to match the mobility distribution of the sample. Another method of eliminating crescent formation is to move the chamber walls mechanically to eliminate pressure-gradient-driven flow profiles in both the axial and the lateral (electroosmosis) directions.[12]

As Equation 3-11 shows, matching the sample and buffer conductivities ($R = 1$) eliminates initial EHD flows. However, the matching will not be exactly maintained throughout the separation. Although attention is confined to EHD effects in CFE, these EHD effects are even more important in isoelectric focusing or isotachophoresis in free fluids, since these two techniques generate far steeper conductivity gradients.

REFERENCES

1. **Snyder, R. S., Griffin, R. N., and Johnson, A. J.,** Free fluid electrophoresis on Apollo 16, *Sep. Purif. Methods,* 2, 259, 1973.
2. **Allen, R. E., Rhodes, P. H., and Snyder, R. S.,** Column electrophoresis on the Apollo-Soyuz Test Project, *Sep. Purif. Methods,* 6, 1, 1977.
3. **Snyder, R. S., Rhodes, P. H., and Herren, B. J.,** Analysis of free zone electrophoresis of fixed erythrocytes performed in microgravity, *Electrophoresis,* 6, 3, 1985.
4. **Snyder, R. S., Rhodes, P. H., and Miller, T. Y.,** Continuous flow electrophoresis experiments on shuttle flights STS-6 and STS-7, *NASA Tech. Paper,* 2778, 1987.
5. **Rhodes, P. H. and Snyder, R. S.,** Sample band spreading phenomena in ground- and space-based electrophoresis separators, *Electrophoresis,* 7, 113, 1986.
6. **Strickler, A. and Sacks, T.,** Focusing in continuous flow electrophoresis system by electrical control of effective cell wall zeta potentials, *Ann. N.Y. Acad. Sci.,* 209, 497, 1973.
7. **Rhodes, P. H., Snyder, R. S., and Roberts, G. O.,** Electrohydrodynamic distortion of sample streams in continuous flow electrophoresis, *J. Colloid Interface Sci.,* 129, 78, 1989.
8. **Saville, D. A.,** The fluid mechanics of continuous flow electrophoresis in perspective physicochemical hydrodynamics, *Electrophoresis,* 1, 297, 1980.
9. **Saville, D. A. and Ostach, S.,** Final Report NAS8-31349, Princeton University, Princeton, NJ, 1978.
10. **Roberts, G. O.,** Three-dimensional electrophoresis code, Final Report on SBIR Contract NAS8-37342, Roberts Associates, Inc., 1991, 106.
11. **Roberts, G. O., Rhodes, P. H., and Snyder, R. S.,** Computations of electrohydrodynamic effects in continuous flow electrophoresis, in preparation, 1994.
12. **Rhodes, P. H. and Snyder, R. S.,** U.S. Patent, 4,752,372,1989.
13. **Lauer, H. H. and McManigill, D.,** Capillary zone electrophoresis of proteins in untreated fused-silica tubing, *Anal. Chem.,* 58, 166, 1986.
14. **Cheng, S. L., Blanchard, W. C., and Wu, C. T.,** Direct control of the electroosmosis in capillary zone electrophoresis by using an external electric field, *Anal. Chem.,* 62, 1550, 1990.

Chapter 4

The Loading and Unloading of Cells in Electrophoretic Separations

Paul Todd

CONTENTS

0-8493-8918-6/94/$0.00+$.50
© 1994 by CRC Press, Inc.

I. INTRODUCTION

A. ELECTROPHORETIC SEPARATORS

An important part of resolution in the separation of cells by electrophoresis is the fidelity
with which cells can be placed into an electric field and the precision with which they can
be removed from the separator. On the following pages, methods for loading and
unloading cells from a variety of electrophoretic separators are reviewed, one separator
at a time, and a physical and/or statistical evaluation of each is made. Table 4-1 provides
a summarized introduction to each of the separation methods to be discussed.

B. PHYSICAL FACTORS AFFECTING SEPARATION
1. Droplet Sedimentation

The diffusion coefficients of small molecules, of macromolecules, and of whole cells and
particles are in the range 10^{-6} to 10^{-5} cm^2/s, 10^{-7} to 10^{-6} cm^2/s, and 10^{-12} to 10^{-9} cm^2/s,
respectively. If a small zone, or droplet, of radius R contains n particles of radius a inside,
the diffusivity of which is much less than that of solutes outside, then rapid diffusion of
solutes in and slow diffusion of particles out of the droplet (with conservation of mass)
leads to a locally increased density of the droplet:

$$\rho_D = \rho_o + \frac{a^3}{R^3} n \left(\rho - \rho_o \right) \tag{4-1}$$

where ρ_D = density of droplet containing cells or particles, ρ_o = solution density, and ρ
= cell or particle density. If $\rho_D > \rho_o$ then the droplet falls down; if $\rho_D < \rho_o$ it is buoyed
upward.[1] Droplet, or zone, sedimentation is usually controlled by the loading of cell
samples at sufficiently low concentration that it does not occur and by choosing a loading
geometry that avoids excessive cell densities. In column electrophoresis droplet sedimentation cannot be overcome,[2] and the only way to scale column electrophoresis is to
increase column cross-sectional area.[3] Under conditions of droplet sedimentation particles

Table 4-1 **Summary of preparative cell electrophoresis methods**

Name	Abbreviation	Brief Description
Free zone electrophoresis	FZE	Horizontal tube, rotating in the laboratory, static in low gravity
Density gradient electrophoresis	DGE	Vertical cylinder with density gradient
Density gradient isoelectric focusing	DGI	Vertical cylinder with density gradient and pH gradient
Reorienting gradient electrophoresis	RGE	Short, vertical cylinder; reorients to widen zones for easier separation
Continuous flow electrophoresis	CFE	Vertically flowing buffer curtain with transverse electric field

still behave individually unless the ionic environment also permits aggregation.[4,5] In the case of erythrocytes there is sufficient electrostatic repulsion among cells to permit the maintenance of stable dispersions up to at least 3×10^8 cells per milliliter.[6,7] Droplet sedimentation is an avoidable initial condition in zone processes, but it can be created during processing when separands become highly concentrated as in isoelectric focusing or interfacial extraction (see below). This general rule will be seen to apply to all methods of free electrophoresis in which a sample zone in involved.

2. Natural Convection

Droplet sedimentation (or buoyancy) is a special case of convection, and its occurrence is governed by the same Rayleigh-Taylor stability rules as convection, in which convection sets in when the density gradient

$$\frac{d\rho}{dx} > 67.94 \frac{vD}{gR^4}$$

(4-2)

where v = kinematic viscosity, D = diffusion coefficient, and g = gravitational acceleration at the earth's surface. The density of loaded sample zones relative to the density of the carrier buffer must be low enough to prevent reaching a critical gradient.

C. EVALUATION OF SEPARATION METHODS

1. Resolution

The quantitation of resolution can follow the paradigm of differential extraction or adsorption chromatography:

$$R = \frac{\Delta t}{2(\sigma_1 + \sigma_2)}$$

(4-3)

where R = resolution of separation, Δt is actual separation (measured as the time between the arrival of the two separands at a collection point or as the distance between two separands), and σ_1 and σ_2 are the standard deviations of the values of t_1 and t_2 (respectively) used to calculate Δt.

The inverse view is also useful. Equation 4-3 applies to the separation between specific *pairs* of separands, and without a value for Δt resolution cannot be calculated. The denominator of Equation 4-3, however, is often a characteristic of the method. In sedimentation, for example, σ is the standard deviation of the distance sedimented, and in electrophoresis it depends on the standard deviation of mobility, but can include other factors related to sample loading and unloading, as discussed in detail below. In the following discussion separation statistics shall be characterized by using the coefficient of variation (CV) or the relative standard deviation, which is the ratio of the standard deviation to the mean (CV $\equiv$ standard deviation/mean).

There are usually method-related physical constraints on the CV, some of which are established by the methods of inserting and harvesting cells, as will be seen below. The CV can never be less than the reciprocal of the maximum number of fractions that can be collected. This measure of "geometrical resolution" should be used as a design criterion for separator outlets. Any method that divides the outflow into n equal fractions will have an intrinsic minimum *geometrical coefficient of variation* (CV) of $1/n$. Uniform cell populations have intrinsic CVs around 0.05, or 5% of their mean mobility,[8,9] so every collected fraction should migrate at least 20 fractions to take full advantage of the resolution offered by free electrophoresis.

Figure 4-1 The free zone electrophoresis (FZE) principle. Trajectory followed by a particle in FZE in a rotating tube. In the presence of gravity the center of the spiral is below the center of rotation.[10] (From Todd, P., in *Low Gravity Fluid Mechanics and Transport Phenomena,* Koster, J. N. and Sani, R. L., Eds., American Institute of Aeronautics and Astronautics, Washington, D.C., 1991, 539. Copyright© 1991 AIAA — Reprinted with permission.)

2. Purity

Purity is defined as the fraction of cells in a separated volume that consists of the desired product. In every method of cell electrophoresis, cell separation occurs inside a chamber, and separated fractions are unloaded using a method that causes some remixing of separated cells. This translates into removing cell fractions at low Reynolds number (Re):

$$Re = \frac{\rho_o \ell v}{\eta}$$

(4-4)

where ℓ = characteristic length of flowing system, v = velocity, and η = viscosity. For dilute aqueous solutions through a 1-mm channel it is necessary to use an exit fluid velocity of less than 100 cm/s (volumetric flow rate = 0.7 cm³/s) to avoid turbulent mixing and to stay within the laminar flow regime, assuming the exit channels have smooth walls. This limit, in turn, translates into a flow rate of 40 ml/s in a continuous-flow separator with, for example, 50 channels.

II. LOADING AND UNLOADING CELLS USING VARIOUS ELECTROPHORETIC METHODS

A. FREE ZONE ELECTROPHORESIS
1. Brief Description of the Method

Horizontal column electrophoresis has been performed in rotating tubes of 1 to 3 mm I.D.[10] The small diameter minimizes convection distance and facilitates heat rejection, while rotation counteracts all three gravity-dependent processes: convection, zone sedimentation, and, where applicable, particle sedimentation. The concept is illustrated in Figure 4-1, which shows that the total vertical velocity vector oscillates as a sample zone moves in the electric field, so that spiral motion results with a radius vector that can be derived from equations of motion in which the sample zone is treated as a solid particle. In actuality, the sample zone is more like a sedimenting droplet, which can be treated as a particle with density ρ_D (see Section I.B.1 above).

In an open system, in which fluid flow in the direction of the applied electric field is possible, the net negative charge on the tube wall results in plug-type electroendosmotic flow (EEO) that enhances total fluid transport according to the Helmholtz relationship[11] integrated over the double layer at the chamber wall:

$$v = \frac{\zeta \varepsilon}{4\pi\eta} E$$

(4-5)

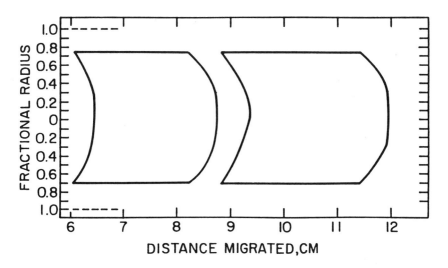

DISTANCE MIGRATED,CM

Figure 4-2 Contour lines of sample bands predicted in simulation of human (right) and rabbit (left) erythrocyte migration, from left to right, in a cylindrical column in low gravity, using simulation method of Vanderhoff and Micale.[15] Assumptions were τ = 1.0 h, μ_{eo} = -0.06, μ_e(human) = -2.05, μ_e(rabbit) = -1.05 × 10^{-4} cm²/V-s, E = 18.6 V/cm, sample width = 0.75 × chamber diameter. Courtesy of Dr. F. J. Micale. (From Todd, P., in *Low Gravity Fluid Mechanics and Transport Phenomena*, Koster, J. N. and Sani, R. L., Eds., American Institute of Aeronautics and Astronautics, Washington, D.C., 1991, 539. Copyright© 1991 AIAA — Reprinted with permission.)

where ζ = zeta potential, ε = absolute dielectric constant, and E = electric field strength. In a closed system, such as is used in almost all practical cases, flow at the walls must be balanced by return flow in the opposite direction along the center of the cylindrical tube, so that the flow profile is a parabola:[12-14]

$$v = \frac{\zeta\varepsilon}{4\pi\eta}E\left(r^2/R^2 - 1/2\right) \tag{4-6}$$

Thus, if the electroosmotic mobility μ_{eo} (coefficient of E in Equation 4-6) is high (on the order of 10^{-4} cm²/V-s), then sample zones will be distorted into parabolas as shown in the simulation of Figure 4-2. It also has been shown that an additive parabolic pattern is superimposed if there is a radial temperature gradient that causes a viscosity gradient. High temperature in the center of the chamber results in low viscosity and higher velocity due to both EEO return flow and faster migration of separands. Increased temperature also increases conductivity. Excessive power input under typical operating conditions can result in 25% higher separand velocity in the center of the tube even in the absence of EEO.[15] This undesirable, strictly thermal effect is independent of gravity, and it can be expected in low-gravity operations and in rotating tubes with inadequate heat rejection. In such small-bore tubes with closed ends, scrupulous care must be taken to suppress EEO by using a nearly neutral, high-viscosity coating,[10,16] and rapid heat rejection and/or low current density are necessary to minimize radial thermal gradients.

Low gravity represents an improvement over the rotating tube in terms of vertical sample zone stability; in low gravity it is not necessary for the tube to rotate, so experiments have been performed with static columns.[17,18] The earliest electrophoresis experiments in low gravity were performed in a static cylindrical column on Apollo 14 and Apollo 16 lunar missions.[19-21] The same free zone electrophoresis (FZE) apparatus was used to separate cells on the Apollo-Soyuz mission in 1974[18] and on Space Shuttle flight STS-3 in an experiment designated "EEVT". The latter experiment was designed to test the hypothesis that very high concentrations of test particles (fixed erythrocytes) can be separated in low gravity owing to the lack of particle and zone sedimentation. Gravity-independent effects such as cell-cell interaction, reduced conductivity, and electrostatic repulsion[22] could be sought in low-gravity experiments at high cell concentrations.[23]

2. Loading of Cells

If electroosmosis is present, it can be seen from the example of Figure 4-2 that a method of loading the sample near the center, away from the tube wall, results in improved resolution relative to samples loaded in zones that encompass the full radius of the tube. The method of adding sample in rotating FZE is shown in Figure 4-3A. The sample is added quickly while the tube rotates at about 70 rpm, and ideally the sample remains in a bolus having dimensions less than the diameter of the rotating tube — typically 3.0 mm — as shown in Figure 4-3B. It is possible to inject several samples for a single separation run, as shown in Figure 4-3C. The starting position of each sample is defined using a ruler that fits precisely alongside the tube when it is not submerged in the thermostated bath (Figure 4-3B).

Recommended maximum sample load is 30 μl containing 10^6 mammalian or 10^8 bacterial cells per milliliter. Experiments have been performed in which larger samples have been injected (200 μl, for example), and resolution has not been adversely affected. Higher loading than recommended is possible when all other conditions of FZE are met: zero electroosmotic backflow at the tube wall, rapid electrophoresis (>0.1 cm/min), and migration over the full length of the tube. When electroosmotic backflow is absent (i.e., negligible relative to electrophoretic velocity), allowing cells to become close to the tube wall does not appear to have serious consequences. Figure 4-3B shows two samples of densely suspended fixed erythrocytes loaded in volumes of 200 and 30 μl. Although the former sample was spread to the tube wall, resolution was not sacrificed; this can be seen in Figure 4-4, where the mobilities and standard deviations were determined for populations of erythrocytes ranging from 3×10^7 to 9×10^8 cells loaded. There is thus an inverse relationship between the magnitude of electroosmosis and the number of cells that can be loaded in FZE.

This is also true in low-gravity FZE. In early experiments, this rule was not followed and serious parabolic spreading resulted.[21] In later experiments, cells were loaded as a frozen disk, about 2 mm thick and the same diameter as the column.[23,24] Several cell types were separated on the Apollo-Soyuz flight in an experiment designated "MA-011". To minimize sample processing on orbit, frozen cells were as a disk in a delrin "key" that could be inserted into the cathode end of electrophoresis columns (up to eight disks and eight columns). Columns were 6 mm in diameter and 15 cm long, to accommodate particles with $\mu = -1.0$ to -3.0×10^{-4} cm²/V-sec during a 1-h separation.

3. Unloading of Cells

In rotating-tube FZE the normal procedure for unloading cells is exactly the reverse of cell loading. First all of the cell-free buffer through which the cells have migrated is removed while the tube is still rotating after switching off the current and removing the tube from the thermostated bath. Then the sample is withdrawn quickly while the tube still rotates at about 70 rpm. These operations are accomplished using a 30-cm-long stainless

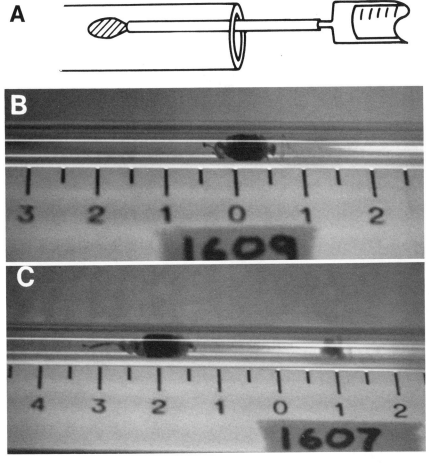

Figure 4-3 The loading of cell suspensions in rotating-tube free zone electrophoresis. Sample is added while the tube is rotating, before it is submerged in the thermostated bath. **(A)** A small volume is added using a syringe. **(B)** Photo showing sample zone just after loading. **(C)** A small number of separate samples can be injected in different zones before the initiation of electrophoresis.

steel syringe needle. With a ruler in place it is possible to harvest fractions that have specified volume and specified migration distance. The process is illustrated diagrammatically in Figure 4-5. It is possible to harvest several fractions from a single separation run and to determine the cellular composition of fractions, as shown in Figure 4-6.

The unloading method chosen for the harvesting of fractions from low-gravity FZE separations was the slicing of sections from a frozen tube using the extruding device shown in Figure 4-7. To freeze columns on orbit, temperature control was provided by a thermoelectric cooler operated in two modes: thermostating at 25°C during electrophoresis and freezing the entire column after completion of each run. Each column was stored frozen for return to earth with separated cells at different positions in the column to be sliced (like a loaf of bread) while still frozen for the collection of separated cell subpopulations. Despite severely attenuated viability after flight, it was possible to obtain electrophoretic mobility distributions and unique subpopulations of cultured human kidney cells, for example.[18,25]

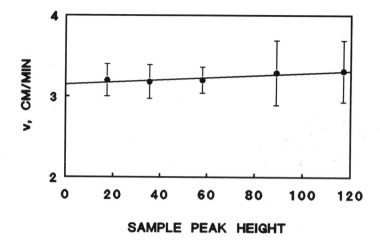

Figure 4-4 Effect of cell load on migration and band spread in free zone electrophoresis. Migration velocity increased by 3% as cell concentration increased from 3 $\times 10^7$ to 9×10^8 cells per milliliter. The abscissa is optical density of the loaded sample determined at the start of electrophoresis.

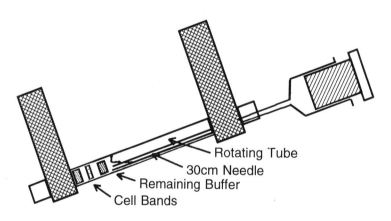

Figure 4-5 Removal of separated fractions from rotating-tube free zone electrophoresis. The tube is removed from the water bath and tilted while still rotating after the current has been switched off. A 30-cm syringe needle is used to sample cell-containing fluid from chosen positions along the tube using a stationary ruler as guide.

4. Effects of Loading and Unloading Procedures on Separation Quality
a. Rotating-Tube Free Zone Electrophoresis
From the above discussion it can be deduced that, to date, only manual methods have been applied to sample loading and unloading in FZE. The fractions harvested from the experiments described in Figure 4-6 had the following purities: fraction 1: 99.7% rabbit RBC, 0.3% chicken RBC; fraction 3: 97.8% chicken RBC, 2.2% rabbit RBC. Although it is a manual method, the sample unloading procedure described in Figure 4-5 yields high-purity products.

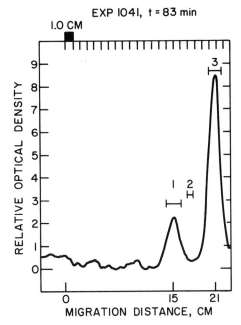

EXP 1041, t = 83 min

Figure 4-6 Electrophoretic separation of chicken and rabbit erythrocytes (fixed with glutaraldehyde) by free zone electrophoresis. Sampling regions 1, 2, and 3 were used to determine purity, and data were used to determine CV. (Reprinted with permission from Todd, P., in *Cell Separation Science and Technology,* Kompala, D. S. and Todd, P., Eds., p. 216. Copyright 1991, American Chemical Society, Washington, D.C.)

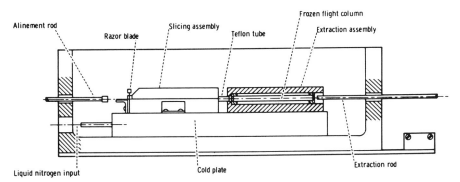

Figure 4-7 Sectioning system for frozen free zone electrophoresis columns used in low gravity. A plunger extrudes the frozen column of fluid out of the glass tube from right to left. At the left end, sections are guillotined at specified intervals. The resulting disks are then thawed and subjected to analysis or subsequent culture.

b. Low-Gravity Free Zone Electrophoresis

Thanks to in-flight photography, the progress of erythrocyte samples could be tracked through the course of a 1-h separation on the STS-3 mission of the U.S. Space Shuttle in 1983.[23] It was generally found that particles (fixed erythrocytes) separated at high concentration on orbit separated in the same fashion as low-concentration cells in normal gravity. However, a single parabolic pattern of human and rabbit red cells was seen migrating when two distinct bands were expected. The leading and trailing edges corresponded to the μ values of human and rabbit red cells, respectively.[23,26] Simulation experiments in a rotating tube indicated that the observed pattern could occur if the human:rabbit cell ratio was 2:1 and a significant (but undetermined) EEO was present.[5]

Narrower bands (lower CV, higher resolution) would have resulted had the frozen sample disk that was loaded been smaller in diameter than the column (see Figure 4-2). Due to loss of cryogenic liquid after sample return, there was no opportunity to slice separated cell subpopulations from the column.[27] Thus, the preservation of the position of the zones of cells by freezing proved to be detrimental to data collection on the Apollo-Soyuz Test Project (ASTP) and fatal on STS-3. Another approach to sample harvesting — namely, capturing separated cells in a reversible gel[28,29] — would have avoided such a costly loss. It was subsequently established that similar separation experiments would be feasible in a colloidal sol that could be regelled after separation and remain solid without freezing or continued refrigeration.[29,30]

B. DENSITY GRADIENT ELECTROPHORESIS
1. Brief Description of the Method

Density gradient electrophoresis (DGE) has been used for protein and virus separation[31] and has been shown to separate cells on the basis of electrophoretic mobility as measured by microscopic electrophoresis.[25,32] Methods and applications of this technique have been reviewed.[2,3] Downward electrophoresis in a commercially available device utilizing a Ficoll gradient was first used to separate immunological cell types.[33,34] Similar Ficoll gradient and phosphate-buffer conditions are employed in the Boltz-Todd apparatus,[2,35] in which electrophoresis of cells and other separands is usually performed in the upward direction. The use of a density gradient counteracts three effects of gravity: (1) particle sedimentation; (2) convection; and (3) zone, or droplet, sedimentation. The migration rates of cells in a density gradient are, however, quantitatively different from those observed under typical analytical electrophoresis conditions,[14] in which physical properties are constant. In DGE the following are also variable:

1. The sedimentation component of the velocity vector changes as the cells migrate upward and the density of the suspending fluid decreases.
2. The viscosity of the suspending Ficoll solution decreases rapidly with increasing migration distance in the chamber as cells migrate upward.[35]
3. The conductivity of the suspending electrolyte increases as the cells migrate upward, consistent with the viscosity reduction.
4. Carbohydrate polymers typically used as solutes to form density gradients, including Ficoll, increase the zeta potential of cells.[36,37]
5. Cells applied to the density gradient in high concentrations undergo droplet sedimentation,[1,2] thereby increasing the magnitude of the sedimentation component of the migration velocity (see Equation 4-1, above).[32,35,37-40]

A simple model of vertical column electrophoresis can satisfactorily predict experimentally observed migration distance vs. time plots at least for times on the order of 2 to 3 h without using fitted parameters.[37,39,41]

Cell separation by electrophoresis in a Ficoll density gradient has been accomplished in a variety of configurations, such as a 2.2×7.0 cm water-jacketed glass column with or without a central cooling pipe using an isotonic gradient from 1.7 to 6.2% (w/v) Ficoll,[2] a flat cylindrical chamber 1.5 cm high and 8 cm in diameter translated in and out of an interelectrode space or a reorienting chamber of the same dimensions,[3] and commercial concentric units such as the Buechler Polyprep 200[42] or LKB isoelectric focusing column.[43] It is found theoretically and experimentally that sedimentation, in addition to electrophoretic migration, plays a significant role in cell separation.[2,3,39]

2. Loading of Cells

Cells are always loaded in DGE by placing a small volume of cells into a thin band at the top or bottom of the density gradient or at some point within it. In the simplest approach,

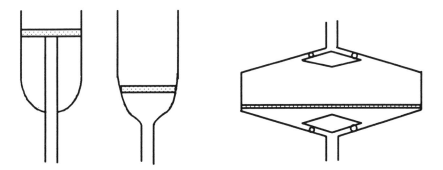

Figure 4-8 Three cell-sample inlet configurations for density gradient electrophoresis. Center: parabolic funnel in which the cell sample band (shaded area) is injected from below after the gradient (above the band) has been added and before the "floor" solution (below the band). Left: fixed insert tube for layering cell band between "floor" (below shaded area) and gradient (above shaded area). Right: inlet with conical flow deflector to prevent pluming as cell sample enters gradient from below.[2,3,35]

Bronson and Van Oss[44] layered cell suspensions by pipetting them onto the top of the appropriate solution in a vertical cylindrical column, while Boltz et al. used two other approaches, shown in Figure 4-8, for carefully layering a cell suspension on top of a dense "floor" beneath a density gradient, through which cells would migrate upward.[2,35] To prevent pluming of the cell sample volume admitted from below, Tulp[3] used a conical flow deflector, also illustrated in Figure 4-8 and also used at the exit at the top of his apparatus.

The range of sample layer thicknesses used in published research varies from 0.30 to 6.7 mm, with only 4 cases out of 22 cited by Tulp being greater than 5 mm and the minimum achievable layer thickness being 0.3 mm.[3]

3. Unloading of Cells
Typically, fractions are collected by pumping "floor" (high-density solution) into the bottom of the column, thereby displacing the gradient and cells through the outlet at the top. Separands are typically collected in 0.5-ml fractions, so that in a typical configuration (20- to 100-ml gradient) this is a low-volume, batch process. Broadening of the bands of separands is expected to occur during the collection procedure. A calculation of the Reynolds number for typical flow rates in the outlet tube leads to values less than 20 so that laminar flow is expected, leading to some dispersion of the band. For this reason, an inverted parabolic funnel or a tapered outlet with a conical deflecter (inverse of those in Figure 4-8) is used in the collection of fractions at the top of the DGE apparatus. Because no fluid flows are present during the separation process, sample removal is the only source of significant band dispersion after separation.

4. Effects of Loading and Unloading Procedures on Separation Quality
From the data given above, a range of minimum CVs can be estimated using typical input sample layer thicknesses (1 to 4 mm) and typical column lengths (1.5 to 10 cm). A 1-mm sample layer in a column with a 10-cm active region will result in CV = 1% and 2% for fractions migrating to the top and midpoints of the column, respectively, while a 4-mm sample layer in a 1.5-cm column (an extreme that has never existed) corresponds to a CV of up to 50%. In other words, short columns require thin sample layers — the "miniaturization principle".[3]

The minimum geometrical CV is established by the volume collected per fraction while pumping the gradient out the top of the column. By collecting 0.25-ml fractions,

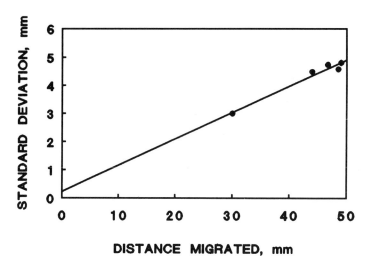

Figure 4-9 Plot of fraction bandwidth as a function of distance migrated by ribosomes in density gradient electrophoresis. The slope is the coefficient of variation of the sample particles, and the intercept is the spreading due to other causes, including initial sample layer thickness and flow distortion during unloading.[41,45]

Boltz and Todd[2] obtained a histogram with a class width of 0.01 mobility units (10^{-4} cm^2/V-s) per fraction, which was 1.1% of the mean in a test separation of mixed erythrocytes. This number is determined from CV_{min} = (volume per fraction)/(volume traversed by cells). It is desirable to keep this value below the CV of the electrophoretic mobility of the sample, in this case about 5.5%.

Figure 4-9 is a plot of standard deviation as a function of distance migrated in DGE.[45] The fitted slope corresponds to the CV of the mobility of the separands under study (bacterial ribosomes), which is $10.3 \pm 0.6\%$. The intercept is roughly equal to the starting sample layer thickness, indicating that the procedure for removing fractions contributes negligibly to fraction bandwidth.[41,45]

When sample loads are too high, additional band broadening is caused by droplet sedimentation or "streaming",[2] a gravity-dependent phenomenon that adversely affects all free electrophoresis processes in which the sample is applied as a zone.[1] Tulp[3] has presented plots of the equation

$$n_c = (1/3\pi a)(g\nabla\rho/\eta D)^{1/2} \qquad (4\text{-}7)$$

for critical cell load n_c as a function of particle radius a in centimeters and density gradient $\nabla\rho$ in grams per cm^4 (other definitions are as in Equation 4-1). For example, the critical concentration for streaming of a suspension of 10-μm cells is about 2×10^6 cells per milliliter in a density gradient of 1 mg/cm^4. For reference, density gradients typically used range from 0.67 for long (10 cm) columns to 20 mg/cm^4 for short (1.5 cm) columns.[3] In experiments over three orders of magnitude of n, Boltz[2,46] showed that final bandwidth increases during electrophoresis due to this phenomenon, as indicated in Table 4-2. Extrapolation of these data to zero bandwidth gives $n_c = 4 \times 10^6$ cells per milliliter, in reasonable agreement with Equation 4-7. Therefore, in density gradient electrophoresis, the most critical variable related to loading and unloading is the cell concentration in the loaded sample.

Table 4-2   **Final bandwidth as a function of cell load,
using fixed rat erythrocytes as test particle (Boltz)[46]**

Cell Load n (cells/ml)	Migration Distance (cm)	Final Band-width (mm)	CV (%)
1×10^7	5.0	2.1	4.0
5×10^7	5.0	5.6	11.0
1×10^8	5.0	7.5	15.0
5×10^8	5.0	12.0	24.0

C. DENSITY GRADIENT ISOELECTRIC FOCUSING
1. Brief Description of the Method
Isoelectric focusing is the movement of a separand through a pH gradient to a pH at which it has zero net charge, at which point it ceases to move through the separating medium. Isoelectric focusing is considered an *equilibrium* process. The steady state condition that occurs when a separand has reached equilibrium can be characterized quantitatively on the basis of the dissociation constants of water, ampholytes (amphoteric electrolytes used to establish the pH gradient), and the separands; this has been done in some detail by Palusinski et al.[47] The ampholytes normally used to establish natural pH gradients are expensive and often harmful to separands, especially living cells. Nonamphoteric, harmless buffer systems have been developed to create artificial pH gradients.[48,49] This separation method is of limited use in the purification of biological cells, most of which are isoelectric at a lower pH than they can tolerate.[43,50,51] Nevertheless, cell separation by isoelectric focusing was studied in the 1970s;[43,50,52,53] most investigators found evidence for the separation of viable cells, but viability was usually low. Jüst and Werner[53] used the continuous flow electrophoresis method and allowed the pH gradient to form in ampholytes in the presence of cells; this caused lysis of living erythrocytes. Most of the research was performed using density gradient columns.

2. Loading of Cells
The highest levels of viability in density gradient isoelectric focusing (DGI) were obtained when cells were inserted in preformed pH gradients at a location close to their isoelectric point. Sherbet et al.[54] designed a vertical cylindrical column with a rubber septum near its midpoint. A cell sample could be injected into the column at mid-level after the pH gradient was formed, thereby limiting the duration of the exposure of living cells to toxic ampholytes to the time required for the cells to migrate electrophoretically to their focusing pH. This concept was improved in the work of Boltz et al.[2,50] by adding an inner conduit to the cooling finger of the column with four outlets midway down the cooling finger, which could be positioned vertically. A sample is withdrawn through this conduit from the prefocused gradient and its pH is measured. If the measured pH of the gradient sample is very close to that of the focusing pH of the desired cells in a mixture, the cell mixture is added to the gradient sample and injected through this conduit. Since the cooling finger can be positioned vertically, the cell mixture can always be injected at a preselected pH. The conduit has four outlets to the gradient so that the cells are injected radially, uniformly, from the center of the column, thereby forming a stable sample band the initial thickness of which is determined by the injected sample volume. This technique is illustrated in Figure 4-10.

3. Unloading of Cells
As in the case of DGE, the absence of flow results in sharp, well-separated bands of cells, and these are pumped out through the top or drained from the bottom of the column just

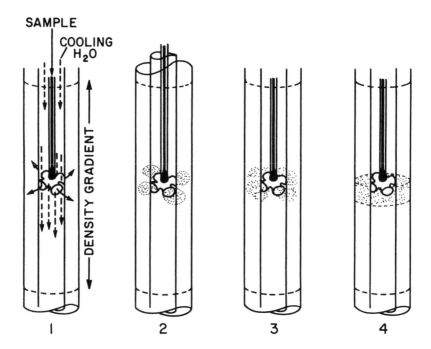

SAMPLE
COOLING
H₂O

DENSITY GRADIENT

1   2   3   4

Figure 4-10 Mechanism for loading cells into a prefocused pH gradient in density gradient isoelectric focusing. Upon injection into the center of the cooling tube cells emerge from four openings at 90° from each other and fill a 3- to 4-mm band in the gradient. (From Boltz, R. C., Jr. and Todd, P., in *Electrokinetic Separation Methods,* Righetti, P. G., van Oss, C. J., and Vanderhoff, J., Eds., Elsevier/North-Holland, Amsterdam, 1979. With permission.)

as in DGE. However, once flow is initiated for harvesting there is a very small loss of resolution due to the broadening effect of Poiseuille flow. When a cooling pipe is used in a vertical column, as is usually the case in DGI, the exit flow path is asymmetric, as indicated in the cross-section drawing of Figure 4-11.

4. Effects of Loading and Unloading Procedures on Separation Quality
Most workers have loaded cell samples by injection into a preformed pH gradient at a predetermined position[50,54] because living cells will not withstand lengthy exposure to ampholytes and low ionic strength at low pH. The method of sample loading is therefore a compromise and, as can be envisioned in Figure 4-10, leads to broad initial sample zones (up to 4 mm). However, the focusing effect of isoelectric focusing substantially counteracts this initial sample layer broadness, and, at steady state, fractions can occupy less than one half of the initial cell layer thickness.

The asymmetry of flow during cell removal (see Section II.C.3 and Figure 4-11) does not appear to compromise resolution of separation, since samples that focus 0.15 pH units apart can be distinguished.[55]

D. REORIENTING GRADIENT ELECTROPHORESIS
1. Brief Description of the Method
Reorienting gradient electrophoresis (RGE) is the electrophoretic counterpart of reorienting gradient sedimentation, introduced by Bont and Hilgers[56] as a rectangular column and

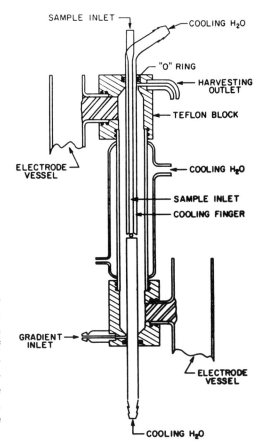

SAMPLE INLET
COOLING H₂O
"O" RING
HARVESTING OUTLET
TEFLON BLOCK
ELECTRODE VESSEL
COOLING H₂O
SAMPLE INLET
COOLING FINGER
GRADIENT INLET
ELECTRODE VESSEL
COOLING H₂O

Figure 4-11   Drawing of a density gradient isoelectric focusing column showing sample exit path at the top. The exit path from the top of the column to the "harvesting outlet" is asymmetric in this application of the Boltz-Todd column since it must bypass the cooling tube, which occupies the center of the column.

refined by Tulp et al.[57] as a flat cylinder, 1.5 cm high by 8 cm in diameter. In the electrophoresis version the top and bottom surfaces of the cylinder are the electrode membranes. Initially, the cylinder is tilted at about 60° and completely filled with "floor" solution (dense Ficoll in electrophoresis buffer) through an inlet located at the lowest point. The density gradient is allowed to flow in from the highest point and displace the "floor" downward until the entire gradient is loaded. Cell sample is added to the top of the gradient, followed by an electrophoresis buffer "ceiling". The chamber is then rotated slowly to its horizontal position, and a field is applied so that the cells migrate downward in the electric field.[3] A sketch of the procedure is shown in Figure 4-12.

2. Loading of Cells

The cell sample layer, when added to the top of the tilted gradient, is typically 5 ml in volume and on the order of 1 cm thick. After return of the gradient to the horizontal position, it is 1 mm thick, so migration distances, and therefore times, can be very short, on the order of 10 min at 5 to 6 V/cm. As in Tulp-column DGE (see Figure 4-8), the density gradient that supports this sample band is about 20 mg/cm⁴.

3. Unloading of Cells

After the electric field has been switched off, the chamber is slowly rotated back to the same 60° tilt as during loading. "Floor" (heavy Ficoll) solution is pumped into the chamber at its lowest point, and fractions are collected from the outlet, which is at the

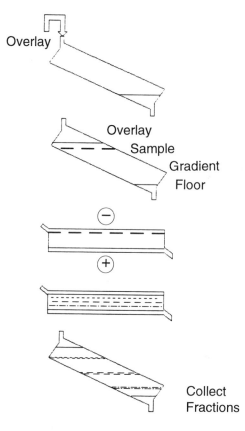

Overlay

Overlay
Sample
Gradient
Floor

Collect
Fractions

Figure 4-12 Reorienting density gradient electrophoresis.[14,29] The flat cylindrical chamber (1.5 to 3.0 cm high × 8 to 20 cm in diameter) is tilted for filling with "floor" at the bottom corner, with sample layered at the top of a linear density gradient added at the top. The field is applied with the chamber in the horizontal position. After electrophoretic separation the chamber is reoriented for the collection of fractions. (Courtesy of R. M. Stewart. See References 3 and 30.)

highest point of the tilted chamber. As with DGE, the minimum CV is established by the volume, and hence the number, of fractions collected. In a test RBC separation, Tulp[3] was able to collect eight to ten fractions across the full width of each cell peak. The unloading statistics of DGE and RGE thus seem to be very similar.

4. Effects of Loading and Unloading Procedures on Separation Quality

The inlet and outlet geometries of RGE, while appearing to be an incidental outcome of the chamber geometry and not the product of a deliberate design like the examples in Figure 4-8, are nevertheless effective in facilitating the loading of thin sample layers and the recovery of resolved fractions. Inlet and outlet geometry does not appear to be a substantial factor in separation quality as long as the flow through them is laminar. As in DGE, the most significant factor is cell concentration in the sample load.

E. CONTINUOUS FLOW ELECTROPHORESIS
1. Brief Description of the Method

Other chapters in this volume describe continuous flow electrophoresis (CFE) in considerable detail.[58-61] The CFE process itself has several characteristics that place constraints on the manner in which cell samples are loaded and the manner in which fractions are collected. These are discussed briefly below in the context of the elements of CFE theory.

Several investigators have constructed mathematical models of CFE.[62-73] The main ingredients of such a model include the following variables:[66] electrophoretic migration of separands, Poiseuille flow of pumped carrier buffer, electroendosmosis at the front and back chamber walls, horizontal and vertical thermal gradients, conductivity gradients,

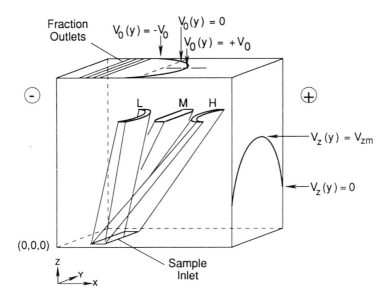

Figure 4-13 Schematic diagram (not to scale) of an upward-flowing free-flow electrophoresis chamber. A coordinate system with (0,0,0) at the lower front left corner is shown. Sample stream is shown as a slit (circular sample streams are commonly used). Anodal migration of low, medium, and high (L, M, H) mobility separands results in crescent-shaped bands due to Poiseuille retardation of particles near the walls [where $v_z(y)$ is low] in the case of high mobility and electroosmotic flow near the walls [where $v_o(y)$ is high] in the case of low mobility. These two parabolic distortions balance in the case of medium-mobility separands. (From Todd, P., in *Low Gravity Fluid Mechanics and Transport Phenomena*, Koster, J. N. and Sani, R. L., Eds., American Institute of Aeronautics and Astronautics, Washington, D.C., 1991, 539. Copyright© 1991 AIAA — Reprinted with permission.)

diffusion, sample input configuration, and fraction collection system. Figure 4-13 indicates the path followed by three categories of separands — low, medium, and high mobility — and the effects of Poiseuille flow and electroendosmosis when sample is injected at the bottom, electrophoresis is to the right, and fractions are collected at the top. A coordinate system is shown in which the z axis is the direction of pumped fluid (buffer) flow, the y axis is the chamber thickness, and the x axis is the chamber width and direction of migration of separands in the applied, horizontal electric field. Separands will arrive at the outlet end of the chamber distributed in nested crescents in which high-mobility separand particles nearest the front and back walls migrate the farthest, as indicated in Figures 4-13 and 4-14. However, the y dependencies can be minimized by using a thick chamber and limiting the sample stream to a central zone near the midpoint of the chamber in the y direction. A thicker chamber is allowed in low gravity where thermal distortions of flow presumably do not occur (ignoring Marangoni flow due to a small thermal gradient in the z direction and the viscosity gradient mentioned above).

2. Loading of Cells

The method of loading cells in CFE is more critical to the quality of results than is usually appreciated. There are three aspects of loading that are important, and all relate to the fact that CFE is a continuous process requiring the confluence of flow streams. The first problem is the suspension of cells in the inlet pump, the second is the geometry of the

Figure 4-14 A thicker free flow electrophoresis chamber or a thinner sample band allows the separation of separands in narrower bands, resulting in higher output purity. Preventing the sample stream from approaching the front and back chamber walls results in less band distortion due to Poiseuille and electroosmotic flow. Dashed rectangle represents a cross section of the original

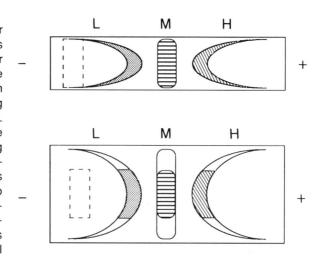

sample stream at the inlet. (From Todd, P., in *Low Gravity Fluid Mechanics and Transport Phenomena,* Koster, J. N. and Sani, R. L., Eds., American Institute of Aeronautics and Astronautics, Washington, D.C., 1991, 539. Copyright© 1991 AIAA — Reprinted with permission.)

nozzle through which cell suspension enters the column, and the third is the matching of input sample conductivity density to flowing-buffer density (closely related to the first).

At a feed rate on the order of 10 ml/h and at cell concentrations of <10^6 (animal) cells per milliliter, the processing of a cell sample requires several hours, during which a steady feed rate must be maintained. This is normally achieved with a syringe (positive pressure) pump or a peristaltic pump. In either case there is a cell-sample reservoir in which cells will settle if they are not properly mixed. If care is not taken to suspend cells vigorously they will be too concentrated early or late in the separation run if pumping is downward or upward, respectively. Consequently, the inlet cell concentration could exceed the droplet sedimentation limit[1] at the beginning or the end of a separation, and substantial fractions of the separands could be lost. Worse yet, a given separation could be subjected to time-varying migration rates (see below).

Figure 4-14 indicates how sharper bands can be achieved using a thicker chamber or by injecting the sample stream into a narrow region midway between the front and back walls. Sample injection ports of CFE equipment are designed to accomplish this, as indicated in Figure 4-13. However, such bands can be unsharpened by electrohydrodynamic phenomena when the conductivity of the injected sample zone is not properly matched to that of the carrier buffer, resulting in E(x,y) not being constant because k(x,y) is not constant.[74,76] This phenomenon is also independent of gravity, although any effect of g on heat transfer will also affect k(x,y), since k depends linearly on temperature. The general relationship for field distortion is derived from Taylor's "electrohydrodynamic" concept applied to a spherical drop;[77] it assumes a sample zone of circular cross section small compared to the y dimension of the chamber and having different conductivity k_s, viscosity η_s, and dielectric constant ε_s from that of the carrier buffer. The surface stress S is given by the equation

$$S = \frac{E^2\varepsilon\left[\varepsilon/\varepsilon_s\left\{\left(k_s/k\right)^2 + \left(k_s/k\right) + 1\right\} - 3\right]}{4\pi\left\{\left(k_s/k\right)^2 + 1\right\}^2} \qquad \textbf{(4-8)}$$

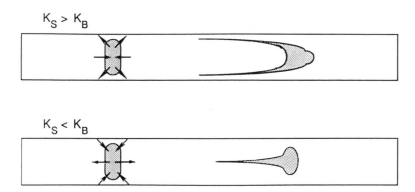

Figure 4-15 Electrohydrodynamic distortion of sample bands in continuous flow electrophoresis. From Equation 4-8 the stress vector at the "surface" of a high-conductivity sample stream is outward perpendicular to the electric field and inward parallel to the field lines (top diagram) and vice versa for a sample stream with lower conductivity than that of the carrier buffer (lower diagram). (From Todd, P., in *Low Gravity Fluid Mechanics and Transport Phenomena,* Koster, J. N. and Sani, R. L., Eds., American Institute of Aeronautics and Astronautics, Washington, D.C., 1991, 537. Copyright© 1991 AIAA — Reprinted with permission.)

where θ is the angle subtended by the stress vector and the applied electric field lines. Thus, at any angle with respect to the direction of migration, the sign of the quantity in square brackets and that of the cosine function combine to determine whether the sample stream is pulled outward or compressed inward. This function predicts that under conditions of $k_s > k$ the sample stream will be pulled outward at $\theta = 90°$ (toward the chamber walls) and compressed in the perpendicular direction, as in the upper example of Figure 4-15, and vice versa as in the lower example. Discontinuities in *buffer* ion concentrations have been shown to produce electrohydrodynamic distortions of sample bands that seriously compromise separations, as illustrated in Figure 4-15. Soluble separands that add significantly to the conductivity of sample streams at high concentrations would presumably have the same effect. It remains to determine whether or not particulate separands (cells) would, at high concentration, also cause such field distortions by adding to the internal conductivity, k_s.

The consequences of loading cells in sample streams with inappropriate densities in CFE are severe. Figure 4-16 illustrates the four possible combinations of sample stream density and flow direction. Critical cell concentrations are comparable to those discussed above in Section II.B.4.

3. Unloading of Cells
At the heart of resolution in CFE is the manner in which cells are unloaded in fractions at the outlet. Material balance and flow continuity require that all fluid entering the chamber via buffer and sample flow must exit the chamber. Although Poiseuille flow occurs in the chamber (Figure 4-13), the parabolic flow profile in the y direction is very shallow, and the rate of accumulation of volume in the outlets is treated as constant for all outlets, especially since outlets near the edges of the chamber are never used.

a. Effect of Number of Outlets
Table 4-3 is a summary of the number of outlets in four different published CFE devices and gives the minimum CV at the maximum fraction and central fraction in each case.

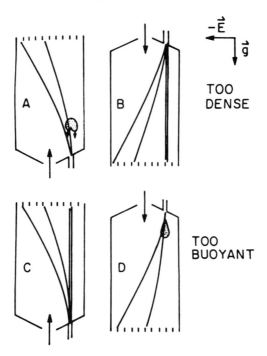

Figure 4-16 Sketch of the consequences of sample zones that are too dense (**A** and **B**) or too buoyant (**C** and **D**) in continuous flow electrophoresis with flow upward (A and C) or downward (B and D). (From Todd, P., in *Cell Electrophoresis,* Schütt, W. and Klinkmann, H., Eds., Walter de Gruyter, Berlin, 1985, 3. With permission.)

Table 4-3 **Number of outlets and resulting minimum geometric coefficients of variation for various continuous flow electrophoresis units**

Name of Unit	Number of Outlets	Full Chamber CV (%)	Half Chamber CV (%)
Beckman CPE	25	4	8
Desaga FF44	44	2.2	4.4
Bender & Hobein VAP	90	1.1	2.2
McDonnell-Douglas CFES	198	0.5	1.0

The range of CVs spans the range of measured CVs for cells as determined by analytical electrophoresis.

b. Broad-End Unloading

One means of achieving higher resolution of collected fractions is the broadening of the column at the top while maintaining constant cross section. The minimum width of an outlet that can be machined might be 1 mm; this allows 60 fractions over a 6-cm-wide column. If the column is widened to 18 cm, then 180 fractions could be collected, thereby increasing theoretical resolution threefold and reducing the minimum CV to $1/N = 1/180 = 0.5\%$ of maximum mobility, which exceeds the accuracy of any electrophoretic measurement. For low-mobility particles that cross one sixth of the column width, $1/N = 1/30 = 3.3\%$ CV, a useful increase compared to $1/10 = 10\%$ CV.

Figure 4-17 Mobilities of polystyrene test particles separated by continuous flow electrophoresis (CFE). Lower pair of curves gives electrophoretic migration patterns (optical density [OD] at 700 nm vs. fraction number) in CFE. Migration was from left to right, starting at fraction 51, 0.8 mm per fraction. Two batches of particles with identical mobilities, 4.4×10^{-4} cm^2/V-s, but different diameters, 0.297 μm (squares) and 0.865 μm (triangles), were used. Upper pair of curves gives mobilities of particles harvested from all seven fractions. All seven fractions collected over the entire peak in both cases contained particles with the same electrophoretic mobility. (From Todd, P., Kurdyla, J., Sarnoff, B. E., and Elsasser, W., in *Frontiers in Bioprocessing,* Sikdar, S. K., Bier, M., and Todd, P., Eds., CRC Press, Boca Raton, FL, 1989, 223. With permission.)

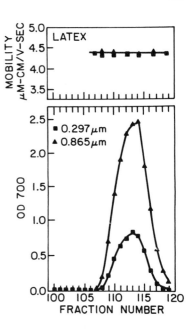

4. Effects of Loading and Unloading Procedures on Separation Quality

While it is very important to hold the inlet sample stream to a diameter less than one half the chamber thickness, it is more important yet to control the concentration of solutes and particles in the sample stream so that density overturns and electrohydrodynamic overturns do not occur. The number of outlets, if less than about 20, can play a significant role in separation statistics. Otherwise sample band spreading in CFE is mainly due to the "crescent phenomena" depicted in Figure 4-14. An experiment with suspensions of latex microspheres was performed to illustrate this point.

Two sizes of microspheres having identical electrophoretic mobilities, $4.44 \pm 0.07 \times 10^{-4}$ cm^2/V-s, were electrophoresed in a 198-outlet chamber 1.5 mm thick in matched-conductivity buffer and below the critical droplet concentration.[25] As seen in Figure 4-17, they migrated 62 fractions with a standard deviation of 2.45 fractions, which corresponds to CV = 4.0%, compared to 1.6% as measured by laser light-scattering analytical electrophoresis[78] and 1/62 =: 1.6%, the minimum geometrical CV. As the figure also shows, the particles in all fractions collected had the same mobility, so band spreading was not due to intrinsic electrophoretic heterogeneity of the sample. In summary, sample loading and unloading, when correctly practiced in CFE, do not limit the quality of separation by CFE.

III. SUMMARY AND CONCLUSIONS

The quantitative effects of cell loading and unloading procedures on the quality of cell separations by five electrophoretic methods have been reviewed. In free zone electrophoresis (rotating tube or low gravity), methods of loading and harvesting have less impact on the quality of separation than does the control of electroosmosis (EEO) at the chamber walls. When significant EEO is present, a loading procedure that keeps cells away from the chamber walls is required. In density gradient electrophoresis (DGE), loading cells at excessive concentrations must be avoided to prevent droplet sedimentation, while the geometrical resolution achieved by current methods of fraction collection exceeds that required on the basis of the CVs of cell mobilities. Although narrow starting sample bandwidths are difficult to achieve in isoelectric focusing, the final fraction bandwidths are narrower due to the focusing process itself. Band spreading during

96

harvesting does not appear to interfere with resolution. Thick sample zones can be loaded in reorienting gradient electrophoresis (RGE) because these are narrowed after reorientation of the chamber for electrophoresis. The major cause of zone broadening in continuous flow electrophoresis (CFE) is the conspiracy of Poiseuille and electroosmotic flows. One antidote is the injection of very narrow sample streams that prevent cells from approaching the chamber walls. As in DGE and RGE, droplet sedimentation due to high cell concentrations can result in band spreading; in the case of CFE, however, it also can prevent sample migration in the electric field.

ACKNOWLEDGMENTS

It is a pleasure to thank Stellan Hjertén, Karin Ellenbring, Inge Johansson, Jeffrey Kurdyla, Debra Hawker, Lindsay D. Plank, Robert C. Boltz, Jr., M. Elaine Kunze, Dennis R. Morrison, Robert S. Snyder, Percy Rhodes, W. C. Hymer, Marian Lewis, John L. Sloyer, Jr., William Elsasser, and Scott R. Rudge for their generous assistance during various parts of the 20-year course of performing the research on which this chapter is based.

NOMENCLATURE

a	=	cell or particle radius (cm or μm)
D	=	diffusion coefficient (cm^2/s)
$d\rho/dx$	=	density gradient (g/cm^4)
E	=	electric field strength (V/m or V/cm)
g	=	gravitational acceleration at the earth's surface (980 cm/s^2)
k	=	buffer conductivity ($[ohm-cm]^{-1}$)
k_s	=	conductivity of suspended-cell sample zone ($[ohm-cm]^{-1}$)
ℓ	=	characteristic length in flowing system
n	=	number of cells per milliliter
n_c	=	critical cell concentration for droplet sedimentation
R	=	radius of cylindrical container
R^3	=	cube of droplet radius
R	=	resolution of separation (Equation 4-3)
r	=	radial distance from the center of cylindrical tube (m or cm)
Re	=	Reynolds number (ratio of inertial to viscous force)
S	=	electrohydrodynamic surface stress vector
Δt	=	physical separation between two separands
v	=	velocity (cm/s or m/s)
ε	=	absolute dielectric constant in MKS (SI) units
ε_s	=	dielectric constant of suspended-cell sample zone
η	=	viscosity (Poise or Pascal-s)
η_s	=	viscosity of suspended-cell sample zone
ζ	=	zeta potential, potential at a hydrodynamic surface (V)
μ_{eo}	=	electroosmotic mobility (cm^2/V-s)
υ	=	kinematic viscosity
ρ	=	cell or particle density
ρ_D	=	density of droplet containing cells or particles
ρ_o	=	solution density
$\nabla\rho$	=	density gradient (g/cm^4)
σ_1	=	standard deviation of the position of separand #1
θ	=	angle between electrohydrodynamic and electric field vectors

REFERENCES

1. **Mason, D. W.,** A diffusion-driven instability in systems that separate particles by velocity sedimentation. *Biophys. J.,* 16, 407, 1976.
2. **Boltz, R. C., Jr. and Todd, P.,** Density gradient electrophoresis of cells in a vertical column, in *Electrokinetic Separation Methods,* Righetti, P. G., van Oss, C. J., and Vanderhoff, J., Eds., Elsevier/North-Holland, Amsterdam, 1979, 229.
3. **Tulp, A.,** Density gradient electrophoresis of mammalian cells, *Methods Biochem. Anal.,* 30, 141, 1984.
4. **Todd, P. and Hjertén, S.,** Free zone electrophoresis of animal cells. I. Experiments on cell-cell interactions, in *Cell Electrophoresis,* Schütt, W. and Klinkmann, H., Eds., Walter de Gruyter, Berlin, 1985, 23.
5. **Todd, P.,** Microgravity cell electrophoresis experiments on the space shuttle: a 1984 overview, in *Cell Electrophoresis,* Schütt, W. and Klinkmann, H., Eds., Walter de Gruyter, Berlin, 1985, 3.
6. **Snyder, R. S., Rhodes, P. H., Herren, B. J., Miller, T. Y., Seaman, G. V. F., Todd, P., Kunze, M. E., and Sarnoff, B. E.,** Analysis of free zone electrophoresis of fixed erythrocytes performed in microgravity, *Electrophoresis,* 6, 3, 1985.
7. **Omenyi, S. N., Snyder, R. S., Absolom, D. T., Neumann, A. W., and van Oss, C. J.,** Effects of zero van der Waals and zero electrostatic forces on droplet sedimentation, *J. Colloid Interface Sci.,* 81, 402, 1981.
8. **Schütt, W., Hashimoto, N., and Shimuzu, M.,** Application of cell electrophoresis for clinical diagnosis, in *Cell Electrophoresis*, Bauer, J., Ed., CRC Press, Boca Raton, FL, 1994, chap. 13.
9. **Bauer, J., Ed.,** *Cell Electrophoresis,* CRC Press, Boca Raton, FL, 1994.
10. **Hjertén, S.,** *Free Zone Electrophoresis,* Almqvist and Wiksells, Uppsala, Sweden, 1962.
11. **Abramson, H. W., Moyer, L. S., and Gorin, M. H.,** *Electrophoresis of Proteins,* Reinhold, New York, 1942.
12. **Bangham, A. D., Flemans, R., Heard, D. H., and Seaman, G. V. F.,** An apparatus for microelectrophoresis of small particles, *Nature,* 642, 642, 1958.
13. **Brinton, C. C., Jr. and Laufer, M. A.,** The electrophoresis of viruses, bacteria, and cells and the microscope method of electrophoresis, in *Electrophoresis*, Bier, M., Ed., Academic Press, New York, 1959, 427.
14. **Seaman, G. V. F.,** Electrophoresis using a cylindrical chamber, in *Cell Electrophoresis*, Ambrose, E. J., Ed., Little, Brown, Boston, 1965, 4.
15. **Vanderhoff, J. W. and Micale, F. J.,** Influence of electroosmosis, in *Electrokinetic Separation Methods,* Righetti, P. G., Van Oss, C. J., and Vanderhoff, J. W., Eds., Elsevier/North-Holland, Amsterdam, 1979, 81.
16. **Harris, J. M., Brooks, D. E., Boyce, J. F., Snyder, R. S., and Van Alstine, J. M.,** Hydrophilic polymer coatings for control of electroosmosis and wetting, in *Dynamic Aspects of Polymer Surfaces: Proc. 5th Rocky Mountain American Chemical Society Meet.,* Andrade, J. D., Ed., American Chemical Society, Washington, D.C., 1987, 111.
17. **Todd, P.,** Separation physics, in *Low Gravity Fluid Mechanics and Transport Phenomena,* Koster, J. N. and Sani, R. L., Eds. American Institute of Aeronautics and Astronautics, Washington, D.C., 1991, 539.
18. **Barlow, G. H., Lazer, S. L., Rueter, A., and Allen, R.,** Electrophoretic separation of human kidney cells at zero gravity, in Bioprocessing in Space, Morrison, D. R., Ed., NASA TM X-58191, L. B. Johnson Space Center, Houston, 1977, 125.
19. **Snyder, R. S.,** Electrophoresis Demonstration on Apollo 16, Rep. NASA TMX-64724, National Aeronautics and Space Administration, Washington, D.C., 1972.

20. **Snyder, R. S., Bier, M., Griffin, R. N., Johnson, A. J., Leidheiser, H., Micale, F. J., Ross, S., and van Oss, C. J.,** Free fluid particle electrophoresis on Apollo 16, *Sep. Purif. Methods,* 2, 258, 1973.

21. **McKannan, E. C., Krupnick, A. C., Griffin, R. N., and McCreight, L. R.,** Electrophoretic Separation in Space — Apollo 14, Rep. NASA TMX-64611, National Aeronautics and Space Administration, Washington, D.C., 1971.

22. **McGuire, J. K. and Snyder, R. S.,** Operational parameters for continuous flow electrophoresis of cells, in *Electrophoresis '81,* Allen, R. C. and Arnaud, P., Eds., Walter de Gruyter, Berlin, 1981, 947.

23. **Snyder, R. S., Rhodes, P. H., Herren, B. J., Miller, T. Y., Seaman, G. V. F., Todd, P., Kunze, M. E., and Sarnoff, B. E.,** Analysis of free zone electrophoresis of fixed erythrocytes performed in microgravity, *Electrophoresis,* 6, 3, 1985.

24. **Allen, R. E., Rhodes, P. H., Snyder, R. S., Barlow, G. H., Bier, M., Bigazzi, P. E., van Oss, C. J., Knox, R. J., Seaman, G. V. F., Micale, F. J., and Vanderhoff, J. W.,** Column electrophoresis on the Apollo-Soyuz Test Project, *Sep. Purif. Methods,* 6, 1, 1977.

25. **Todd, P., Kurdyla, J., Sarnoff, B. E., and Elsasser, W.,** Analytical cell electrophoresis as a tool in preparative cell electrophoresis, in *Frontiers in Bioprocessing,* Sikdar, S. K., Bier, M., and Todd, P., Eds., CRC Press, Boca Raton, FL, 1989, 223.

26. **Sarnoff, B. E., Kunze, M. E., and Todd, P.,** Analysis of red blood cell electrophoresis experiments on Space Shuttle flight STS-3, in Proc. Conf. Manufacturing in Space, O'Neill, R., Ed., *Adv. Astronaut. Sci.,* 53, 139, 1983.

27. **Morrison, D. R. and Lewis, M. L.,** Electrophoresis tests on STS-3 and ground control experiments: a basis for future biological sample selections, in 33rd Int. Astronautical Federation Congr., Paris, 1983, Paper No. 82-152.

28. **Todd, P., Szlag, D. C., Plank, L. D., Delcourt, S. G., Kunze, M. E., Kirkpatrick, F. H., and Pike, R. G.,** An investigations on gel forming media for use in low gravity bioseparations research, *Adv. Space Res.,* 9(11), 97, 1989.

29. **Plank, L. D., Kunze, M. E., Gaines, R. A., and Todd, P.,** Density gradient electrophoresis of cells in a reversible gel, *Electrophoresis,* 9, 647, 1988.

30. **Todd, P.,** Comparison of methods of preparative cell electrophoresis, in *Cell Separation Science and Technology,* Kompala, D. S. and Todd, P., Eds., American Chemical Society, Washington, D.C., 1991, 216.

31. **Poulson, A. and Cramer, R.,** Zone electrophoresis of type 1 poliomyelitis virus, *Biochim. Biophys. Acta,* 29, 187, 1958.

32. **Boltz, R. C., Jr., Todd, P., Gaines, R. A., Milito, R. P., Docherty, J. J., Thompson, C. J., Notter, M. F. D., Richardson, L. S., and Mortel, R.,** Cell electrophoresis research directed toward clinical cytodiagnosis, *J. Histochem. Cytochem.,* 24, 16, 1976.

33. **Griffith, A. L., Catsimpoolas, N., and Wortis, H. H.,** Electrophoretic separation of cells in a density gradient, *Life Sci.,* 16, 1693, 1975.

34. **Platsoucas, C. D., Good, R. A., and Gupta, S.,** Separation of human T-lymphocyte subpopulations (Tμ; Tγ) by density gradient electrophoresis. *Proc. Natl. Acad. Sci. U.S.A.,* 76, 1972, 1979.

35. **Boltz, R. C., Jr., Todd, P., Streibel, M. J., and Louie, M. K.,** Preparative electrophoresis of living mammalian cells in a Ficoll gradient, *Prep. Biochem.,* 3, 383, 1973.

36. **Brooks, D. E. and Seaman, G. V. F.,** The effect of neutral polymers on the electrokinetic potential of cells and other charged particles, *J. Colloid Interface Sci.,* 43, 670, 1973.

37. **Todd, P., Hymer, W. C., Plank, L. D., Marks, G. M., Hershey, M., Giranda, V., Kunze M. E., and Mehrishi, J. N.,** Separation of functioning mammalian cells by density gradient electrophoresis, in *Electrophoresis '81,* Allen, R. C. and Arnaud, P., Eds., Walter de Gruyter, New York, 1981, 871.

38. **Gaines, R. A.,** A Physical Evaluation of Density Gradient Cell Electrophoresis, M.S. Thesis, The Pennsylvania State University, University Park, PA, 1981.

39. **Plank, L. D., Hymer, W. C., Kunze, M. E., and Todd, P.,** Studies on preparative cell electrophoresis as a means of purifying growth-hormone producing cells of rat pituitary. *J. Biochem. Biophys. Methods,* 8, 273, 1983.

40. **Rudge, S. R. and Todd, P.,** Applied electric fields for downstream processing, in *Protein Purification from Molecular Mechanisms to Large-Scale Processes,* Ladisch, M. R., Willson, R. C., Painton, C. D. C., and Builder, S. E., Eds., American Chemical Society, Washington, D.C., 1990, 257.

41. **Hawker, D. T. L., Todd, P., Davis, R. H., Lawson, R. C., and Rudge, S. R.,** Electrokinetic isolation of vesicles and ribosomes derived from *Serratia marcescens, Biotechnol. Prog.,* 8, 429, 1992.

42. **Catsimpoolas, N. and Griffith, A. L.,** Preparative density gradient electrophoresis and velocity sedimentation at unit gravity of mammalian cells, in *Methods of Cell Separation,* Vol. 1, Catsimpoolas, N., Ed., Plenum Press, New York, 1977, 1.

43. **Sherbet, G. V.,** *The Biophysical Characterisation of the Cell Surface,* Academic Press, London, 1978.

44. **Van Oss, C. J. and Bronson, P. M.,** Vertical ascending cell electrophoresis, in *Electrokinetic Separation Methods,* Righetti, P. G., van Oss, C. J., and Vanderhoff, J. W., Eds., Elsevier/North-Holland, Amsterdam, 1979, 251.

45. **Hawker, D. T. L.,** Electrokinetic and Extractive Isolation of Bacterial Ribosomes and Vesicles, Thesis, University of Colorado, Boulder, 1992.

46. **Boltz, R. C., Jr.,** The Development and Use of Whole Cell Density Gradient Electrophoresis and Isoelectric Focusing for Detection of Differences in the Ionogenic Composition of Mammalian Cell Surfaces, Ph.D. thesis, The Pennsylvania State University, University Park, 1980.

47. **Palusinski, O. A., Allgyer, T. T., Mosher, R. A., Bier, M., and Saville, D. A.,** Mathematical modelling and computer simulation of isoelectric focusing of simple ampholytes, *Biophys. Chem.,* 14, 389, 1981.

48. **Troitsky, G. V. and Azhitsky, G. Yu.,** *Isoelectric Focussing of Proteins in Natural and Artificial pH Gradients,* Kiev Nauka Dumka, Kiev, 1984.

49. **Boltz, R. C., Jr., Miller, T. Y., Todd, P., and Kukulinsky, N. E.,** A citrate buffer system for isoelectric focusing and electrophoresis of living mammalian cells, in *Electrophoresis '78,* Catsimpoolas, N., Ed., Elsevier/North-Holland, Amsterdam, 1978, 345.

50. **Boltz, R. C., Jr., Todd, P., Hammerstedt, R. H., Hymer, W. C., Thompson, C. J., and Docherty, J. J.,** Initial studies on the separation of cells by density gradient isoelectric focusing, in *Cell Separation Methods,* Bloemendal, H., Ed., Elsevier/North-Holland, Amsterdam, 1977, 145.

51. **McGuire, J. K., Miller, T. Y., Tipps, R. W., Snyder, R. S., and Righetti, P. G.,** New experimental approaches to isoelectric fractionation of cells, *J. Chromatogr.,* 194, 323, 1980.

52. **Leise, E. and LeSane, F.,** Isoelectric focusing of peripheral lymphocytes, *Prep. Biochem.,* 4, 395, 1974.

53. **Jüst, W. W. and Werner, G.,** Cell separation by continuous-flow electrophoresis and isoelectric focusing, in *Electrokinetic Separation Methods,* Righetti, P. G., van Oss, C. J., and Vanderhoff, J. W., Eds., Elsevier/North-Holland, Amsterdam, 1979, 143.

54. **Sherbet, G. V., Lakshmi, M. S., and Rao, K. V.,** Characterization of ionogenic groups and estimation of the net negative electric charge on the surface of cells using natural pH gradients, *Exp. Cell Res.,* 70, 113, 1972.

55. **Thompson, C. J., Docherty, J. J., Boltz, R. C., Jr., Gaines, R. A., and Todd, P.,** Electrokinetic alterations of the surfaces of herpes simplex virus infected cells, *J. Gen. Virol.,* 39, 449, 1978.

56. **Bont, W. S. and Hilgers, J. H. M.,** Ramol separation of cells at unit gravity, *Prep. Biochem.,* 7, 45, 1977.

57. **Tulp, A., Timmerman, A., and Barnhoorn, M. G.,** in *Electrophoresis '82,* Stathokos, D., Ed., Walter de Gruyter, Berlin, 1973, 317.

58. **Grateful, T. M., Athalye, A. M., and Lightfoot, E. N.,** Numerical description of zone electrophoresis in the continuous flow electrophoresis device, in *Cell Electrophoresis,* Bauer, J., Ed., CRC Press, Boca Raton, FL, 1994, chap. 2.

59. **Bauer, J.,** The negative surface charge density of cells and their actual state of differentiation or activation, in *Cell Electrophoresis,* Bauer, J., Ed., CRC Press, Boca Raton, FL, 1994, chap. 14.

60. **Rhodes, P. H. and Snyder, R. S.,** Theoretical and experimental studies on the stabilization of hydrodynamic flow in free-fluid electrophoresis, in *Cell Electrophoresis,* Bauer, J., Ed., CRC Press, Boca Raton, FL, 1994, chap. 3.

61. **Cohly, H. P. and Das, S. K.,** Cell electrophoresis using antibodies and antigens as ligands, in *Cell Electrophoresis,* Bauer, J., Ed., CRC Press, Boca Raton, FL, 1994, Chap. 5.

62. **Hannig, K., Wirth, H., Neyer, B., and Zeiller, K.,** Theoretical and experimental investigations of the influence of mechanical and electrokinetic variables on the efficiency of the method, *Hoppe-Seyler's Z. Physiol. Chem.,* 356, 1209, 1975.

63. **Saville, D. A.,** The fluid mechanics of continuous flow electrophoresis, *Physicochem. Hydrodynamics,* 2, 893, 1977.

64. **Deiber, J. A. and Saville, D. A.,** Flow structure in continuous flow electrophoresis chambers, in *Materials Processing in the Reduced Gravity Environment of Space,* Rindone, G. E., Ed., North-Holland, New York, 1982, 217.

65. McDonnell-Douglas Astronautics Company, Feasibility of Space Manufacturing — Production of Pharmaceuticals, Vol. 2, Technical Analysis, National Aeronautics and Space Administration Contractor Rep. NASA CR-161325, McDonnell-Douglas, St. Louis, MO, 1978.

66. **McCreight, L. R.,** Electrophoresis for biological production, in Bioprocessing in Space, Morrison, D. R., Ed., NASA TM X-58191, L. B. Johnson Space Center, Houston, 1977, 143.

67. **Giannovario, J. A., Griffin, R., and Gray, E. L.,** A mathematical model of free flow electrophoresis. *J. Chromatogr.,* 153, 329, 1978.

68. **Rhodes, P. H.,** High resolution continuous flow electrophoresis in the reduced gravity environment, in *Electrophoresis '81,* Allen, R. C. and Arnaud, P., Eds., Walter de Gruyter, New York, 1981, 919.

69. **Babsky, V. G., Zhukov, M. Y., and Yudovich, V. I.,** *Mathematical Theory of Electrophoresis,* Naukova Dumka, Kiev, 1983.

70. **Boese, F. G.,** Contributions to a mathematical theory of free flow electrophoresis, *J. Chromatogr.,* 438, 145.

71. **Biscans, B., Alinat, P., Bertrand, J., and Sancez, V.,** Influence of flow and diffusion on protein separation in a continuous flow electrophoresis cell: computation procedure, *Electrophoresis,* 9, 84, 1988.

72. **Clifton, M. J. and Marsal, O.,** Heat transfer design of an electrophoresis experiment, *Acta Astronaut.,* 1989.

73. **Bello, M. S. and Polazhaev, V. I.,** Fluid flow in free flow electrophoresis chamber in microgravity, *Microgravity Sci. Technol.,* 3, 3, 1990.

74. **Miller, T. Y., Williams, G. P., and Snyder, R. S.,** Effect of conductivity and concentration on the sample stream in the transverse axis of a continuous flow electrophoresis chamber, *Electrophoresis,* 6, 377, 1985.

75. **Rhodes, P. H., Snyder, R. S., and Roberts, G. O.,** Electrohydrodynamic distortion of sample streams in continuous flow electrophoresis, *J. Colloid Interface Sci.,* 129, 78, 1989.

76. **Snyder, R. S. and Rhodes, P. H.,** Electrophoresis experiments in space, in *Frontiers in Bioprocessing,* Sikdar, S. K., Bier, M., and Todd, P., Eds., CRC Press, Boca Raton, FL, 1989, 245.

77. **Taylor, G. I.,** Studies in electrohydrodynamics. I. The circulation produced in a drop by an electric field, *Proc. R. Soc. London,* A291, 159, 1966.

78. **Goetz, P.,** System 3000 Automated Electrokinetic Analyzer for biomedical applications, in *Cell Electrophoresis,* Schütt, W. and Klinkmann, H. Eds., Walter de Gruyter, Berlin, 1985, 261.

Chapter 5

Electrode Compartment-Separating Membranes for Cell Electrophoresis

Jörn Heinrich and Horst Wagner

CONTENTS

I. INTRODUCTION

Continuous flow electrophoresis (CFE) is a versatile and mild method for the separation of sensitive biological samples. The separation process is performed within a laminarly flowing buffer film into which the sample — i.e., the cell suspension — is injected as a fine filament. Separation inside a free buffer solution has many advantages, but also has the drawback that it is sensitive to convective interferences. All convective fluxes which are not parallel to the forced, slow buffer flow lead to a decrease of the resolution. An undisturbed flow pattern of the separation buffer is of eminent importance, especially during the preparative long-time fraction collection.

In commonly used CFE equipment the electrode compartments are separated from the sample processing space by ion-permeable diaphragms. These membranes allow the electric current to pass while preventing gas bubbles from the electrode reactions from coming into the separation space. The membranes have a low hydrodynamic permeability. This ensures that the fast-flowing electrode buffers do not disturb the laminar flow of the separation buffer.

Under the influence of the electric current, the surfaces of the membranes generate concentration shifts inside the separation buffer solution and thereby cause density gradients. With *in situ* measurements of the conductivity and field strength inside the separation chamber and with computer simulation of the flow field based on the measurement data, the influence of the membrane-induced density gradients was evaluated for process parameters and membrane materials commonly used for CFE separations.[1] The results of this study indicate that the membranes are a main source of resolution decrease. It is very difficult to evaluate the exact magnitude of this decrease because the resulting density gradients are dependent on many factors: the residence time of the buffer in the electric field, the buffer flow velocity, the current, the arrangement of the membranes, the membrane material, the cooling properties of the electrophoresis chamber walls, and the concentrations and compositions of both separation buffer and electrode solution.

0-8493-8918-6/94/$0.00+$.50
© 1994 by CRC Press, Inc.

II. MEMBRANE-INDUCED CONCENTRATION SHIFTS

It has been known for a long time that concentration shifts occur inside electrolyte solutions if electrical current is applied to nonuniform systems. Hittorf already described such effects in porous diaphragms.[2] Later, Nernst and Riesenfeld observed concentration shifts at the boundary between a phenolic and an aqueous salt solution. They also gave the first theoretical description of the phenomenon.[3,4] More recently, studies were performed on the concentration shifts near membranes inside CFE equipment. Thereby, equations were deduced from a phenomenological model. They allow the quantitation of the zones of changed concentration inside the separation chamber.[5]

The separation process of CFE may be affected by these concentration changes in several ways. They lead to gradients of buffer density, conductivity, and joule heat which decrease the effectivity of the method. Due to the local conductivity change near the surface of the membrane the electrical field strength around the suspended cells may decrease, thus preventing a complete separation within the residence time of the sample inside the field. Furthermore, a zone with changed pH and buffer composition may be generated near the membrane and may spread into the sample zone, causing loss of cell vitality.

The conditions for the solvated ions inside the membranes are different from those in a free electrolyte solution. The uncharged membrane matrix and membrane-fixed ions interact with the ions of the electrolyte. The influences of the membrane material on the different types of electrolyte ions are unequal. This unequal interaction is the reason why different ions have different transference numbers. It causes the concentration change within the electrolyte near the boundary to the current-conducting membrane. The concentrations shifts resulting in the CFE chamber are presented in Figure 5-1. Five possible arrangements of membranes commonly used in the CFE apparatus are shown. In each the concentration of the separation buffer, after the passage through the electrical field (y axis), is plotted against the distance in field direction (x axis). The concentration data were obtained by analyzing the buffer in the fractions of the corresponding CFE experiment, which were performed using an ELPHO-VAP 21 (Bender and Hobein, Munich, Germany).

Figure 5-1 is based on numerous experimental results obtained with a variety of different electrolytes. The magnitude of the concentration shift changes with the kind of ions dissolved in the electrolyte. However, an inversion of the electrolyte concentration shift pattern was not observed when the commonly used buffer substances (and mixtures of them) were tested (Tris, ammonia, triethanolamine, sodium, potassium, nitrate, chloride, phosphate, carbonate, amino acids, carbonic acids, Good's buffers). Depending on the kind of membrane and on the electrolyte concentration, the original electrolyte cation may be replaced by membrane-generated H_3O^+ ions. They cause a zone of higher conductivity within the regime of decreased electrolyte cation concentration. In the following section the different sources of the concentration shifts and their effects on the CFE process are discussed in detail.

A. WEAKLY CHARGED MEMBRANES

Weakly charged membranes (cellulose acetate or polyethersulfone membranes) produced the smallest concentration shift among the membranes tested. Only a small number of charged groups are found in the matrix of weakly charged membranes. The effect of the charged groups on the difference in the transference numbers and, thus, on the concentration shifts depends on the electrolyte concentration. The lower the electrolyte concentration, the higher the concentration shifts caused by the membrane fixed charges. An influence of the membrane-fixed charges in weakly charged membrane material is detectable only if the electrolyte concentration is in the range of the membrane charge

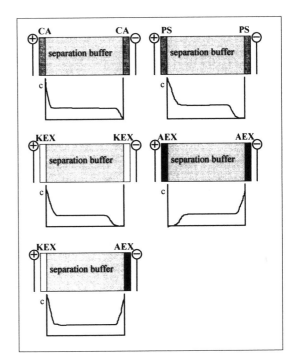

Figure 5-1 Qualitative schemes of the concentration shifts caused by the different possible arrangements of commonly used membranes. AEX, anion exchange membrane; KEX, cation exchange membrane; CA, cellulose acetate membrane; PS, polyethersulfone membrane; c = concentration.

concentration (Figure 5-2). In the case of the cellulose acetate membrane, a remarkable concentration shift increase can be detected only beyond the electrolyte concentration of 2 mmol/l. With increasing electrolyte concentration the magnitude of the concentration shift reaches a constant value.

The difference in the transference numbers is caused by the unequal interaction of the uncharged membrane matrix on the various types of ions of an electrolyte. The cations migrate faster through the membrane than the anions compared to their migration in free solution. Different velocities of different kinds of ions passing through the uncharged membrane matrix can be due to molecular sieving, based on the hydrated ion diameter, to different adsorption coefficients of the anions and cations on the matrix, or to selective solvation of the ions in the matrix. However, it is important to note that the magnitude of the differences in the transference numbers cannot be decreased by a change in the electrolyte concentration within the working range of cell electrophoresis buffer systems.

Weakly charged or nearly uncharged membranes are often used in cell electrophoresis since they have the advantage of producing the smallest concentration shifts and letting all kinds of electrolyte ions migrate through at very similar rates. They have the disadvantage of producing a depletion zone near the membrane sealing the cathode compartment. This has to be taken into account when selecting the parameters of an electrophoresis experiment.

B. CHARGED MEMBRANES

In electrophoresis experiments with charged membranes a portion of the electrolyte ions is replaced by membrane-fixed ions. In the commercially available, hydrated ion exchange membranes which were used in the authors' experiments, the concentration of membrane-fixed ions is in the range of 1 mol/l. Mannecke and Laupenmühlen showed

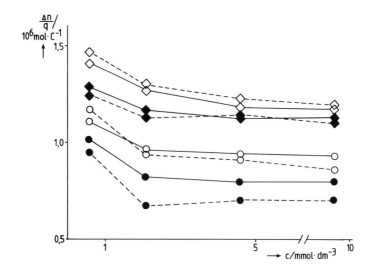

Figure 5-2 Influence of the electrolyte concentration on the enrichment and depletion equivalents (particles per charge) obtained with the weakly charged cellulose acetate membrane. Residence time, 300 s; open diamond, Tris/acetic acid; closed diamond, Tris/nitric acid; closed circle, sodium nitrate; open circle, sodium acetate/acetic acid; all buffer solutions pH 7, current densities from 30 to 130 A/m²; broken line, accumulation equivalent; normal line, depletion equivalent; electrode compartment electrolyte concentrations equal to the chamber electrolyte concentration.

experimentally that an ion exchanger membrane is free of ions with the same charge sign as those fixed in the membrane matrix if it is immersed into an electrolyte with an ion concentration ten times lower than the concentration of membrane-fixed ions.[6] If the cation exchange membrane is mounted so as to seal the cathodic electrode compartment, the buffer solution is greatly depleted during an electrophoresis experiment: the cations move into the membrane and the anions move away from the boundary between the membrane and the buffer without being replaced. In this way the charge equivalents of the electrolytes are depleted at the boundary according to the transference number of the anions. The same amount of electrolyte is accumulated in the electrode compartment. The accumulation (or depletion) equivalent for common electrolytes such as sodium chloride, phosphate buffers, or acetate buffers ranges from 20 to 80% of the overall charge transport, depending on the composition of the electrolytes and the transference numbers in free solution. If the membrane is a strong cation exchanger containing sulfonic acid groups, for instance, then the conductivity of the buffer at the depletion boundary is decreased, giving rise to a high field strength and temperature (Figure 5-3). The strong cation exchange membranes produce concentration shifts comparable to the weakly charged membranes, but with a larger depletion (or accumulation) equivalent.

If a weak anion exchanger is used to seal both electrode compartments, a process different from mere electrolyte depletion can be seen near the membranes. In this case, the pK value of the ion-exchanging groups is important for the concentration changes within the chamber buffer. In the depleting boundary the buffer cations are replaced by H⁺ ions generated by the hydrolysis equilibrium of the ion-exchanging groups near the membrane surface. If there is no buffering anion present in the chamber electrolyte, a zone of very high conductivity is generated at the anodic membrane which continuously enlarges (Figure 5-4). If an electrophoresis experimented is performed with this membrane arrangement and the buffer is collected at the bottom of the separation chamber,

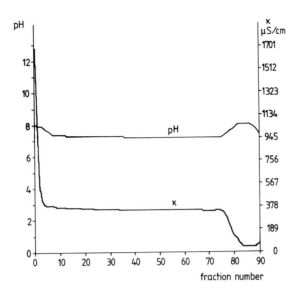

Figure 5-3 Fraction evaluation: zones of changed concentration generated by cation exchange membranes at the anode and cathode electrode compartments. Anode to the left, cathode to the right. Residence time, 300 s; voltage, 360 V; current density, 31 A/m². Buffer: 5 mmol/l Tris and 4.4 mmol/l HNO₃, pH 7.5; anode buffer: 0.3 mol/l Tris; cathode buffer: 0.1 mol/l HNO₃.

more than three quarters of the fractions show a decreased pH value between 2.3 and 3.6. Only a small number have kept the original pH value of about 7. Such a pH change has to be avoided because sensitive cells would suffer (Figure 5-5). Furthermore, the uncontrolled enlargement of a zone of completely changed conductivity and pH produces unstable conditions of temperature and field strength in the rest of the separation space.

The uncontrolled broadening of the zone of lowered pH can be avoided by the addition of a buffering anion such as acetate. In this way the concentration of the H⁺ ion is always kept above a value which would produce a higher effective conductivity for H⁺ than that of the buffer cation. A zone of decreased pH coming from the membrane then enlarges slowly, but is controlled by the self-sharpening properties of the boundary toward the unaltered electrolyte (Figure 5-6). Such systems are well known in the isotachophoretic analysis of cations, the corresponding base of a weak acid being the counterion and the terminating zone with the smallest effective cation conductivity being the buffered H⁺ ion zone.

The most useful arrangement of ion exchanger membranes for cell electrophoresis is to combine an anion exchanger membrane sealing the cathode electrode compartment with a cation exchanger membrane sealing the anode electrode compartment. With this membrane arrangement two accumulation layers are generated during electrophoresis near the membranes. They have a higher density than the rest of the electrolyte, but cause a convective resolution decrease only at low flow velocities in the vertical working mode of the electrophoresis apparatus. Such an arrangement of the membranes is used in the isoelectric focusing of proteins and has an inherent disadvantage for cell electrophoresis buffer systems which contain multivalent ions. They have a higher affinity to the ion exchanger than single charges so that the ions pass through the membrane with different transference numbers. The highly mobile H⁺ ions (and alkaline ions) migrate much faster inside the membrane than the divalent or trivalent ones, and the same is also applicable to the anions.

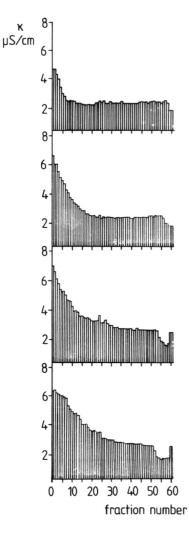

Figure 5-4 *In situ* measurement of the conductivity with a potential gradient conductivity scanner. The data are collected at 50, 100, 150, and 200 mm after the beginning of the electrode compartments. The anode is on the left. Both electrode compartments are sealed with anion exchange membranes. Process conditions: buffer, sodium nitrate, 2.5 mmol/l, pH 6.5; residence time, 300 s; current density, 31 A/m^2, anode buffer, 0.1 mol/l NaOH; cathode buffer, 0.1 mol/l HNO$_3$. K = conductivity.

The risk of a change of the electrolyte composition inside the electrophoresis chamber induced by ion exchange membranes can be minimized if concentrated marginal electrolyte solutions are pumped through the electrophoresis chamber near the membranes. They stop the changes in concentration by the low field strength and by the slow migration rates inside these zones. This special buffer solution arrangement with a lower conducting buffer zone, to which the biological material is applied, between two high conductivity zones is also known as field step electrophoresis.[7]

If systems with a small buffer capacity are used, the highly concentrated, stabilizing marginal solutions can be the source of concentration changes too. This comes from concentration-dependent transference numbers, especially of the buffering ions. Generated at the concentration boundary between the highly concentrated marginal solution and the separation buffer, pH shifts may enlarge continuously into the separation space and change the pH at the location of the sample stream. To overcome this problem, the pH of the marginal solution — that is, the buffer ion concentration(s) — must be adjusted to a value at which the transference numbers of these ions are the same as in the less concentrated and less conductive separation buffer. This problem can be solved easily by

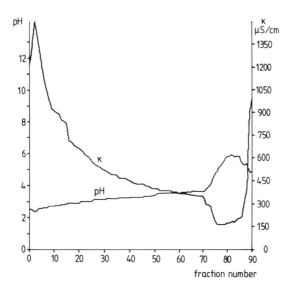

Figure 5-5 Fraction evaluation: generation of a zone of increased H$^+$ concentration at the anodic anion exchange membrane in sodium nitrate buffer, 2.5 mmol/l, pH 6.5. Residence time, 300 s; current density, 31 A/m^2; anode buffer, 0.1 mol/l NaOH; cathode buffer, 0.1 mol/l HNO$_3$.

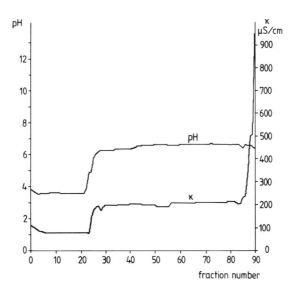

Figure 5-6 Fraction evaluation: generation of a zone of decreased pH at the anodic anion exchanger membrane. The broadening of the zone is controlled by a moving boundary caused by the acetate ion present in the buffer (sodium acetate, 5 mmol/l, acetic acid 0.04 mmol/l); residence time, 300 s, current density, 100 A/m^2; anode buffer, 0.1 mol/l NaOH; cathode buffer, 0.1 mol/l acetic acid.

trial and error in the case of simple systems such as the ammonium acetate saline buffer. In this case either ammonia or acetic acid has to be added to the marginal solution until the pH in the separation buffer remains unchanged during electrophoresis. For more complex systems with more than one pH-adjusting component, only time-consuming experiments or computer simulations can help to find the optimal composition of a marginal solution to replace the migrating separation buffer ions by exactly the amount needed to keep the composition of the buffer unchanged at the sample separation.

III. CALCULATIONS ON CONCENTRATION SHIFTS

With the knowledge of a few apparative and method parameters, the membrane-induced concentration changes in the separation buffer can be calculated in advance, thus helping to start optimization of the method with preoptimized conditions. The knowledge of the width of a depletion zone, together with the expected migration distance of the sample, helps to find the right position for the introduction of the sample stream if the electrophoresis equipment offers the possibility to select it.

The parameters which are needed for the calculation of a depletion or accumulation layer must be evaluated experimentally since they depend on the special geometry of the CFE equipment and the membrane properties, together with the electrolyte used. An experimental concept for the measurement of the relevant quantities is presented which can be adapted to any CFE apparatus and electrolyte system. The first parameter to be evaluated is the difference between the electrolyte transference numbers in free solution and those inside the membrane. The second one is the diffusion velocity of the electrolyte in the membrane. The differences in transference number yield information on the extent to which the charge transport through the membrane leads to an accumulation or depletion of an electrolyte; the diffusion velocity yields information on the possibility of reducing an electrolyte depletion by diffusive concentration balance from the electrode compartment.

A measurement of the differences in transference numbers is carried out in the following way: the electrolyte system used for the cell separation is introduced into the chamber. The electrode compartments contain the same buffers as those present at the corresponding membrane boundary inside the separation space. A medium residence time and current is applied to the equipment, and the fractions are collected after the equilibration time of the process. The fractions near the membranes are examined for a change in concentration, determining the volumes and the ion concentrations of the electrolytes collected in each fraction. They are not equal to the fractions obtained from the center of the separation chamber. The depletion (or accumulation) equivalent is obtained from the number of electrolyte ions depleted (or accumulated) divided by the electrophoresis current within the residence time (Equation 5-1):

$$\Delta T = \frac{\Delta n \cdot F}{i \cdot t} \tag{5-1}$$

where T = transference number (dimensionless), n = number of ions depleted (or accumulated in moles), F = Faraday constant = 96,480 C/mol, i = electrophoresis current (in amps), and t = residence time of the buffer in the electric field (in seconds).

The diffusive transport through the membrane, which may help to minimize a depletion zone, can be estimated if the electrolyte diffusion velocity inside the membrane is known. A series of experiments is necessary: the electrode buffer concentration is varied until no further depletion or even an increased concentration can be detected in the fractions collected near the depleting membrane boundary. The experiment must be carried out with a moderate flow velocity and a relatively small current since it is hardly possible to balance the depletion by a difference in concentration at high currents. If the depletion is exactly balanced by diffusive transport from the electrode compartment, then for a state of a constant concentration difference across the membrane the diffusion velocity is[5]

$$v_{Diff} = \frac{D_M}{l_M} = \phi \cdot \frac{\Delta T}{F \cdot (c_E - c_S)}$$

(5-2)

where v_{Diff} = diffusion velocity (in meters per second), D_M = diffusion coefficient in the membrane (in square decimeters per second), l_M = membrane width (in meters), c_E = electrode buffer concentration (in moles per liter), c_S = separation buffer concentration (in moles per liter), and ϕ = density of the electrical current (in amps per square meter).

An accumulation zone is detectable only very close to the membrane if no convective process leads to a broadening of it. The depletion zone, however, enlarges continuously as the depleted electrolyte comes from inside the separation chamber, except for the amount transported by diffusion through the membrane. In the depletion zone a steep gradient of field strength, temperature, and conductivity causes an intensive mixing of liquid, which accelerates the transport of electrolytes to the membrane. For the estimation of the concentration needed in a marginal solution to prevent a depletion zone from coming into the separation space, it is essential to know the width of the depletion zone after the buffer film has passed through the separation space. The width of the depletion zone is calculated from the differences in transference numbers, the diffusion velocity, the current, the cross section, and the concentrations of the electrolytes on both sides of the membrane:[5]

$$b_D = t \cdot \left[\frac{i \cdot \Delta T}{A \cdot F \cdot c_S} + v_{Diff} \cdot \left| \left(\frac{c_E}{c_S} - 1 \right) \right| \right]$$

(5-3)

where b_D = width of the depletion zone in field direction (in meters) and A = cross section of the separator, i.e., the current flown area (in square meters).

The width of the depletion zone can be reduced by introducing a buffer solution with a higher concentration near the membrane. It is reduced by the factor by which the marginal concentration exceeds the separation buffer concentration. However, if a marginal solution is used, the separation buffer must be adjusted to the viscosity and density of the marginal solution by addition of uncharged substances such as sucrose for higher density and polymers for increased viscosity. This is necessary to maintain the same flow velocities of both marginal solutions and separation electrolyte and, thus, equal fractionation rates, especially in vertically operated equipment.

IV. CONCLUDING REMARKS

In continuous flow electrophoresis (CFE) concentration changes near the electrode compartment diaphragms may cause severe interference with the separation process. The changes in concentration lead to gradients of density and joule heat which may produce a convective flow that reduces the resolution of the method. Furthermore, the field strength at the location of the sample may be altered. This not only leads to incorrect mobility values, if CFE is employed for such measurements, but also severely decreases the efficiency of the method for preparative separations. The choice of the optimal membrane material is a very sensitive point for the methodological optimization. A sophisticated electrolyte arrangement, such as combining separation solutions with optimally adjusted marginal solutions, may help to improve cell electrophoresis results.

REFERENCES

1. **Heinrich, J., Clifton, M.J., and Wagner, H.**, Use of in situ conductivity measurements to calculate the flow field and heat transfer in continuous flow electrophoresis, *Int. J. Heat Mass Transfer*, 36, 3703, 1993.
2. **Hittorf, J.**, Bemerkungen über die Bestimmungen der Ueberführungszahlen der Ionen während der Elektrolyse ihrer Lösungen, *Z. Phys. Chem.*, 39, 613, 1902.
3. **Nernst, W. and Riesenfeld, E.H.**, Ueber elektrolytische Erscheinungen an der Grenzfläche zweier Lösungsmittel, *Ann. Phys.*, 8, 600, 1902.
4. **Nernst, W. and Riesenfeld, E.H.**, Bestimmung der Ueberführungszahl einiger Salze in Phenol, *Ann. Phys.*, 8, 609, 1902.
5. **Heinrich, J. and Wagner, H.**, Influence of the electrode compartment separating membranes on continuous flow electrophoresis, *Electrophoresis*, 14, 99, 1993.
6. **Mannecke, G. and Laupenmühlen, E.O.**, Elektrische Leitfähigkeit von Kationernaustauschermemebranen, *Z. Phys. Chem.*, 2, 336, 1955.
7. **Wagner, H. and Kessler, R.**, Free-flow field step focusing — a new method for preparative protein focusing, in *Electrophoresis '82*, Stathakos, D., Ed., Walter de Gruyter, Berlin, 1982.

Chapter 6

Cell Electrophoresis Using Antibodies and Antigens as Ligands

Hari H. P. Cohly and Suman K. Das

CONTENTS

I. INTRODUCTION

It has been demonstrated that cells in suspension migrate in a characteristic and reproducible pattern when placed in an electrical field. This book focuses entirely on events related to the electrical charge. Any issue related to charge hinges on whether a molecule is positively or negatively charged or is neutral. With this in mind, we may begin with the generalization that cells are charged species and that charge densities on the cell surface membrane may play a pivotal role in explaining our observations.[1] It appears evident from the work of the authors of this chapter and that of others that electrophoretic mobility of cells depends not only on the electrokinetic charge density, but also on the glycocalyx thickness and the friction of the electroosmotic flow within the glycocalyx.[2]

This chapter contains a presentation of experimental findings in the field of immunology and a discussion of their significance in light of current concepts. Important new directions of investigation are also suggested. The main topics include cell (intraepithelial lymphocytes from mouse gut) separation by electrophoresis and double antibody tagging, quantification of antibody tagging by microelectrophoresis, and study of the effect of the presence of antigen on the electrophoretic mobility and proliferating response of T-cells

0-8493-8918-6/94/$0.00+$.50
© 1994 by CRC Press, Inc.

in the continuous flow electrophoresis device of McDonnell-Douglas Aeronautics Corporation.

A. CELL SURFACE

The cell surface was first envisioned as a rigid lipid bilayer membrane model which was seen as a barrier between the extracellular environment and a poorly organized cytoplasm.[3] This perception was altered by Singer and Nicolson,[4,5] who proposed their "fluid mosaic model" of cell membrane structure, in which they indicated that the lipids are in a fluid state, permitting lateral diffusion of the proteins embedded within the membrane. The proteins were envisioned somewhat as serenely floating icebergs in a sea of lipid bilayer. Two kinds of membrane-associated proteins, integral and peripheral, were envisioned in their model.[6]

The current concept of the cell surface is that of a dynamic organelle extending from the extracellular matrix through to the cytoskeleton. With all possible diversity there are still a limited number of functional domains: first, a ligand-binding region on the surface of the cell; second, a transmembrane component; and third, a signal generating system in the cytoplasm.[7-9] Conditions of cell cytoplasm in turn modulate the properties of these receptors. Surface proteins also function in transporting material and information into and out of the cell. Simultaneously, fusion of vesicles regenerates the membrane and implements the laying down of the extracellular matrix; thus, the stable membrane has now gone through a "metamorphosis" and is envisioned as a dynamic, continuously changing membrane structure.[10] To understand this dynamic membrane structure it is essential that we know the mechanism that regulates the timing of appearance as well as the spatial distribution of specific receptors, adhesion mediators,[11] and specialized structures involving the glycocalyx on the cell surface.

The general characteristics of the membrane include the binding of the ligand to its receptor. This ligand recognition is accompanied by a change in the receptor, which is transmitted across the membrane to an intracellular component that acts as a signal generator. The signal itself may be either of two classes: excitation and adaptation.[3] The excitation signal results from a cascade of events initiated by interaction of the intracellular portion of the receptor with another molecule, eventually leading to the stimulation of an effector. This effector function can be due to activation of protein kinase, the generation of second messengers, the opening of an ion channel, or still other factors.[12] The net result of these changes is to amplify the impact of the initial event. Concomitant with this excitation signaling there is a feedback mechanism that depolarizes this impact by modifying the sensitivity or regeneration of the excitation signal. The desensitization may include the removal of receptor, thus down-regulating or modulating the sensitivity to reset the membrane for the incoming stimulus.[3] Hence, as in other systems the events occurring on the cell surface may be envisioned as following a sinusoidal curve, as shown by Secvcikova et al.[13]

Conformational changes can be studied with voltage-gated K^+ channels in a millisecond range of timing, while site-directed mutagenesis provides insight into the role that certain amino acids play in the binding interaction. X-ray crystallography, on the other hand, at a resolution of better than 3 Å makes clear the crucial role that the structure of the receptor plays in regulating the first events.[6] One of the simplest forms of conformational change is the aggregation of individual subunits to form a multimeric structure. The transmembrane conformation change has been envisioned by Stoddard et al.[7] as a pulley, while the analysis of Kim et al.[8] involves the stabilization of a small rotational movement in the molecules that allows the two subunits to pivot around each other.

Lymphocyte receptors, the target cells in immunological studies, can be grouped into three categories: adhesion molecules that interact with antigen-presenting cells, second receptors for cytokines that promote differentiation or proliferation, and, finally, antigen receptors (the T-cell antigen receptor and B-cell-surface immunoglobulin) that associate with multiple signal complexes to permit activation following interactions with ligand.[14]

The late 1970s and early 1980s were devoted to elucidation of the kinetics of endocytosis and recycling. Two components can be recognized during ligand-induced rearrangement. The first component involves the formation of clusters of ligand; the second involves the preferential relocation of ligand to certain areas on the cell, i.e., capping.

Experiments with antibody as ligand shed light on the phenomenon of binding.[15] Cluster formation has been proposed to be from cross-linking of surface sites, whereas the second component of rearrangement must involve more than cross-linking or aggregation of labeled ligands, which is dependent on energy.[16] Cap formation and pinocytosis are ATP-requiring active processes induced by specific antibody; they may involve microfilament activity.[17-19] The observation that Fab anti-immunoglobulin-treated cells cap if they are subsequently treated with anti-Fab antibody[20] suggests that cross-linking by divalent antibody is essential for capping. Whether a given antigenic determinant is immunogenic in relation to a given lymphocyte receptor depends on the differentiation state or maturity of that cell.[20,21] Cap formation is not a prerequisite for pinocytosis. Patching, which is an intermediate step between binding and capping, is an energy-independent but temperature-dependent process.[22]

With respect to receptors and ligands, the triggering event can be envisioned as a transport system in which the import rather than the export sorting system is used.[23] In this import system the membrane structures used inside the cells are endosomes and lysosomes; they are referred to as afferent arms of the general transport apparatus, which nevertheless is in continuous dynamic flux. The two pathways which have emerged from receptor studies are the clathrin-dependent (coated pits) and clathrin-independent endocytic pathways. Clathrin-dependent pathways have been shown to exist for polymeric IgA, growth factors, and plant lectins, involving more of the physiological molecules. Clathrin-independent pathways, which may include the nonselective uptake of solute molecules by surface vesiculation and engulfment of media without utilizing a specific surface binding site,[23] have been shown to occur in tissue culture cells *in vitro,* but may not really represent the *in vivo* situation. Coated pits usually are associated with transmembrane glycoproteins or glycolipid receptors.[24] Clathrin-independent pathways also can be designated as pinocytic pathways.

The capping process with concanavalin A (Con A) or fluorescent conjugate has been shown by fluorescence and peroxidase labeling to be completed in 30 min.[25] Immunofluorescent and ultrastructural studies of COS-7 cells infected with recombinant thrombomodulin have shown that antithrombomodulin antibodies can be internalized in a time- and temperature-dependent manner, with internalization detectable within 10 min. Initiation of endocytosis for 10 to 60 min resulted in a redistribution of gold particles into small clusters in a noncoated pit. Hence, the most recent literature shows that the process of endocytosis of thrombomodulin involves the participation of both clathrin-coated and noncoated pits, but that the latter process predominates.[26] Coated pits may play a role in polarization of fibroblast cells, but noncoated pits appear to play a role in initial attachment, spreading, and formation of focal adhesions.[27] The approach used for the clathrin-independent pathway has been selective inhibition of the coated pathway. Antibodies specific to clathrin have been injected into the cytosol for inhibition of coated pits.[24] To date, no marker has been shown for vesicles other than those of the pits, but van Deurs et al.[23] have proposed the population of small (50 to 100 nm), noncoated, flask-shaped invaginations seen at the surface of many cell types (e.g., the smooth muscle cells, fibroblasts, fat cells, and endothelial cells) as a potential candidate. Other possibilities include the pits which are between 150 and 300 nm and the membrane ruffling and outgrowth of lamellipodia associated with glial cells.

B. CELL SURFACE RECYCLING

The concept of membrane recycling, which is involved in the maintenance of the surface area and the volume of the entire cell, must be addressed. Steinman et al.[28] have observed that remarkable amounts of plasma membrane can actually be internalized during endocytosis with no cell shrinkage or increase in total surface area of intracellular vacuoles within a 3-h period. They conclude that membrane cycles between the cell surface and the intracellular endocytic compartments. The turnover is fairly rapid, and a high fraction of ligand is released by the cells in the first 30 min, continuing thereafter at a much lower rate. [^{14}C]-sucrose has been used as a fluid-phase endocytic marker, and it was shown to return to the extracellular medium in about 20 min.[29]

Although there are some problems in the interpretation of the literature, the general conclusion from these studies is that the uptake rate of fluid-phase markers and plasma membrane may be very high, but a considerable amount of the plasma membrane and the endocytosed fluid is rapidly returned to the exterior of the cell (5 to 30 min for physiologic ligands). Further, more vesicular and tubular appendices or extensions are involved in recycling of membrane and associated receptors on the cell surface. Various studies with nonspecific markers such as horseradish peroxidase (HRP) and conjugated ferritin have established a transcytotic pathway, i.e., a communication between the apical and basolateral epithelial surface in various epithelia. Also, Wilson and King[30] found transcytosis in both cationic and anionic HRP in monkey trophoblastic epithelium, indicating that the charge of the solute really does not make too significant a difference.

Considering the current concepts, it appears that all endocytic molecules reach the cytoplasm of the cell to be recycled; the only point of difference could be the percentage of vesicles that will be recycled which are specific for a given solute or substrate. The process of transcytosis has been well established for ligands involving immunoglobulins of the polymeric IgA class.[31]

In the endosomal compartment, molecules (i.e., receptors and ligands) that are not selected for recycling, for transcytosis, or for the Golgi complex are eventually delivered to the lysosomes, which may represent the default pathway of endocytosis. The distinction between the endosomal and lysosomal compartments is the temperature block. Endocytosed molecules are said to be found in endosomes, but not in lysosomes, when the cells are incubated at 18 to 20°C.[32] Using fluorescein probes and digital analysis, typical multivesicular bodies have been shown to be the immediate precursor of the secondary lysosome. Roederer et al.[33] have shown that hydrolysis of endocytosed molecules begins within a few minutes after uptake at 37°C and that exposure of endocytosed molecules to lysosomal enzymes can take place below 20°C. These data indicate that although most hydrolytic enzymes may be added at a certain late stage in the endocytic cycle, some enzymes have already been added at an early stage, suggesting that endosomes gradually mature into lysosomes. The conclusion from all these studies is that the endosome transforms into a lysosome by receiving lysosomal enzymes via the mannose-6-phosphate receptor system. Hence, the endosome and lysosome probably should be considered as collectives for very early and late endocytic compartments.

In addition, molecules also may be transported from endosomes to the Golgi complex. This process of transport was clearly demonstrated by Snider and Rogers, who followed the fate of asialotransferrin receptors on the cell surface.[34] Since the only site for sialytransferase is the trans Golgi network (TGN),[35] the appearance of asialotransferrin receptors on the cell surface indicated that the cells must have passed through the TGN pathway.

Lectins such as wheat germ agglutinin also have been shown to reach the Golgi complex.[36] Thus, apparently all ligands or ligand conjugates can be traced to the same endosomal structures, regardless of their binding sites, their valency, or their preference for coated or uncoated pits. Also, time sequence studies show that whereas internalized molecules reach the endosomal compartment in minutes at 37°C, it takes at least 30 min for the same probes to reach the Golgi complex in detectable amounts.

Many aspects of the endocytic part of the protein handling system have yet to be resolved.

C. MEMBRANE VESICLES

Patch-clamp analysis and fluorescence microscopy in recent studies involving influenza virus hemagglutinin (HA) have revealed that the HA fusion pore opens before there is substantial lipid mixing.[10] The trigger which activates the fusion protein of viruses that are activated at neutral pH is questionable. The concept of a highly localized fusion event, most likely involving fusion pores that are circumscribed with proteins, is gaining momentum, and the fusion pore seems to be the common element that unites all these cellular fusion events. Kinetic studies with fusion proteins have shown that free, cis-unsaturated fatty acids are most effective in promoting fusion.[37] Recent simulation experiments on the cell membrane interior on the 100-ps time scale indicate that the apparently high viscosity of the cell membrane is more closely related to molecular interactions that occur on the cell surface rather than in the cell interior.[38] Annexin VI appears to be essential for "pinching off" membrane of clathrin-coated vesicles in conjunction with calcium when endocytosis is reconstituted *in vitro* with isolated plasma membranes and cytosolic fractions.[39]

Three approaches that have contributed to the recent progress are genetic screening for translocation components, identification of membrane proteins adjacent to translocating polypeptides by cross-linking, and reconstitution of the translocation components into proteoliposomes after their solubilization and purification.[40]

D. EPITHELIAL CELLS

Both selective targeting and selective retention pathways determine protein distribution at the cell surface in polarized epithelial cells.[41] In Maldin-Darby canine kidney (MDCK) cells, a cell line derived from polarized renal epithelia, proteins are generally sorted into different vesicle populations in the TGN and are recruited directly to the apical- or basal-lateral membrane (a selective targeting pathway). Some proteins are delivered to both membrane domains and sorted subsequently by differential retention and turnover (a selective retention pathway).

In hepatocytes *in situ* and in the intestinal cell line Caco-d2, proteins are first delivered to baso-lateral membrane; basal-lateral proteins are selectively retained, while apical proteins are resorted and selectively targeted to the apical membrane. Differences in protein sorting in these cells may reflect the presence of cell type-specific machinery for protein sorting or the localization of the same sorting machinery to different membrane compartments. An apical membrane protein, dipeptidyl peptidase IV, is sorted to the apical membrane in both hepatocytes and MDCK cells. Sorting of one class of proteins to the apical membrane is determined by a glycosyl phosphatidylinositol (GPI) membrane anchor. In some proteins, a basal-lateral sorting signal includes a tyrosine residue that confers an increased rate of endocytosis from the cell surface.[42] The structural character-istics of this basal-lateral surface suggest that it may be recognized by cytosolic proteins that are related to the family of adaptor proteins (components of the clathrin-coated pit assembly) that regulate signal-mediated endocytosis of receptor proteins at the cell surface.[43]

A family of cell surface glycoproteins, the selectins, is capable of using some of these carbohydrate structures in adhesive mechanisms that transfer leukocytes to regions of inflammation. Data from homing studies pointed to a multistep combinatorial process whereby selectins mediated the initial, low-affinity interaction with the vascular endo-thelium, after which a given leukocyte activator mediated the induction of a higher-affinity interaction determined by the leukocyte integrins.[11] Hence, diverse selectins, leukocyte activation signals, and integrins could be utilized in a combinatorial matrix to induce migration of different leukocyte subpopulations to a given type of tissue during inflammatory episodes.

The early experiments on lymphocyte triggering by antibodies and lectins indicate that membrane phenomena are an important initial part of this program. Extensive study of the membrane phenomena is essential in order to capture the initial events of lymphocyte activation.

Because these events probably occur within minutes on the surface of the cell membrane, a technique using electrophoresis would allow immediate assessment of membrane alterations. Cell electrophoresis provides a probe of membrane interactions involving changes in the surface charge density, and these changes are reflected by an alteration of the cell's electrophoretic mobility. Thus, the specificity of electrophoretic mobility has become a valuable investigational tool in the study of alterations in cell membranes during biological processes.

A number of models have been proposed for the receptor-ligand binding, but the emphasis has been on transmembrane signaling, which actually involves structures that may lie within the membrane or inside the cell cytoplasm.[7,8] Emerging from all of the new technology is the opportunity to design systems that are economical, easy to use, macroscopically visible, and associated with the fewest complications.

II. ELECTROPHORESIS

Electrophoresis is the transport of charged molecules or suspended particles in a homogeneous polar liquid under the influence of an electrical field.[44,45] Each particle has its own net electrical charge. When numerous particles with different charges are mixed, subjecting the mixture to an electric current causes separation of the different particles based on the charge of each particle.

Electrophoretic separation of cells is based on the difference in density of the cell surface charge. This charge is an "apparent" surface charge, since the "true" charge of the cells that reflects the composition of the cell surface is greatly reduced by a layer of ions of opposite charges surrounding the cell and forming an electrical double layer.[46,47] Since mammalian cells are negatively charged at physiologic pH, their isoelectric points (pI) being in the range of 2.1 to 3.4, cations are attracted to cells by electrostatic forces.[48,49] Further removed from the cell membrane, the cells are surrounded by an additional, diffuse region in which ions are distributed according to the electrostatic forces present and random thermal movement. The electrophoretic mobility depends on the zeta potential[44,45] and is at the outer part of the double layer ("surface shear") composed of the cell surface and its surrounding ions, representing the reduced (i.e., apparent) charge of the cell.

A basic characteristic of the surface of the eukaryotic cell is the charge, which is determined by the nature and number of ionogenic groups exposed on the plasma membrane.[50] Anionic charges associated with the cell surface charges are phosphate, sulfate, and carboxyl groups, while the cationic charge comes from the guanidium and ammonium groups.[51] Sialic acids and sulfated proteoglycans are responsible for the anionic charge. Sialic acids comprise several derivatives of neuraminic acids, including N-glycol, N-acetyl, and N,O-acyl derivatives. Phosphates and sulfated glycosaminoglycans are anionically charged carbohydrate molecules that contain carboxylates and sulfates from uronic acids.[52]

Charge has been shown to play a significant role in the degeneration and regeneration of cells. Bush et al. showed that transient anionic fenestrations developed in mouse endoneural vessels in 4-day crush injury measured at pH 2.0 by labeling the glycocalyx moieties such as sialic acid with cationic colloidal gold.[53]

The influence of charge on the cell surface was examined recently through observations of Vargas et al.,[52] who found that sulfated glycosaminoglycans are the main carriers of the surface charge in vascular endothelial cells. This was shown by the suppression of the cell mobility by treatment with chondroitin sulfate lyase. The uptake of poly(L-lysine) conjugates in cultured MDCK epithelial cells has been used as a model for nonabsorptive endocytosis of cationic macromolecules.[54] Neuraminidase pretreatment resulted in a 40% increase, whereas trypsin treatment did not. These results suggest that the negative charges, rather than the glycoprotein matrices, of glycocalyx play an important role in the control of cationic macromolecules in kidney epithelial cells. Sowemimo-Coker and Meiselman showed that lower concentrations of cationic anesthetics such as procaine hydrochloride increased electrophoretic mobility while higher concentrations decreased it.[2] This study clearly showed the increase in electrophoretic mobility to be due to the expansion of the RBC glycocalyx membrane. The pattern of observing first an increase in electrophoretic mobility with low concentrations of cationic ligand and then a decrease with high concentrations is similar to that reported by Blume et al. with Con A on lymphocytes.[55]

There are two types of electrophoresis. In one type, the material to be separated is placed on a gel and is the current applied as in preparative electrophoresis (ascending and descending). In the second type, free flow electrophoresis, the material is placed into a moving stream of buffer solution. As the material passes through an electric field, the individual components congregate into discrete streams in the solution. This method is used to process larger quantities of materials than is possible with the gel method.

Electrophoresis of cells in density gradient columns was introduced simultaneously and independently by Kolin[56] in the U.S., and Svenson and Valmet in Sweden.[57] A dense and a less dense solution are mixed to produce a density steadily increasing downward in the column. A density gradient counteracts heat convection and prevents mixing of the electrophoresed particles.[58]

A. PREPARATIVE ELECTROPHORESIS

Attempts to perform both descending and ascending preparative electrophoresis using density gradients of polymer solutions[59-62] have often yielded inconsistent results and produced undesirable effects on cells and their zeta potential.[63-67]

In ascending density gradient electrophoresis, gradient components and ionic strength of buffers are the most important parameters influencing electrophoretic mobility (u). Generation of joule heat can be reduced by using electric fields of low strength and buffers of low conductivity made isotonic by the addition of nonionic sugars. Furthermore, the procedure can be carried out at 4°C.[61,68]

Heavy water (D_2O) has been used in making gradients.[59] It is biologically, chemically, and physicochemically inert and thus suitable for the separation of living cells. Conductivity decreases with increasing D_2O concentration.[59] At the same time, the viscosity increases approximately 25% when D_2O concentration increases from 0 to 100%. Considering the von Smoluchowski's equation for electrophoretic mobility,[45]

$$u = \frac{7ESe}{88n}$$

where E is the electric field, S the electrokinetic potential, e the dielectric current, and n the viscosity, it is obvious that changes in E and n (which both increase at the same rate with increasing D_2O concentration) roughly cancel each other's effects. In addition, the

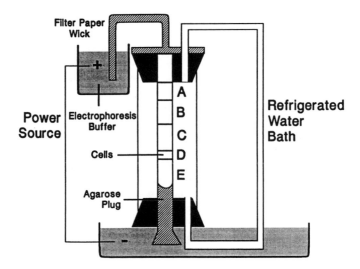

Figure 6-1 Scheme of a preparative ascending gradient cell electrophoresis device. See text for details. (From Gillman, C. F., Bigazzi, E. E., Bronson, P. M., and van Oss, C. J., *Prep. Biochem.*, 4, 457, 1974. With permission.)

dielectric constant for D_2O is approximately 1% lower than that of water. Thus, electrophoretic mobilities in D_2O are rather similar to those in water.

Using a fairly shallow discontinuous D_2O gradient instead of water, ascending electrophoresis has been shown to be affected much less by convective disturbances; it does not participate in biochemical interactions, nor does it lead to cell aggregation.[59] The apparatus used in this study consisted of a rectangular chamber with a jacket through which water was cooled to 4 to 8°C and circulated from a refrigerated water bath (Figure 6-1). The bottom third of the chamber was filled with a 1.5% agarose plug in electrophoresis buffer (Sigma, St. Louis, MO). The electrophoresis buffer (pH 7.0) consisted of 0.679 g NaCl, 0.147 g Na_2HPO_4, 0.043 g KH_2PO_4, 1.0 g EDTA, and 40.0 g dextrose per liter of distilled water. The same buffer was also prepared using D_2O as the solvent. A 1-ml aliquot of solution E, with density = 1.11 g/ml was placed on top of the agarose plug (see Figure 6-1). The cell preparations (percoll-purified intraepithelial lymphocytes or Ficoll/Hypaque (F/H)-purified splenocytes) to be electrophoresed were suspended in solution D, (density = 1.070 g/ml) and layered on top of the D_2O cushion (solution E); this was overlaid with solutions C, B, and A. Thus, a discontinuous gradient was established, decreasing in density from 1.11 g/ml (solution E) at the cushion end to a density of 1.02 g/ml (solution A) at the anodal end of the electrophoresis chamber. The chamber was connected to the electrode vessel by a filter paper wick at the top and through the 1.5% agarose plug at the bottom, making the top the anode and the bottom the cathode. An electric field of 7 to 9 V/cm was applied across the chamber, yielding a current of 7 to 9 mA. After 1.5 h of electrophoresis, 1-ml fractions were removed; the uppermost fraction, corresponding to solution A, was denoted as "fraction 1" and the lowermost fraction, corresponding to solution E, was denoted as "fraction 5".

In order to measure the movements of cells in electrophoresis, different concentrations of D_2O are mixed with varying amounts of electrophoresis buffer in a discontinuous stepwise gradient. Gravity pulls solution downward while heat convection causes mixing and, hence, attempts to alter the discontinuous gradient caused by the heavy-water gradient.[68]

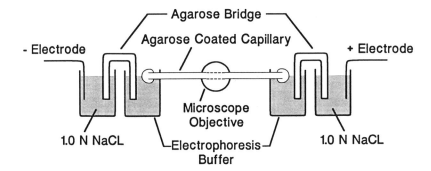

Figure 6-2 Scheme of an analytical microelectrophoresis device. (From van Oss, C. J. and Fike, R. M., in *Electrokinetic Separation Methods,* Righetti, P. G., van Oss, C. J., and Vanderhoff, J. W., Eds., Elsevier/North Holland Publishers, Amsterdam, 1979. With permission.)

B. MICROELECTROPHORESIS

Preparative electrophoresis is the electrophoresis of cells at the global level, while microelectrophoresis is the electrophoresis of cells at the individual level. The classical technique of particle and cell microelectrophoresis has been described by Abramson et al.[69] Seaman[70] further defined the conditions for microelectrophoresis for cylindrical chambers, and Fuhrman and Ruhenstroth-Bauer[71] described the conditions for rectangular chambers. The classical technique is complicated by the effect of osmotic backflow in the vicinity of the inner wall of the electrophoresis chamber, which causes the observed mobilities of particles to vary considerably according to their position in the chamber.

To cope with this effect, the observed mobilities must be obtained by focusing on the "stationary level", provided that the particles remain long enough at the same level. The paraboloid velocity distribution resulting from electroosmotic backflow in closed capillaries can be obviated by reducing the system to a uniform plug flow. The inner surface of glass capillaries is coated with a gel layer[72] after the sample cells or particles are in the capillaries; both ends are plugged with the gel. The observed electrophoretic mobilities of particles are then the same at all points of the capillary tube diameter. Cells may be tracked continuously by preventing sedimentation of very large cells by occasional rotation of the capillary around its cylindrical axis. The extent of electroosmotic backflow in the agarose must be determined for each buffer, and this value must be added to the electrophoretic mobilities measured for the cells or particles.

The microelectrophoresis apparatus was described in detail by van Oss and Fike.[72] Briefly, it consisted of a 1.6 to 1.8×100 mm glass capillary tube coated inside with 1.5% agarose solution. The capillary connected two buffer chambers, in turn connected to Ag/AgCl "reversible" electrodes via two U-shaped agarose bridges (Figure 6-2). The capillary was coated by dipping one end into the heated (80°C) agarose solution, aspirating agarose into the capillary, and then forcing the agarose out by gentle blowing. A diluted suspension of cells (10^6 cells per milliliter, amounting to approximately 3×10^4 cells per capillary) was drawn into the capillary by capillary action. Subsequently, the capillary was plugged at both ends with agarose. All experiments were performed in the electrophoresis buffer. A Microstar microscope (Reichert Instruments, Buffalo, NY) was used to observe the cells. The mobility measurements were taken by using both "normal" and "reverse" polarity of electrodes for each cell measured at the same amperage in the normal and reversed position. The strength of the electric field was determined by direct measurement of the voltage. The electrophoretic mobilities have the dimensions 10^{-4} cm^2 V^{-1} s^{-1}.

The velocity of individual cells (V) was measured by using the equation

$$u = (d/t) \times (D/V)$$

where d = distance (in micrometers) covered by cells during measurements, t = time (in seconds) required by cells to cover the distance d, D = distance between the two agarose plugs (which is approximately the length of the tube), and V = potential applied to the electrodes. The cell mobility was timed in alternate directions to minimize electrode polarization.

C. FREE FLOW ELECTROPHORESIS

A major problem in the system of preparative electrophoresis used in this study was the yield, i.e., the number of cells obtained from the most purified fraction. This issue may be resolved by using free flow electrophoresis, which offers a practical means of achieving a scaled-up electrophoretic procedure for separating cells without thermoconvection-stabilizing additives. This type of electrophoresis apparatus places the cells into a moving stream of buffer solution. As the cells pass through an electric field, the individual cells congregate into discrete streams in the solution. The systems that are marketed at present are based on the free-flow electrophoresis system of Hannig et al.[73-75] The Hannig apparatus ensures separation in a thin, free-flowing system by processing cells at a rate of up to 10^8 cells per hour.[73,74] A device marketed by Harwell (Oxfordshire, U.K.), using a continuous flow electrophoresis system in which laminar flow is stabilized by a gradient of angular velocity, can process up to 10^{11} cells per hour.[76] The various space shuttle flights employing the continuous flow electrophoresis device of McDonnell-Douglas Aeronautics Corporation allow throughput comparable to the system of Harwell.[77] In addition, sterile recovery of cells can be achieved after electrophoresis.

While earth-based electrophoresis processing provides better separation than many other processes, the purity of some cells (e.g., the stem cells for gene therapy) can still be improved. Earth's gravity causes convection currents and sedimentation, which remix the particles.[68] Experiments in microgravity have demonstrated that the free flow electrophoresis of McDonnell-Douglas Aeronautics Corporation is a promising way to separate cells.[77]

III. ANTIBODIES AS LIGANDS

A. EFFECT OF ANTIBODY TREATMENT ON THE ELECTROPHORETIC MOBILITY OF INTRAEPITHELIAL LYMPHOCYTES USING PREPARATIVE ELECTROPHORESIS

In the small intestine, lymphocytes are present in the epithelium — the intraepithelial lymphocytes (IEL) — and in the lamina propria; antigens in the intestinal lumen may be of prime importance in regulating the immune response to food antigens and to the microorganisms inhabiting the intestine. It is well established that IEL are a heterogeneous cell population, with a majority of the cells reacting with T-cell markers. Many IEL contain large, distinct granules in their cytoplasm. The nature and function of IEL remain poorly defined, however, despite extensive research in this field. It appears that at least some of the problems in IEL research in the past have been due to the limitations of isolation and purification methods proposed in the literature.

Pursuing the original intention to characterize both surface markers of IEL and some of their functional aspects, it was soon realized that the claims of high purity of IEL preparations obtained using the established isolation procedures could not be confirmed. Tagliabue et al.,[78] using discontinuous percoll gradients, reported on the isolation of two IEL fractions containing no more than 2% and 1%, respectively, with intestinal epithelial

Table 6-1 **Reactivity with rabbit IgG anti-rat intestinal brush border of untreated (no antibody treatment), single antibody-treated, and double antibody-treated percoll-purified intestinal intraepithelial lymphocytes (IEL) after preparative electrophoresis[a]**

Fraction Number	No Ab Treatment		Single Ab Treatment[b]		Double Ab Treatment[c]	
	% Tot	% ABB	% Tot	% ABB	% Tot	% ABB
1	13.1 ± 4.5	11.5 ± 2.6	25.6 ± 4.5	12 ± 4.0	23.0 ± 6.4	0.6 ± 1.0
2	22.5 ± 10.9	32.5 ± 13.6	23.9 ± 6.7	18 ± 2.4	28.7 ± 5.0	8.2 ± 2.3
3	29.5 ± 9.1	23.1 ± 7.2	24.8 ± 10.5	20 ± 10.2	23.3 ± 4.9	40.0 ± 14.3
4	19.9 ± 1.4	14.0 ± 1.5	12.8 ± 2.0	39 ± 12.2	14.1 ± 3.0	34.0 ± 8.4
5	15.0 ± 4.8	12.7 ± 0.8	12.8 ± 4.0	20 ± 10.2	8.4 ± 2.2	34.0 ± 8.4

Note: Ab = antibody, % Tot = percentage of the total cell number in a given fraction, % ABB = percentage of the cells reacting positively with rabbit IgG anti-rat intestinal brush border in a given fraction.

[a] Mean ± SD; four experiments were done for no Ab treatment and for double Ab treatment, while two experiments were done for single Ab treatment (average number of animals per experiment: 15).
[b] Single Ab treatment = percoll-purified IEL incubated with rabbit IgG anti-rat intestinal brush border in a given fraction.
[c] Double Ab treatment = percoll-purified IEL incubated with rabbit IgG anti-rat intestinal brush border + goat IgG anti-rabbit IgG.

cells as contaminants (at the 55/70% and 70/80% percoll interphase). Subsequently, the same authors[79] reported slightly different results — namely, 3 ± 1% epithelial cells in the 55/70% interphase and 0% in the 70/80% interphase. The presence of an even larger proportion of epithelial cells in the IEL preparations was acknowledged by Petit et al.[80] They reported 7.5 ± 8% and 25 ± 15% intestinal epithelial cells in the 55/70% and 70/80% percoll interphase, respectively. Since this report[80] came from Bienenstock's group, which also participated in the studies of Tagliabue et al.,[81] the explanation for this remarkable change in the performance of their purification method with time might have been expected, but has not yet been addressed.

An attempt was made to use ascending preparative electrophoresis in a discontinuous heavy-water gradient. Using this technique, all the fractions obtained had >10% epithelial cells as detected by indirect immunofluorescence testing. Hence, cell electrophoresis alone would not meet the needs of this study.

A new approach was investigated in which cells were incubated with antibodies. Because mammalian cells have a pI of 2 to 3.3[82] and immunoglobulins have a pI of 5 to 8.5, it was postulated that incubation of cells with an antibody should decrease the negative charge of the cell and, thus, lead to decreased electrophoretic mobility of epithelial cells. IEL from mice were obtained by percoll gradient centrifugation.[83,84] As shown in Table 6-1, all the fractions contained >10% cells staining positively with antibody to brush border (BB). For single antibody treatment, IEL were incubated with rabbit IgG to rat intestinal BB, and for double antibody treatment this was followed by goat IgG to rabbit IgG. Virtually no cells reacted with anti-BB antibody in fraction 1. Fewer than 1% (Table 6-1) of contaminating epithelial cells (EC) were found in the uppermost fraction 1 after double antibody treatment. After antibody treatment, cells were electrophoresed in the ascending preparative electrophoresis system for 1.5 h, and the cells from different fractions were analyzed for reactivity with rabbit IgG to rat intestinal BB. The cells from the same fractions were also tested for their reactivity with fluorescein-conjugated rabbit IgG to mouse brain Thy-1 antigen, a mouse monoclonal IgG to Thy-1, a rat monoclonal IgG to Lyt-1 (L3T4), a fluorescein-conjugated mouse monoclonal IgG to Lyt-1, a rat monoclonal IgG to Lyt-2, a fluorescein-conjugated mouse monoclonal IgG to Lyt-2, and rabbit IgG to asialo-GM1 from bovine brain. The activity in YAC-1 lytic assays was also tested, but results were inconsistent.

In accordance with previously published reports, these experiments also showed >90% of the percoll-purified IEL to react with fluorescein isothiocyanate (FITC)-labeled anti-Thy-1 antibody and <2% to react with anti-total immunoglobulin.[85-87] The binding characteristics of rabbit IgG to brain Thy-1 to the cell membrane were reported to be similar to those of anti-Thy-1 alloantibody.[88-90] Both anti-Thy-1 reagents reacted not only with peripheral T-cells from the spleen and mesenteric lymph nodes and IEL, but also with up to 70% of the epithelial cells. Thus, one may argue that a subpopulation of epithelial cells has Thy-1 antigen on the surface. Unfortunately, the rat monoclonal IgG to Thy-1 reacted only very weakly with IEL.

Purified IEL have features that are fairly similar to the stem cells of the mouse, which have been characterized as follows: Thy1.1 lo, Lin–, Sca1+, Rho123 lo.[91] Fichtelius had predicted that murine small intestine is a first-level lymphoid organ.[92] More recently, Mosley and Klein[93] had suggested that IEL are in part responsible for the initiation and regulation of T-cell development. Hence, it would not be too far fetched to hypothesize a new functional role for IEL, i.e., as a stem cell of the hematopoietic system in the adult immune system when the thymus has involuted. Recently, heterotopic bone formation was mimicked by mixing rat xiphoid and ileum in tissue culture. Here also, IEL have been implicated as the stem cells responsible for activation by a soluble factor from the xiphoid and for laying down the extracellular white ground material.[94]

B. EFFECT OF ANTIBODY TREATMENT ON THE ELECTROPHORETIC MOBILITY OF SPLENIC LYMPHOCYTES USING PREPARATIVE ELECTROPHORESIS

Several investigators have separated T- and B-lymphocyte populations from the spleen using density gradient electrophoresis.[95-97] A number of workers have documented that T-cells are electrophoretically faster-moving cells than B-cells.[98]

Mouse splenocytes were obtained as described by Andersson et al.[98] Briefly, spleen cells were minced on a wire gauze, partially purified by F/H centrifugation, and resuspended in electrophoresis buffer. To study the effect of antibody treatment on spleen B-lymphocytes, single antibody treatment consisted of incubation with a goat IgG to mouse immunoglobulins (specific for IgA, IgG, and IgM) and double antibody treatment consisted of a second incubation with rabbit IgG to goat IgG. Lymphocytes obtained after electrophoresis were analyzed by immunofluorescence microscopy for the number of T- and B-lymphocytes in each fraction. T-lymphocytes were defined by a FITC-conjugated rabbit serum to brain Thy-1 antigen, and B-lymphocytes were labeled by rhodamine isothiocyanate-conjugated goat IgG to mouse total immunoglobulin (specific for IgA, IgG, and IgM).

Spleen cells were electrophoresed, and cells from different fractions were analyzed for T- and B-cell surface markers. Again, in this system, electrophoretic retardation of cells was seen with double antibody treatment, but not with single antibody treatment (data not shown). With no antibody treatment there was 20% B-cell contamination in the uppermost fraction, while after double antibody treatment there was only about 2% B-cell contamination in the uppermost fraction (Table 6-2). Table 6-3 further supports the fact that the double antibody incubation procedure alone does not cause the T-cell enrichment. In particular, Table 6-2 shows that there are only 2.4% B-cells in the uppermost fraction after double antibody treatment, compared to 53.5% B-cells in Table 6-3. In addition, density gradient in conjunction with antibody treatment effects the enrichment of T-cells in the uppermost fraction.

Table 6-2 **Preparative electrophoresis of Ficoll/Hypaque-purified splenocytes from Balb/c mice.**[a]

Fraction Number	No Ab Treatment			Double Ab Treatment[b]		
	% Tot	% T-cells	% B-cells	% Tot	% T-cells	% B-cells
1	29.1 ± 12.3	80.5 ± 6.4	20.7 ± 10.4	31.8 ± 13.9	90.6 ± 4.2	2.4 ± 1.6
2	16.7 ± 7.0	65.6 ± 10.4	35.5 ± 13.4	21.3 ± 4.3	78.2 ± 11.9	19.5 ± 9.7
3	14.1 ± 4.3	70.1 ± 3.2	33.8 ± 8.6	18.2 ± 6.7	67.5 ± 12.5	29.2 ± 15.6
4	13.3 ± 4.6	68.6 ± 4.1	33.9 ± 7.0	12.0 ± 5.2	76.0 ± 10.0	28.3 ± 10.3
5	26.7 ± 18.2	65.5 ± 13.2	38.7 ± 10.7	16.7 ± 9.0	60.8 ± 4.5	44.4 ± 11.1

Note: Ab = antibody, % Tot = percentage of the total cell number, % T-cells = percentage of the lymphocytes in a given fraction reacting with fluorescein-labeled rabbit IgG anti-mouse brain Thy-1, % B-cells = percentage of B-lymphocytes in a given fraction reacting with rhodamine-labeled goat IgG anti-mouse immunoglobulins (specific for IgA, IgG, and IgM).

[a] Mean ± SD of five experiments (average six animals per experiment).

[b] Double Ab treatment = splenocytes incubated with goat anti-mouse immunoglobulins (specific for IgA, IgG, and IgM) + rabbit IgG anti-goat IgG.

The experiments reported have shown that double antibody treatment is an effective method for purification of IEL and splenic T-cells in preparative electrophoresis using a heavy-water gradient. Use of a heavy-water gradient makes cell separation by electrophoresis much more economical than other methods that require expensive instruments.[73-76] Thus, the results reported here are of practical value, even though other researchers had suggested earlier that double antibody treatment could be a promising approach.[67,99,100] The most likely explanation for the decreased electrophoretic mobility induced by antibody appeared to be the addition of positive charges to the cell. Although it was reasonable to expect that a single antibody treatment would suffice for induction of the decrease in mobility and successful electrophoretic separation, experiments did not confirm this expectation. Treatment with a second antibody was necessary. It can be concluded that the single antibody treatment adds only a few positive charges, whereas a second antibody increases the magnitude of these charges sufficiently to induce a

Table 6-3 **Preparative electrophoresis of Ficoll/Hypaque-purified splenocytes from Balb/c mice without D$_2$O**

Fraction Number	No Ab treatment			Double Ab treatment[a]		
	% Tot	% T-cells	% B-cells	% Tot	% T-cells	% B-cells
BE	N.D.	46.7	39.3	N.D.	27.6	53.5
1	25.4	71.2	17.8	26.4	63.9	24.4
2	12.5	64.0	17.5	17.0	67.0	21.4
3	13.6	67.6	14.9	17.7	69.2	17.5
4	6.5	67.4	15.9	10.5	63.1	35.7
5	42.0	57.0	24.3	28.3	54.7	54.7

Note: Ab = antibody, % Tot = percentage of the total cell number, % T-cells = percentage of the lymphocytes in a given fraction reacting with fluorescein-labeled rabbit IgG anti-mouse brain Thy-1, % B-cells = percentage of B-lymphocytes in a given fraction reacting with rhodamine-labeled goat IgG anti-mouse immunoglobulins (specific for IgA, IgG, and IgM).

[a] Double Ab treatment = splenocytes incubated with goat anti-mouse immunoglobulins (specific for IgA, IgG, and IgM) + rabbit IgG anti-goat IgG.

measurable and statistically significant decrease in electrophoretic mobility. If this were the only mechanism operative in the experiments, a moderate reduction of cells after a single antibody treatment would be expected. However, no changes in electrophoretic mobility were detectable after single antibody treatment. From these results one cannot support the original hypothesis, that is, that addition of a second antibody simply adds further positive charges to the cell.

C. EFFECT OF ANTIBODY TREATMENT USING MICROELECTROPHORESIS

The electrophoretic mobility is reduced after treatment with antibody, as was first demonstrated by Coulter.[101] Erythrocytes of the A and B blood groups have a reduced mobility after reaction with the appropriate antibody. Using antisera to Rh blood groups, however, no change in mobility was seen. With the addition of Coomb's reagent (i.e., double antibody treatment), an increase in mobility was reported.[102] With nucleated cells, Zerial and Wilkins[103] showed that addition of specific antibodies to human lymphocytes produced a 24.4% reduction in mobility, whereas addition of nonimmune sera produced only a 2.5% reduction. Similarly, Bert et al.[104] used circulating lymphocytes to demonstrate a 30% reduction in mobility after incubation of cells with antilymphocyte serum; under similar experimental conditions, Bona et al.[105] showed a 45% reduction in mobility.

Furthermore, this technique was used to formulate a hypothesis on the precise mechanism involved in the reduction of antibody-treated cells. A system was needed in which the primary antibody reacted with all, rather than some, of the cells present in the suspension; i.e., a homogeneous cell population was required. As reported by Cohly et al.,[106] three homogeneous cell populations were studied: intestinal epithelial cells from two different sources of mice and F/H-enriched human mononuclear cells. The intestinal epithelial cells were obtained from 10- to 12-day-old mice because IEL are not present in the intestines at this age.[107,108] The availability of an antibody reacting with 95 to 99% of the mononuclear cells from human blood made the use of such cells in microelectrophoresis desirable. In addition, some data are available regarding microelectrophoresis of mononuclear cells.[67,102-105]

Microelectrophoresis of intestinal epithelial cells and human mononuclear cells showed that single antibody treatment caused no significant change in mobility, while double antibody treatment produced a 36 to 38% reduction in mobility (Table 6-4). There was no reduction in electrophoretic mobility when the primary antibody used was not specific for the antigenic determinants of the cell in question. Therefore, the effect is not caused by nonspecific coating of the cells with immunoglobulin, but is dependent on specific immunoglobulin-cell surface determinant interactions.

In contrast, it seems reasonable to postulate that, owing to its length, the first antibody is unable to project beyond the glycocalyx of the cell and, therefore, does not cause a

Table 6-4 **Reduction in electrophoretic mobility of intestinal epithelial cells (EC) and human mononuclear cells (HMNC) by single antibody treatment and double antibody treatment in microelectrophoresis**

Cells	Single Antibody Treatment		Double Antibody Treatment	
	% Reduction	p Value	% Reduction	p Value
C57BL/6Ja EC	3.1 (a)	<0.005	37.5 (a + b)	<0.001
CBA/LAC EC	2.5 (a)	<0.001	38.3 (a + b)	<0.01
HMNC	1.1 (c)	<0.001	36.3 (c + b)	<0.01

Note: p value obtained by Student's *t*-test; a = rabbit IgG anti-rat intestinal brush border, b = goat IgG anti-rabbit IgG, and c = fluorescein isothiocyanate (FITC)-rabbit IgG anti-HMNC.

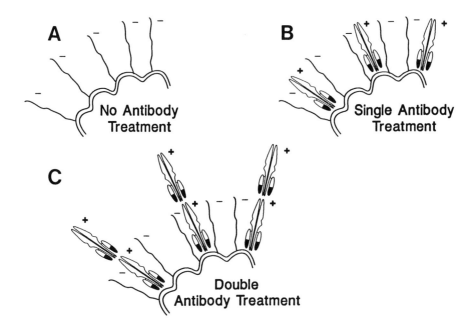

Figure 6-3 Schematic representation of the effect of double antibody treatment on the effective charge of cells. (**A**) A cell with its negatively charged glycoprotein strands projecting. (**B**) A cell that has an antibody sticking on the cell surface where the height of the antibody is similar to the height of the glycoprotein strands. (**C**) A cell in which the second antibody is on top of the first antibody and extends beyond the glycocalyx. Only in the last case will an effective decrease in electrophoretic mobility occur.

significant effective decrease in negative charge. The addition of a second antibody, however, allows the positive charge to extend beyond the glycocalyx and, therefore, results in an effective charge reduction responsible for retardation of cells in an electric field. This hypothesis is explained diagrammatically in Figure 6-3.

To study quantitatively the effect of antibody treatment on the electrophoretic mobility of human red blood cells (RBC), microelectrophoresis was performed. The cells included human RBC of NN, A_1, and D+ blood groups (Table 6-5). For human RBC of blood group NN, monoclonal primary antibody treatment consisted of B010 in one experiment and monoclonal antibody AH7 in another. For the A_1 blood group a rabbit IgG to human A_1 was used. For the blood group D antigen for single antibody treatment, human IgG to Rh blood group D was used. A goat IgG to rabbit IgG was used for double antibody treatment for D antigen. The donor erythrocytes for Rh blood group had a genotype CDe/cde (R^1/r).

The use of erythrocytes in microelectrophoresis allowed comparison with published work on the effect of antibody treatment on cell mobility and further testing of the hypothesis proposed above. The reported quantitative data on electrophoretic mobility with erythrocytes were in the same range as the data reported here.[102] The fact that the three blood group determinants used in this test are localized at different levels of the cell surface was used to test our hypothesis. Blood group N and A_1 antigens are part of the glycocalyx, single antibody treatment resulted in a reduction in electrophoretic mobility of 27 to 31% and 18%, respectively (Table 6-5).

In contrast, blood group D antigen is cell membrane associated and does not extend far into the glycocalyx;[109] single antibody treatment indeed caused no change in electrophoretic

Table 6-5 **Reduction in electrophoretic mobility of human red blood cells of N, A₁, and D blood groups by single antibody treatment and double antibody treatment in microelectrophoresis**

Cells	Single Antibody Treatment		Double Antibody Treatment	
	% Reduction	p Value	% Reduction	p Value
NN	27.0 (a)	<0.001	N.D.	N.D.
	30.7 (b)	<0.001	N.D.	N.D.
A₁	17.8 (c)	<0.001	N.D.	N.D.
D+	1.7 (d)	<0.005	18.3 (d + e)	<0.01

Note: N.D. = not done; *p* value obtained by Student's *t*-test; a = mouse monoclonal IgG anti-N (B010), b = mouse monoclonal IgG anti-N (AH7), c = rabbit IgG anti-human blood group A₁, d = human IgG anti-D, and e = goat IgG anti-human IgG.

mobility. Addition of a second antibody directed against the Fc portion of the primary antibody, however, caused a 20% reduction in electrophoretic mobility (Table 6-5).

These experiments also suggest that the reduction in electrophoretic mobility by antibody is larger with mammalian nucleated cells than with nonnucleated cells. This may be explained by the difference in zeta potentials of the nucleated[110] and nonnucleated cells.[47] When slightly anionized antibodies were used the retarding effect of double antibody treatment was weak.[99,100] With strongly anionic antibodies the effect of antibodies could be reversed.[67] Since it was postulated that the relatively positive charge of the antibody causes the reduction in cellular electrophoretic mobility, it appeared logical to test the effect of cationized antibodies on the cell mobility. To our knowledge, nothing has been published on the use of cationized antibodies in this context.

The carbodiimide used for cationization of antibodies was 1-ethyl-3(3-dimethyl-amino-propyl)carbodiimide hydrochloride (EDAC), a water-soluble compound. The reaction steps involved were the activation of carboxyl groups and the displacement by the nucleophile (i.e., ethylene diamine), releasing the EDAC as a soluble urea derivative. The cationization[111] was carried out in a 25-ml flask on ice. Ethylene diamine (0.84 ml; Sigma) was added to 10.7 ml of 0.1 *M* sodium acetate buffer (pH 4.75), and the pH was adjusted to 4.75 with 5 *N* HCl. The total volume was brought to 15.9 ml with sodium acetate buffer. Antibodies were dissolved in 14.3 ml of sodium acetate buffer, 0.48 g of EDAC was added, and the reaction mixture was maintained at pH 4.75 for 30 min. The antibodies were then dialyzed against 0.2 *M* sodium borate buffer and 0.15 *M* NaCl (pH 8.0). For electrophoresis the antibodies were dialyzed against phosphate-buffered saline (PBS) containing 0.1% sodium azide.

Rabbit IgG to rat intestinal BB, goat IgG to rabbit IgG, and rabbit IgG to goat IgG were cationized. The second antibody used for double antibody incubation (goat IgG to rabbit IgG) was a heterogeneous population of molecules with respect to charge. Isoelectric focusing rendered 25 to 30 bands between pH 5.2 and 8.5. In contrast, the pI of all three cationized antibodies was found to be ≥9.3. Microelectrophoresis of human mononuclear cells (HMNC) was performed with the cationized antibody (goat IgG to rabbit IgG) as the second antibody. Microelectrophoresis of untreated HMNC resulted in a û of 1.41 ± 0.10, single antibody (rabbit IgG to HMNC) treated cells had a û of 1.45 ± 0.1, and double antibody (rabbit IgG to HMNC followed by cationized goat IgG to rabbit IgG with pI ≥ 9.3) treated cells had a û of 0.53 ± 0.06. Thus, cationizing the second antibody led to a 62.4% decrease in electrophoretic mobility (p < 0.001; Table 6-6).

To establish whether cationization of the primary antibody might also reduce electrophoretic mobility, EC from 10- to 12-day-old C57Bl/6Ja mice were treated with cationized rabbit IgG to rat intestinal BB. Untreated EC from C57Bl/6Ja mice had a û of 1.15 ± 0.18,

Table 6-6   **Reduction in electrophoretic mobility of intestinal epithelial cells (EC) and human mononuclear cells (HMNC) by single antibody treatment and double antibody treatment in microelectrophoresis**

Cells	Single Antibody Treatment		Double Antibody Treatment	
	% Reduction	p Value	% Reduction	p Value
HMNC	−2.8 (a)	<0.001	62.4 (a + b)	<0.001
C57Bl/6Ja EC	4.3 (c)	<0.001	19.1 (c + d)	<0.01
			26.1 (c + e)	<0.01
			77.4 (c + b)	<0.001

Note: p value obtained by Student's *t*-test; a = fluorescein-labeled rabbit IgG anti-human mononuclear cells, b = cationized goat anti-rabbit immunoglobulin IgG, c = cationized absorbed rabbit IgG anti-rat intestinal brush border, d = rhodamine-labeled goat IgG anti-rabbit IgG, and e = unconjugated goat IgG anti-rabbit IgG.

and those treated only with cationized first antibody had a û of 1.10 ± 0.16. Treatment first with cationized and then with native antibody or rhodamine-labeled antibody resulted in a û of 0.85 ± 0.14 or 0.93 ± 0.17, respectively. In contrast, cells treated with cationized second antibody had a û of 0.26 ± 0.03.

Thus, the use of only cationized first antibody did not lead to an appreciable change in electrophoretic mobility, but treatment with either native or rhodamine-labeled second antibody reduced the electrophoretic mobility by 26.1% and 19.1%, respectively. Treatment with cationized first and second antibody reduced the electrophoretic mobility by 77% (Table 6-6).

Cationizing the second antibody resulted in a 62% reduction in electrophoretic mobility of HMNC and a 77% reduction in that of intestinal EC (Table 6-6). These observations further confirm the importance of the positive charge of the second antibody. Furthermore, cationization of the first antibody did not alter electrophoretic mobility significantly, substantiating the need for the electrical charge to extend from the glycocalyx to affect cell movement in an electric field. These results further corroborate the importance of effective charge in achieving the effect on electrophoretic mobility.

Even though agglutination was not observed in the presence of electrophoresis buffer when cationized antibodies were used, extensive agglutination did occur in the presence of heavy water. Thus, cationized antibody cannot be used yet in preparative electrophoresis involving heavy water. The use of double antibody electrophoresis in conjunction with cationized antibody, however, should be explored further in other density gradient systems such as percoll or in other electrophoresis systems that do not require density gradients for stabilization. Should these attempts be successful, cationized antibodies would produce a significant increase in the efficiency of cell purification in electrophoresis because cationization of the second antibody increases retardation of cells by antibody incubation by a factor of two.

The reduction in electrophoretic mobility of erythrocytes tested with antibodies to N antigen was much greater than that obtained with antibodies to A_1. The primary antibodies to N blood group were monoclonal. Thus, increased reduction in electrophoretic mobility by monoclonal antibodies possibly could have been caused by higher pI. Indeed, preliminary data showed that the two monoclonal antibodies used had pIs of 7.2 to 8.0.

From the quantitative results obtained in microelectrophoresis with double antibody treatment a reduction in electrophoretic mobility of roughly one third was observed. In preparative electrophoresis the uppermost fifth of the electrophoresis pathway covered by the fastest moving cells was lacking cells reactive with antibodies. These two observations correlate well. Use of cationized antibody could render at least the two fastest moving fractions devoid of all contaminating cells.

D. EFFECT OF ANTIBODY TREATMENT USING CONTINUOUS FLOW ELECTROPHORESIS

Hansen[100] used double antibody treatment in conjunction with continuous free-flowing electrophoresis in an attempt to separate human lymphocyte populations. He based his separation procedure on the fact that immunoglobulin bears a much lower negative charge than the cell surface and, therefore, the electrophoretic mobility of cells toward the anode can be reduced by reaction with appropriate antibodies. In this system the antibody-labeled cells are separated on the basis of their antigen density.

It was suggested by Hansen[100] that the differences in cell surface charge induced by bound antibody could be amplified by coupling neutralized groups such as DNP or charged groups such as poly-L-aspartic acid or poly-L-glutamic acid to immunoglobulins without interfering with antibody specificity.

IV. ANTIGENS AS LIGANDS

A. EFFECT OF PLANT LECTINS ON THE ELECTROPHORETIC MOBILITY OF T-CELLS

The interaction of lectins with cells triggers quiescent, nondividing lymphocytes into a state of growth and proliferation. In contrast to stimulation by antigens, in which specific clones of lymphocytes are induced to proliferate, lectins activate multiple lymphocyte clones irrespective of their antigen specificity, so that the percentage of cells responding is high (up to 70 to 80% of an appropriate population).[112] The earliest changes detectable when a mitogen is added to the cell are in the cell surface membrane, a phenomenon still not clearly understood.

It is generally accepted that the initial step is the binding of lectin to the cell-surface sugar; this binding generates a certain message that, if stimulatory, activates at low ionic concentrations. Binding does not necessarily mean stimulation. Bessler et al.[113] showed that in order for a lectin to be mitogenic it must be able to induce interleukin 2 (IL-2) production and secretion. Mitogenic stimulation could be resolved into at least two separately regulated phases.[114] The first is mediated by lectin-receptor interaction and represents the transition of a cell from a resting state to G1 phase. This initial stage has been an issue of great controversy.

Blume et al.[55] studied the effect of different concentrations of Con A on the electro-phoretic mobility of murine T-cells at physiologic temperatures. A concentration of 10^{-17} g ml^{-1} of Con A was associated with an optimum increase in electrophoretic mobility, while 10^{-5} g ml^{-1} was associated with a maximum decrease in electrophoretic mobility. In other reports[115,116] a concentration of 10^{-11} g ml^{-1} was associated with the release of IL-2.[115,116] Concomitantly, an optimum blastogenic dose also occurs at 10^{-5} g ml^{-1}.[117]

When Con A is used as a ligand, the labelling is specific for Con A binding sites since it can be removed and prevented with alpha-methyl mannoside (α-MM), a hapten inhibitor of Con A binding.[118] In a time course study in which 0.01 M inhibitor was used, the entire label from the cell surface was removed after about 10 min and microvilli and ruffles reappeared. Con A-treated cells are committed to DNA replication after approximately a 24-h incubation with mitogens. On the other hand, addition of α-MM treated with a high (supraoptimal), nonmitogenic concentration of Con A elicits a marked blastogenic response.[119] In sharp contrast to these effects, the addition of α-MM to low (suboptimal) or optimal concentrations of Con A results in severe inhibition of [^{3}H]-thymidine incorporation. Addition of optimum Con A in the presence of α-MM causes A.E7, a CD4+ T-cell hybridoma clone, to be nonresponsive to the antigen. Furthermore, tolerance induced by antigen receptor stimulation in the absence of costimulation by Con A were demonstrated to have a defect in antigen-induced transcription of the IL-2 gene.[120]

Increase in electrophoretic mobility was associated with one type of receptor followed by redistribution with the second type of receptor associated with the decrease in electrophoretic mobility.[121,121a] Two sets of Con A receptors on splenocytes react sequentially with the mitogenic lectin; it is possible that they may have different functional roles. Studies by McClain and Edelman[119] and Stenzel et al.[122] have indicated a mechanism involving a first activating (positive) signal and a second inhibiting (negative) signal to explain the nonmitogenicity of supraoptimal doses of Con A to human lymphocytes.

Con A interaction produces short-range lateral movements that are required for Con A-induced cell-cell binding and agglutination[123] and large lateral movements that result in cap formation.[124,125] An interesting study by Wang et al. showed that Con A receptor(s) possibly interact(s) with at least two kinds of GTP-binding proteins in murine thymocytes.[126]

The ionized groups of the thymocyte membrane are drastically modified during an increase in electrophoretic mobility. This is accompanied by (1) a 61% diminution in the density of sialic carboxyl groups, (2) a 40% decrease in the density of phosphate groups and a 60% decrease in the density of amino groups, and (3) a 20-fold increase in density of unidentified negatively charged groups.[127]

A key observation by Wioland et al. was that the electrophoretic mobility did not decrease with varying amounts of Con A. This could be interpreted as evidence of not having the appropriate receptor for Con A.[127] This receptor probably is the IL-2 receptor itself. The initial response could be the high-affinity response, which is coupled with an increase in electrophoretic mobility, while the second response — the expression of low-affinity receptor — takes place when the threshold goes beyond the critical concentration of the mitogen. In the case of EL4, the latter does not take place since this may be the defect in this transformed cell line. Further study of this process by Szamel et al. revealed that a short-term activation of protein kinase C is sufficient for synthesis and expression of IL-2 receptors, while long-term activation of the enzyme is necessary for IL-2 synthesis in human lymphocytes.[128] IL-2 beta unit is associated with the high-affinity IL-2 and is responsible for IL-2 signal transduction.[129] That the lack of IL-2 receptor plays a role in the growth of malignant T-cells was demonstrated by Debatin et al.[130] Three classes of IL-2 receptors have been identified: high, intermediate, and low affinity. It is possible that the high-affinity receptor may correspond to an increase in electrophoretic mobility, the intermediate may correspond to the point of inflection or the time period or concentration range between 10^{-6} and 10^{-5} g ml^{-1} of Blume et al.'s study, and the low-affinity receptor may correspond to a decrease in electrophoretic mobility. If true, this would make electrophoresis an indispensable adjunct for mitogenic studies.[131]

The issue of whether the reaction one sees with Con A and lymphocytes is charge related is highly debatable in the context of the recent work on synaptosomal membranes. Con A, which is envisioned as a cationic mitogen, would react with anionic species only, but the Con A binding proteins in this particular instance resolved up to 40 spots, ranging in pI from 4.5 to 8.0, indicating broad heterogeneity.[132] There appear to be other long-range attractive forces existing the molecular origin of which is not well understood.[133] Baker et al.[134] have observed that proteolytic enzymes such as pronase treatment and/or sialidase neuraminidases reduce electrophoretic mobility, but the mechanism may involve nonelectrostatic forces.

Furthermore, binding of mitogen stimulates a receptor tyrosine kinase that activates Ras, Raf, mitogen-activated protein, (MAP) kinase kinase, MAP kinase, and several transcription factors in the phosphorylation cascade.[135] MAP kinases have been postulated to have a pivotal role in promoting the G0 to G1 transition.[14]

A recent review of the literature shows that there has been extensive study in the field of the cell cycle, which involves the G0-G1-S-G2-M-G0 phases of the cell. Cell cycle progression is regulated at two principal points, one that occurs before DNA replication[136] and the other that occurs before mitosis.[137] Both transitions are controlled by the activation of specialized essential protein kinases called cyclin-dependent kinases. The early events or the triggering, which (when seen in the context of the virgin G0 cell and/or the activated cell at G1) both occur at the cell surface, and their consequences can be detected easily and economically by the cell electrophoresis procedure. Investigations would include whether the mobility of the cell has been increased or decreased. The method which is currently being used to study the state of the cell division is centrifugal elutriation.[138]

Receptor-mediated endocytosis has been shown to be inhibited during mitosis of eukaryotic cells, and work on A431 cells has demonstrated that one of the sites inhibited was invagination of the coated pits.[139,140] Furthermore, cancer cells of the gastric mucosa in M phase displayed uniform distribution of Con A receptor complex with strong fluorescence.[141]

The focus of research has recently shifted from soluble intracellular second signal messengers to enzymes that activate transcription factors in the nuclear region.[135] Also, in this domain lies the difference between using electrophoresis or another system as an indicator to detect the initial change. Receptors appearing on the cell surface can be evaluated by electrophoresis, while the receptors that lie in the intracellular domain of the cytoplasm and nucleus, like that of the mineralocorticoids and glucocorticoids, require other equipment and strategies.

B. EFFECT OF PROTEIN ANTIGENS ON THE ELECTROPHORETIC MOBILITY OF B-CELLS

Sundaram et al.[143] showed that incubation of nonstimulated cells with antigen altered the electrophoretic mobility by mere adsorption. This alteration in mobility was reversible since the mobility change could be removed by washing. However, antigen-stimulated lymph node cells with two different antigens had lowered their electrophoretic mobility by 0.43 u/s/V/cm for TAB (*Salmonella typhi, Salmonella paratyphi* A and B) and 0.29 u/s/V/cm for *Vibrio cholerae* as antigen.

A series of metabolic events culminating in activation of B-cells and their entry into the cell cycle by antigen or by anti-immunoglobulin induces the phosphorylation of a number of proteins on tyrosines, including PLC gamma-2. The mechanism of the B-cell interaction with its ligand has yet to be fully elucidated.

Cross-linking of surface immunoglobulin increases *in vitro* kinase activity of p53/56[lyn] along with phosphate inositol-3 kinase activity,[144] while stimulation of B-cells with lipopolysaccharide, a B-cell mitogen, has little effect on p53/56[lyn].[145]

Wheat germ agglutinin, which exerts its stimulatory effect predominantly on B-cells, increases the levels of phosphotyrosines in splenocytes as well as the activity of the cytosylic protein p72[syk].[146,147] This route of B-cell activation also needs further study.

C. EFFECT OF PROTEIN ANTIGENS ON THE ELECTROPHORETIC MOBILITY OF T-CELLS

The T-cell receptor, with its sole monovalent alpha and beta chains, is associated with CD3 subunits, giving rise to the TCR receptor complex which is closely associated with the CD4/CD8 coreceptor.[14] Although the T-cell has been studied extensively, several controversial aspects remain. The first events taking place on the cell surface are very debatable, with no clear-cut resolution on issues as simple as the ability of T-cells to recognize shape or conformation of protein structures. Much attention has been directed toward second or possibly third messages associated with the T-cells.

With respect to second and third messages, evidence is accumulating which indicates that p56[lck] may indirectly participate in the antigen-induced T-cell activation by physical association with CD4 and CD8 coreceptors.[148,149] Furthermore, the possibility of mice bearing a targeted disruption of the *lck* gene manifests a severe defect in T-cell development.[150] Infection of an insulin-specific mouse T-hybridoma cell line with retroviruses encoding an activated form of p56[lck] markedly increased the sensitivity of cells to antigen as measured by IL-2 production.[151] It appears that p56[lck] serves as a signal amplifier rather than as an obligatory activator.

Hence, the mysteries of the T-cell still perplex immunologists from the initial interaction to the final episode. Finally, a recent publication demonstrates that a particular T-cell with alpha beta gene products can have two types of V alpha proteins on the surface, while the T-cell with gamma delta gene products can have dual alleles of the V gamma proteins. These two observations challenge the old dogma of "one cell, one receptor".[152,153]

Early stages of antigen presentation and recognition by T-cells[154-156] were further investigated by generating antigen-specific T-cell lines and then separating these lines in the presence and absence of antigen using the continuous flow electrophoresis system (CFES) device designed by the McDonnell-Douglas Aeronautics Corporation for NASA. Preliminary data from this study showed that incubation of antigen-specific T-cell lines for myoglobin and lysozyme at its optimum antigen dose of proliferative assay without antigen-presenting cells altered the electrophoretic mobility by 20%. This change in electrophoretic mobility suggests that antigen alone interacts with the T-cell receptor for 45 min of incubation at 4°C and causes a change in the glycocalyx of the membrane. Furthermore, after passage through the CFES device the cells show functional activity of T-cells by incorporation of tritiated thymidine in a proliferative response to the antigen. Additional studies are currently underway.

V. FUTURE

Meshing with other fields of investigation is progressing at a rapid pace, with flexibility and simplicity of concepts being the order of the day.[157] The possibility that similar events are taking place in other systems and, hence, a universality of laws may be emerging from this trend gives rise to the potential of addressing issues that may bridge the gap between not only the body and the mind, but also the spirit.[158]

Cellular studies in immunology are highly dependent on the ability to isolate the appropriate cell lines and clones. The procedures presently employed have been quite effective, but new and promising technologies are emerging. The ability to perform cell separations in microgravity using the continuous flow electrophoresis system of McDonnell Douglas Aeronautics Corporation may prove to be one of the most important biologically related spinoffs of the space program. Cell electrophoresis is the future arena of research activities of the 1990s. Double antibody cell separation and purification using double antibody treatment were done 8 years ago at The State University of New York at Buffalo, but were published only in thesis form pending further investigation. Continuation of my Ph.D. work was conducted at NASA's Johnson Space Center and the Baylor University School of Medicine at Houston. Current investigations on double antibody and antigen labeling of IEL for use as a source of hematopoietic stem cells are in progress at the University of Mississippi Medical Center.

REFERENCES

1. **Seaman, G. V. F. and Swank, R. L.,** Surface characteristics of dog chylomicra and some lipid models, *J. Physiol. (London),* 168, 118, 1963.
2. **Sowemimo-Coker, S. O. and Meiselman H. J.,** Effect of procaine hydrochloride on the electrophoretic mobility of human red blood cells, *Cell Biophys.,* 15, 235, 1989.
3. **Simmon, M. I.,** Summary: the cell surface regulates information flow, material and transport, and cell identity, *Cold Spring Harbor Symp. Quant. Biol.,* 57, 673, 1992.
4. **Singer, S. J. and Nicolson, G. L.,** The fluid mosaic model of the structure of cell membrane, *Science,* 175, 720, 1972.
5. **Singer, S. J.,** Architecture and topography of biologic membranes, *Hosp. Pract.,* 8, 81, 1973.
6. **Singer, S. J.,** The structure and function of membranes — a personal memoir, *J. Membr. Biol.,* 129, 3, 1992.
7. **Stoddard, B. L., Biemann, H. P., and Koshland, D. E.,** Receptors and transmembrane signalling, *Cold Spring Harbor Symp. Quant. Biol.,* 57, 1, 1992.
8. **Kim, S. H., Prive, G. G., Yeh, J., Scott, W. G., and Milburn, M. V.,** A model for transmembrane signalling in a bacterial chemotaxis receptor, *Cold Spring Harbor Symp. Quant. Biol.,* 57, 17, 1992.
9. **Hunter, T., Lindberg, R. A., Middlemas, D. S., Tracy, S., and van Der Geer, P.,** Protein receptor tyrosine kinases and phosphatases, *Cold Spring Harbor Symp. Quant. Biol.,* 57, 25, 1992.
10. **White, J. M.,** Membrane fusion, *Science,* 258, 917, 1992.
11. **Lasky, L. A.,** Selectins: interpreters of cell-specific carbohydrate information during inflammation, *Science,* 258, 964, 1992.
12. **Creutz, C. E.,** The annexins and exocytosis, *Science,* 258, 924, 1992.
13. **Secvcikova, H., Marek, M., and Muller, S. C.,** The reversal and splitting of waves in an excitable medium caused by an electric field, *Science,* 257, 951, 1992.
14. **Perlmutter, R. M., Levin, S. D., Appleby, M. W., Anderson, S. J., and Alberola-Ila, J.,** Regulation of lymphocyte function by protein phosphorylation, *Annu. Rev. Immunol.,* 11, 451, 1993.
15. **Loor, F. and Roelants, G. E.,** Immunofluorescence studies of a possible prethymic T-cell differentiation in congenitally athymic (nude) mice, *Ann. N.Y. Acad. Sci.,* 254, 226, 1975.
16. **Unanue, E. R., Engers, D. H., and Karnovsky, M.,** Antigen receptors on lymphocytes, *Fed. Proc.,* 32, 44, 1973.
17. **Inbar, M., Ben-Bassat, H., and Sachs, L.,** Difference in the mobility of lectin sites on the surface membrane of normal lymphocytes and malignant lymphoma cells, *Int. J. Cancer,* 12, 93, 1973.
18. **Rutishauser, U. and Sachs, L.,** Receptor mobility and the mechanism of cell-cell binding induced by concanavalin A, *Proc. Natl. Acad. Sci. U.S.A.,* 71, 2456, 1974.
19. **Brown, S. S. and Revel, J. P.,** Reversibility of cell surface label rearrangement, *J. Cell Biol.,* 68, 628, 1976.
20. **Rapin, A. M. C. and Burger, M. M.,** Tumor cell surfaces: general alterations detected by agglutinins, *Adv. Cancer Res.,* 20, 1, 1974.
21. **Raff, M. C. and Petris, S. D.,** Movement of lymphocyte surface antigens and receptors: the fluid nature of the lymphocyte plasma membrane and its immunological significance, *Fed. Proc.,* 32, 48, 1973.
22. **Petris, S. D.,** Concanavalin A receptors, immunoglobulins, and ϕ antigen of the lymphocyte surface: interactions with concanavalin A and with cytoplasmic structures, *J. Cell Biol.,* 65, 123, 1975.

23. **van Deurs, B., Peterson, O. W., Olsnes, S., and Sandvig, K.,** The ways of endocytosis, *Int. Rev. Cytol.,* 117, 131, 1989.
24. **Heuser, J.,** Effects of cytoplasmic acidification on clathrin lattice morphology, *J. Cell Biol.,* 108, 401, 1989.
25. **Priester, D. W., Baker, A., and Lamers, G.,** Capping of ConA receptors and actin distribution are influenced by disruption of microtubules in *Dictyostelium discoideum, Eur. J. Cell Biol.,* 51, 23, 1990.
26. **Conway, E. M., Boffa, M. C., Nowakowski, B., and Steiner-Mosonyi, M.,** An ultrastructural study of thrombomodulin endocytosis: internalization occurs via clathrin-coated and noncoated pits, *J. Cell. Physiol.,* 151, 604, 1992.
27. **Altankov, G. and Grinell, F.,** Depletion of intracellular potassium disrupts coated pits and reversibly inhibits cell polarization during cell polarization, *J. Cell Biol.,* 120, 1449, 1993.
28. **Steinman, R. M., Brodie, S. E., and Cohn, Z. A.,** Membrane flow during pinocytosis A stereologic analysis, *J. Cell Biol.,* 51, 722, 1976.
29. **Gibbs, E. M. and Lienhard, G. E.,** Fluid-phase endocytosis by isolated rat adipocytes, *J. Cell. Physiol.,* 121, 569, 1984.
30. **Wilson, J. M. and King, B. F.,** Transport of horseradish peroxidase across monkey trophoblastic epithelium in coated and uncoated vesicles, *Anat. Rec.,* 211, 174, 1985.
31. **Geuze, H. J., Slot, J. W., Strouz, G. J. A. M., Peppard, J., von Figura, K., Haslik, A., and Schwartz, A. L.,** Intracellular receptor sorting during endocytosis: comparative immunoelectron microscopy of multiple receptors in rat liver, *Cell,* 37, 185, 1984.
32. **van Deurs, B., Peterson, O. W., Olsnes, S., and Sandvig, K.,** Delivery of internalized ricin from endosomes to cisternal Golgi elements is a discontinuous temperature-sensitive process, *Exp. Cell Res.,* 171, 137, 1987.
33. **Roederer, M., Bowser, R., and Murphy, R. F.,** Kinetics and temperature dependence of exposure of endocytosed material to proteolytic enzymes and low pH: evidence for a maturation model for the formation of lysosomes, *J. Cell. Physiol.,* 131, 200, 1987.
34. **Snider, M. D. and Rogers, O. C.,** Intracellular movement of cell surface receptors after endocytosis: resialylation of asialotransferrin receptor in human erythroleukemia cells, *J. Cell Biol.,* 100, 826, 1985.
35. **Roth, J., Taatjes, D. J., Lucoq, J. M., Weinstein, J., and Paulson, J. C.,** Demonstration of an extensive trans-tubular network continuous with the Golgi apparatus stack that may function in glycosylation, *Cell,* 43, 287, 1985.
36. **Mezitis, S. G. E., Stieber, A., and Gonatas, N. K.,** Quantitative ultrastructural, autoradiographic evidence for the magnitude and early involvement of the Golgi apparatus complex in the endocytosis of wheat germ agglutinin by cultured neuroblastoma, *J. Cell. Physiol.,* 132, 401, 1987.
37. **Creutz, C. E.,** Cis-unsaturated fatty acids induce the fusion of chromaffin granules aggregated by synexin, *J. Cell Biol.,* 91, 247, 1981.
38. **Venable, R. M., Zhang, Y., Hardy, B. J., and Pastor R. W.,** Molecular dynamics simulations of a lipid bilayer and of hexadecane: an investigation of membrane fluidity, *Science,* 262, 223, 1993.
39. **Lin, H. C., Sudhof, T. C., and Anderson, R. G. W.,** Annexin VI required for budding of clathrin-coated pits, *Cell,* 70, 283, 1992.
40. **Rapoport, T. A.,** Transport of proteins across the endoplasmic reticulum membrane, *Science,* 258, 931, 1992.
41. **Nelson, W. J.,** Regulation of cell surface polarity from bacteria to mammals, *Science,* 258, 948, 1992.

42. **Bivic, A. L., Sambuy, Y., Patzak, A., Patil, N., and Chao, M.,** An internal deletion in the cytoplasmic tail removes the apical localization of human NGF receptor in transfected MDCK cells, *J. Cell Biol.,* 115, 607, 1991.

43. **Pearce, B. M. F. and Robinson, M.,** Clathrin, adaptors, and sorting, *Annu. Rev. Cell Biol.,* 6, 151, 1990.

44. **van Oss, C. J.,** The influence of the size and shape of molecules and particles on their electrophoretic mobility, *Sep. Purif. Methods,* 4, 167, 1975.

45. **van Oss, C. J.,** Electrokinetic separation methods, *Sep. Purif. Methods,* 8, 119, 1979.

46. **Schwartz, A. M. and Rader, C. A.,** Surface chemistry —utilization in analysis, in *Treatise on Analytical Chemistry,* 2nd ed., Vol. 2, Sect. V-B and V-C, Koltoff, I. M., Elving, P. J., and Grushka, E., Eds., Marcel Dekker, Amsterdam, 1955.

47. **Sherbet, G. V.,** Cell electrophoresis, in *The Biophysical Characterization of the Cell Surface,* Sherbet, G. V., Ed., Academic Press, New York, 1978, 36.

48. **Thomson, A. E. R. and Mehrishi, J. N.,** Surface properties of normal human circulating small lymphocytes and lymphocytes in chronic lymphocytic leukemia: separation and adhesiveness and electrokinetic properties, *Eur. J. Cancer,* 5, 195, 1969.

49. **Vassar, P. S., Hards, J. M., and Seaman, G. V. F.,** Surface properties of human lymphocytes, *Biochim. Biophys. Acta,* 291, 107, 1973.

50. **Singh, A. K., Kasinath, B. S., and Lewis, E. J.,** Interaction of polycations with cell-surface negative charges of epithelial cells, *Biochim. Biophys. Acta.,* 1120, 337, 1992.

51. **Teixeira, L. A., Figueiredo, A. M. S., Ferreira, B. T., Alves, V. M., Nagao, P. E., Alviano, C. S. J., Angluster, J., Silva-Filho, F. C., and Benchetrit, L. C.,** Sialic acid content and surface hydrophobicity of group B streptococci, *Epidemiol. Infect.,* 110, 87, 1993.

52. **Vargas, F. F., Osorio, H. M., Basilio, C., De Jesus, M., and Ryan, U. S.,** Enzymatic lysis of sulfated glycosaminoglycans reduces the electrophoretic mobility of vascular endothelial cells, *Membr. Biochem.,* 9, 83, 1990.

53. **Bush, M. S., Reid, A. R., and Alt, G.,** Blood-nerve barrier: Ultrastructural and endothelial surface charge alterations following nerve crush, *Neuropathol. Appl. Neurobiol.,* 19, 31, 1993.

54. **Persiani, S. and Shen, W. C.,** Increase of poly(L-lysine) uptake but not fluid phase endocytosis in neuraminidase pretreated Madin-Darby canine kidney (MDCK) cells, *Life Sci.,* 45, 2605, 1989.

55. **Blume, P., Malley, A., Knox, R. J., and Seaman, G. V. F.,** Electrophoretic mobility as a sensitive probe of lectin-lymphocyte interaction, *Nature,* 271, 378, 1978.

56. **Kolin, A.,** Isoelectric spectra and mobility spectra: a new approach to electrophoretic separation, *Proc. Natl. Acad. Sci. U.S.A.,* 41, 101, 1955.

57. **Svensson, H. and Valmet, E.,** Density gradient electrophoresis. A new method of separating electrically charged compounds, *Sci. Tools,* 2, 11, 1955.

58. **van Oss, C. J., Fike, R. M., Good, R. J., and Reinig, J. M.,** Cell microelectrophoresis simplified by the reduction and uniformization of the electroosmotic backflow, *Anal. Biochem.,* 60, 242, 1975.

59. **Bronson, P. M. and van Oss, C. J.,** Preparative cell electrophoresis with D_2O as a stabilizing agent, *Prep. Biochem.,* 9, 61, 1979.

60. **Boltz, R. C., Todd, P., Streibel, M. J., and Louie, M. K.,** Preparative electrophoresis of living mammalian cells in a stationary Ficoll gradient, *Prep. Biochem.,* 3, 383, 1973.

61. **Boltz, R. C. and Todd, P.,** Density gradient electrophoresis of cells in a vertical column, in *Electrokinetic Separation Methods,* Righetti, P. G., van Oss, C. J., and Vanderhoff, J. W., Eds., Elsevier, Amsterdam, 1979, 229.

62. **Ku, D. N.,** Cell electrophoresis in percoll density gradients, *Sep. Purif. Methods,* 11, 71, 1982.

63. **Brooks, D. E.,** The effect of neutral polymers on the electrokinetic potential of cells and other charged particles. III. Experimental studies on the dextran/erythrocyte system, *J. Colloid Interface Sci.,* 43, 700, 1973.

64. **Brooks, D. E.,** The effect of neutral polymers on the electrokinetic potential of cells and other charged particles. IV. Electrostatic effects in dextran-mediated cellular interactions, *J. Colloid Interface Sci.,* 43, 714, 1973.

65. **Brooks, D. E. and Seaman, G. V. F.,** The effect of neutral polymers on the electrokinetic potential of cells and other charged particles. I. Models for the zeta potential increase, *J. Colloid Interface Sci.,* 43, 670, 1973.

66. **Gillman, C. F., Bigazzi, P. E., Bronson, P. M., and van Oss, C. J.,** Preparative electrophoresis of human lymphocytes. I. Purification of non-immunoglobulin bearing lymphocytes by electrophoretic levitation, *Prep. Biochem.,* 4, 457, 1974.

67. **Overbeek, J. T. G. and Wiersema, P. H.,** Interpretation of electrophoretic mobilities, in *Electrophoresis,* Bier, M., Ed., Academic Press, New York, 1967, 1.

68. **Amato, I.,** Microgravity materials science strives to stay in orbit, *Science,* 257, 882, 1992.

69. **Abramson, H. A., Moyer, L. S., and Gorin, M. H.,** Surface chemistry of cells, in *Electrophoresis of Proteins and the Chemistry of Cell Surfaces,* Hafner, New York, 1964, 303.

70. **Seaman, G. V. F.,** Electrophoresis using a cylindrical chamber, in *Cell Electrophoresis,* Ambrose, E. J., Ed., Little, Brown, Boston, 1965, 4.

71. **Fuhrman, G. F. and Ruhenstroth-Bauer, R.,** Cell electrophoresis employing a rectangular measuring cuvette, in *Cell Electrophoresis,* Ambrose, E. J., Ed., Little, Brown, Boston, 1965, 22.

72. **van Oss, C. J. and Fike, R. M.,** Simplified cell microelectrophoresis with uniform electroosmotic backflow, in *Electrokinetic Separation Methods,* Righetti, P. G., van Oss, C. J., and Vanderhoff, J. W., Eds., Elsevier/North-Holland, Amsterdam, 1979, 111.

73. **Hannig, K., Wirth, H., Meyer, B. H., and Zeiller, K.,** Free flow electrophoresis. I. Theoretical and experimental investigations of the influence of mechanical and electrokinetic variables on the efficiency of the method, *Hoppe-Seyler's Z. Physiol. Chem.,* 356, 120, 1975.

74. **Hannig, K. and Heidrich, H. G.,** Electrophoretic methods, continuous free-flow electrophoresis and its application in biology, in *Cell Separation Methods,* Biomendal, H., Ed., Elsevier/North-Holland, Amsterdam, 1977, 93.

75. **Hannig, K., Wirth, H., Schindler, R. K., and Spiegel, K.,** Free flow electrophoresis. III. An analytical version for a rapid, quantitative determination of electrophoretic parameters, *Hoppe-Seyler's Z. Physiol. Chem.,* 358, 753, 1977.

76. **Mattock, P., Aitchison, G. F., and Thomson, A. R.,** Velocity gradient stabilized, continuous free-flow electrophoresis, *Sep. Purif. Methods,* 9, 1, 1980.

77. **Snyder, R. S.,** Review of the NASA Electrophoresis Program, in *Electrophoresis '81,* Allen, R. C. and Arnaud, P., Eds., Walter de Gruyter, Berlin, New York, 1981, 883.

78. **Tagliabue, A., Luni, W., Soldateschi, D., and Boraschi, D.,** Natural killer activity of gut mucosal lymphoid cells in mice, *Eur. J. Immunol.,* 11, 919, 1981.

79. **Tagliabue, A., Befus, A. D., Clark, D. A., and Bienenstock, J.,** Characteristics of natural killer cells in intestinal epithelium and lamina propria, *J. Exp. Med.,* 155, 1785, 1982.

80. **Petit, A., Ernst, P. B., Befus, A. D., Clark, D. A., Rosenthal, K. L., Ishizaka, T., and Bienenstock, J.,** Murine intestinal intraepithelial lymphocytes. I. Relationship of a novel Thy-1 –, Lyt-1 –, Lyt-2 +, granulated subpopulation to natural killer cells and mast cells, *Eur. J. Immunol.,* 15, 211, 1985.

81. **Tagliabue, A., Befus, A. D., Clark, D. A., and Bienenstock, J.,** Characteristics of natural killer cells in intestinal epithelium and lamina propria, *J. Exp. Med.,* 155, 1785, 1982.

82. **Thomson, A. E. R. and Mehrishi, J. N.,** Surface properties of normal human circulating small lymphocytes and lymphocytes in chronic lymphocytic leukemia: separation and adhesiveness and electrokinetic properties, *Eur. J. Cancer,* 5, 195, 1969.

83. **Cohly, H. H. P.,** Cell Separation by Preparative Electrophoresis and Double Antibody Tagging, Ph.D. thesis, State University of New York, Buffalo, 1986.

84. **Cohly, H. H. P., van Oss, C. J., Weiser, M. M., and Albini, B.,** Purification of intestinal intraepithelial lymphocytes by preparative electrophoresis, in *Cell Electrophoresis,* Schutt, W. and Klinkmann, H., Eds., Walter de Gruyter, Berlin, 1985, 603.

85. **Davies, M. D. J. and Parott, D. M. V.,** Preparation and purification of lymphocytes from the epithelium and lamina propria of murine small intestine, *Gut,* 22, 481, 1981.

85a. **Guy-Grand, D., Griscelli, C., and Vassalli, P.,** The gut associated lymphoid system: nature and properties of the large dividing cells, *Eur. J. Immunol.,* 4, 435, 1974.

86. **Guy-Grand, D., Griscelli, C., and Vassalli, P.,** The mouse gut T lymphocyte, a novel type of T cell, *J. Exp. Med.,* 148, 1661, 1978.

87. **Guy-Grand, D. and Vassali, P.,** Gut mucosal lymphocyte subpopulations and mast cells, in *Regulation of the Immune Response. 8th Int. Convoc. Immunology,* Ogra, P. L. and Jacobs, D. M., Eds., S. Karger, Basel, 1982, 122.

88. **Loor, F. and Roelants, G. E.,** Immunofluorescence studies of a possible prethymic T-cell differentiation in congenitally athymic (nude) mice, *Ann. N.Y. Acad. Sci.,* 254, 226, 1975.

89. **Golub, E. S.,** Brain associated theta antigen: reactivity of anti-mouse brain with mouse lymphoid cells, *Cell. Immunol.,* 2, 353, 1971.

90. **Raff, M. C.,** Two distinct populations of peripheral lymphocytes in mice distinguishable by immunofluorescence, *Immunology,* 19, 637, 1970.

91. **Spangrude, G. J. and Johnson, G. R.,** Resting and activated subsets of mouse multipotent hematopoietic stem cells, *Proc. Natl. Acad. Sci. U.S.A.,* 87, 7433, 1990.

92. **Fichtelius, K. E.,** The gut epithelium: a first level lymphoid organ?, *Exp. Cell Res.,* 49, 87, 1967.

93. **Mosley, R. L. and Klein, J. R.,** Peripheral engraftment of fetal intestine into athymic mice sponsors T cell development: direct evidence for thymopoietic function of murine small intestine, *J. Exp. Med.,* 176, 1365, 1992.

94. **Cohly, H. H. P., Buckley, R. C., and Das, S. K.,** Experimental model of heterotopic bone formation, *J. Invest. Surg.,* 6, 351, 1993.

95. **Andersson, L. C., Nordling, S., and Hayry, P.,** Fractionation of mouse T and B lymphocytes by preparative cell electrophoresis, *Cell Immunol.,* 8, 235, 1973.

96. **Platsoucas, C. D.,** Separation of cells by preparative density gradient electrophoresis, in *Cell Separation: Methods and Selected Applications,* Pretlow, T. G. and Pretlow, T. P., Eds., Academic Press, New York, 1985, 145.

97. **Shortman, K.,** Equilibrium density gradient separation and analysis of lymphocyte populations, in *Modern Separation Methods: Macromolecules and Particles,* Gerritsen, T., Ed., Interscience, New York, 1982, 167.

98. **Andersson, L. C., Nordling, S., and Hayry, P.,** Fractionation of mouse T and B lymphocytes by preparative cell electrophoresis, *Cell Immunol.,* 8, 235, 1973.

99. **Hansen, E. and Hannig, K.,** ASECS: antigen specific electrophoretic cell separation, *Electrophoresis '82,* Vol. 2, Stathokos, D., Ed., Walter de Gruyter, Berlin, 1982, 313.

100. **Hansen, E.,** Preparative free-flow electrophoresis of lymphoid cells, in Proc. Int. Meet. Cell Electrophoresis, Rostock, Germany, 1984, 217.

101. **Coulter, C. B.,** The isoelectric point of red blood cells and its relation to agglutination, *J. Gen. Physiol.,* 3, 309, 1920.

102. **Sachtleben, P.,** The influence of antibodies on the electrophoretic mobility of red blood cells, in *Cell Electrophoresis,* Ambrose, E. J., Eds., J. & A. Churchill, London, 1965, 100.

103. **Zerial, A. and Wilkins, D. J.,** The electrophoretic properties of some human blood cells, *Vox Sang.,* 26, 14, 1974.

104. **Bert, G., Cossano, L., Pecco, D., and Mazzei, D.,** Effect of ALS and prednisone on the electrophoretic mobility of lymphocytes, *Lancet,* 1, 365, 1970.

105. **Bona, C., Anteunis, A., Robineaux, R., and Halpern, B.,** Structure of the lymphocyte membrane. III. Chemical nature of the guinea pig lymphocyte membrane macromolecules reacting with heterologous ALS, *Clin. Exp. Immunol.,* 12, 377, 1972.

106. **Cohly, H. H. P., Albini, B., Weiser, M. M., and van Oss, C. J.,** Microelectrophoresis of epithelial cells and lymphocytes, in *Cell Electrophoresis,* Schutt, W. and Klinkmann, H., Eds., Walter de Gruyter, Berlin, 1985, 611.

107. **Ferguson, A. and Parott, D. M. V.,** The effect of antigen deprivation on thymus-dependent and thymus-independent lymphocytes in the small intestine of the mouse, *Clin. Exp. Immunol.,* 12, 477, 1972.

108. **Ferguson, A.,** Immunology, in *Scientific Basis of Gastroenterology,* Duthie, H. L. and Wormsley, K. G., Eds., Churchill Livingstone, Edinburgh, 1979, 49.

109. **Lorusso, D. J. and Green, F. A.,** Solubilized human erythrocyte membranes and the Rh system, in *A Seminar on Antigens on Blood Cells and Body Fluids,* Bell, C. A., Ed., American Association of Blood Banks, Washington, D.C., 1990, 226.

110. **Thomson, A. E. R. and Mehrishi, J. N.,** Surface properties of normal human circulating small lymphocytes and lymphocytes in chronic lymphocytic leukemia: separation and adhesiveness and electrokinetic properties, *Eur. J. Cancer,* 5, 195, 1969.

111. **Reith, E. J. and Ross, M. H.,** Digestive system, in *Atlas of Descriptive Histology,* 3rd ed., Harper & Row, New York, 1970, 142.

112. **Hume, D. M. and Wiedemann, M. J.,** *Mitogenic Lymphocyte Transformation,* Elsevier/North-Holland, Amsterdam, 1980.

113. **Bessler, W. G., Kraut, H., Busing, D., Muller-Hermes, W., and Peters, H.,** Membrane Alterations and induction of responsiveness to interleukin-2 in lymphocytes by lima bean lectins, in *Lectins,* Vol. 3, Bog-Hanson, T. C. and Spengler, G. A., Eds., Walter de Gruyter, Berlin, 1983, 45.

114. **Klaus, G. G. and Hawrylowicz, C. M.,** Cell-cycle control in lymphocyte stimulation, *Immunol. Today,* 5, 15, 1984.

115. **Frelinger, J. A., Neiderhuber, J. E., and Shreffler, D. C.,** Effect of anti-Ia sera on mitogenic responses. III. Mapping the genes controlling the expression of Ia determinants on ConA reactive cells to the I-J subregion of the H-2 gene complex, *J. Exp. Med.,* 144, 1141, 1979.

116. **Sainis, K. B. and Phondke, G. P.,** Confirmatory evidence for the existence of two types of receptor sites for Con A on mouse lymphocytes, *Immunology,* 40, 1, 1980.

117. **McClain, D. A. and Edelman, G. M.,** Analysis of stimulation inhibition paradox exhibited by lymphocytes exposed to concanavalin A, *J. Exp. Med.,* 144, 1494, 1976.

118. **Goldstein, I. J., Hollerman, C. E., and Merrick, M. M.,** Protein-carbohydrate interaction. I. The interaction of polysaccharide with concanavalin A, *Biochim. Biophys. Acta,* 97, 68, 1965.

119. **McClain, D. A. and Edelman, G. M.,** Analysis of stimulation inhibition paradox exhibited by lymphocytes exposed to concanavalin A, *J. Exp. Med.,* 144, 1494, 1976.

120. **Kang, S.-M., Beverly, B., Tran, A.-C., Borson, K., Schwartz, R. H., and Lenardo, M. J.,** Transactivation by AP-1 is a molecular target of T cell clonal anergy, *Science,* 257, 1138, 1992.

121. **Sainis, K. B., Bhisey, A. N., Sundaram, K., and Phondke, G. P.,** Cell electrophoretic studies on ConA treated lymphocytes of AKR mice: possible existence of two types of receptors for concanavalin A, *Immunology,* 37, 355, 1979.

121a. **Sainis, K. B. and Phondke, G. P.,** Confirmatory evidence for the existence of two types of receptor sites for ConA on mouse lymphocytes, *Immunology,* 40, 1, 1980.

122. **Stenzel, K. H., Rubin, A. L., and Novogrodsky, A.,** Lymphocyte commitment to DNA replication induced by concanavalin A, *Exp. Cell Res.,* 115, 285, 1978.

123. **Mintz, U. and Sachs, L.,** Changes in the surface membrane of lymphocytes from patients with chronic lymphocytic leukemia and Hodgkin's disease, *Int. J. Cancer,* 15, 253, 1975.

124. **Yahara, I. and Edelman, G. M.,** The effects of Concanavalin A on the mobility of lymphocyte surface receptors, *Exp. Cell Res.,* 81, 143, 1973.

125. **Sachs, L.,** Regulation and membrane changes, differentiation and malignancy in carcinogenesis, *Harvey Lect.,* 68, 1, 1974.

126. **Wang, C. P., Matsumoto, N., Toyoshima, S., and Osawa, T.,** Concanavalin A receptor(s) possibly interacts with at least two kinds of GTP-binding proteins in murine thymocytes, *J. Biochem.,* 105, 4, 1989.

127. **Wioland, M., Donner, M., and Neauport-Sautes, C.,** Modifications of the thymocyte membrane during redistribution of concanavalin A receptors, *Eur. J. Immunol.,* 6, 273, 1976.

128. **Szamel, M., Kracht, M., Krebs, B., Hubner, U., and Resch, K.,** Activation signals in human lymphocytes: interleukin 2 synthesis and expression of high affinity interleukin 2 receptors require differential signalling for the activation of protein kinase C, *Cell. Immunol.,* 126, 117, 1990.

129. **Tsudo, M., Kitamura, F., and Miyasaka, M.,** Characterization of the interleukin 2 receptor beta chain using three distinct monoclonal antibodies, *Proc. Natl. Acad. Sci. U.S.A.,* 86, 1982, 1989.

130. **Debatin, K. M., Woodroofe, C., Lahm, H., Fischer, J., Falk, W., Brandeis, W. E., and Krammer, P. H.,** Lack of interleukin-2 (IL-2) dependent growth of TAC positive T-ALL/NHL cells is due to the expression of only low affinity receptors for IL-2, *Leukemia,* 3, 566, 1989.

131. **Minami, Y., Kono, T., Miyazaki, T., and Taniguchi, T.,** The IL-2 receptor complex: its structure, function, and target genes, *Annu. Rev. Immunol.,* 11, 245, 1993.

132. **Nashioka, N. and Kurioka, S.,** Analysis of ConA-binding glycoproteins in synaptosomal membranes, *Neurochem. Res.,* 17, 1011, 1992.

133. **Tsao, Y.-H., Evans, D. F., and Wennerstrom, H.,** Long-range attractive force between hydrophobic surfaces observed by atomic force microscopy, *Science,* 262, 547, 1993.

134. **Baker, A. J., Coakley, W. T., and Gallez, D.,** Influence of polymer concentration and molecular weight and of enzymic glycocalyx modification on erythrocytes interaction in dextran solutions, *Eur. Biophys. J.,* 22, 53, 1993.

135. **Tan, Y. H.,** Yin and yang of phosphorylation in cytokine signalling, *Science,* 262, 376, 1993.

136. **Reed, S. I.,** The role of p34 kinases in the G_1 to S-phase transition, *Annu. Rev. Cell Biol.,* 8, 529, 1992.

137. **Forsburg, S. L. and Nurse, P.,** Cell cycle regulation in the yeasts *Saccharomyces cervisiae* and *Schizosaccharomyces pombe*, *Annu. Rev. Cell Biol.,* 7, 227, 1991.

138. **Koff, A., Giordano, A., Deasi, D., Yamashita, K., Harper, J. W., Elledge, S., Nishimoto, T., Morgan, D. O., Franza, B. R., and Roberts, J. M.,** Formation and activation of a cyclin E-cdk2 complex during the G1 phase of the human cell cycle, *Science,* 257, 1689, 1992.

139. **Pyapert, M., Lucocq, J. M., and Warren, G.,** Coated pits in interphase and mitotic A431 cells, *Eur. J. Cell Biol.,* 45, 23, 1987.

140. **Pyapert, M., Mundy, D., Souter, E., Labbe, J. C., and Warren, G.,** Mitotic cytosol inhibits invagination of coated pits in broken mitotic cells, *J. Cell Biol.,* 114, 1159, 1991.

141. **Wang, D. S., Liang, Y. Y., and Shi, Y. J.,** Distribution and lateral mobility of ConA receptors in the course of cell cycle of human gastric cancer cells, *Acta Biol. Exp. Sin.,* 22, 407, 1989.

142. **Funder, J. W.,** Mineralocorticoids, glucocorticoids, receptors and response elements, *Science,* 259, 1132, 1993.

143. **Sundaram, K., Phondke, G. P., and Ambrose, E. J.,** Electrophoretic mobilities of antigen-stimulated lymph node cells, *Immunology,* 12, 21, 1967.

144. **Yamanashi, Y., Kakiuchi, T., Mizuguchi, J., Yamamoto, T., and Toyoshima, K.,** Association of B cell antigen receptor with protein tyrosine kinase, *Science,* 251, 192, 1991.

145. **Burkhardt, A. L., Brunswick, M., Bolen, J. B., and Mond, J. J.,** Anti-immunoglobulin stimulation of B lymphocytes activates src-related protein-tyrosine kinases, *Proc. Natl. Acad. Sci. U.S.A.,* 88, 7410, 1991.

146. **Yamada, T., Taniguchi, T., Nagai, K., Saitoh, H., and Yamamura, H.,** The lectin wheat germ agglutinin stimulates a protein-tyrosine kinase activity of p72syk in porcine splenocytes, *Biochem. Biophys. Res. Commun.,* 180, 1325, 1991.

147. **Hutchcroft, J. E., Harrison, M. L., and Geahlen, R. L.,** Association of the 72-kDa protein-tyrosine kinase PTK72 with the B cell antigen receptor, *J. Biol. Chem.,* 267, 8613, 1992.

148. **Berry, N., Ase, K., Kishimoto, A., and Nishizuka, Y.,** Activation of resting human T cells requires prolonged stimulation of protein kinase C, *Proc. Natl. Acad. Sci. U.S.A.,* 87, 2294, 1990.

149. **Turner, J. M., Brodsky, M. H., Irving, B. A., and Levin, D. R.,** Interaction of the unique N-terminal region of the tyrosine kinase p56[lck] with the cytoplasmic domains of CD4 and CD8 is mediated by cysteine motifs, *Cell,* 60, 755, 1990.

150. **Molina, T. J., Kishihara, K., Siderovski, D. P., van Ewijk, W., Narendran, A., Timms, E., Wakeham, A., Paige, C. J., Hartmann, K. U., Veillette, A., Davidson, D., and Mak, T. W.,** Profound block in thymocyte development in mice lacking p56[lck], *Nature,* 357, 161, 1992.

151. **Abraham, N., Miceli, M. C., Parnes, J. R., and Veillette, A.,** Enhancement of T-cell responsiveness by the lymphocyte-specific tyrosine protein kinase p56[lck], *Nature,* 350, 62, 1991.

152. **Padovan, E., Casorati, G., Dellabona, P., Meyer, S., Brockhaus, M., and Lanzavecchia, A.,** Expression of two T cell receptor alpha chains: dual receptor T cells, *Science,* 262, 422, 1993.

153. **Davodeau, F., Peyrat, M.-A., Houde, I., Hallet, M.-M., Libero, G. D., Vie, H., and Bonneville, M.,** Expression of two distinct functional antigen receptors of human gamma delta T cells, *Science,* 260, 1800, 1993.

154. **Cohly, H. H. P., Morrison, D. R., and Atassi, M. Z.,** Antigen presentation by nonimmune B cell hybridoma clones: presentation of synthetic antigen sites reveals clones that exhibit no specificity and clones that present only one epitope, *Immunol. Invest.,* 18, 987, 1989.

155. **Cohly, H. H. P., Morrison, D. R., and Atassi, M. Z.,** Presentation of antigen to T lymphocytes by nonimmune B cell hybridoma clones: evidence for specific and nonspecific presentation, *Immunol. Invest.,* 18, 651, 1989.

156. **Cohly, H. H. P., Morrison, D. R., and Atassi, M. Z.,** Conformation dependent recognition of a protein by T-lymphocytes: apomyoglobin specific T cell clone recognizes conformational changes between apomyoglobin and myoglobin, *Immunol. Invest.,* 17, 337, 1988.

157. **Koshland, D. E.,** The two component pathway comes to eukaryotes, *Science,* 262, 532, 1993.

158. **Takahashi, J. S.,** Circadian clocks a la CREM, *Nature,* 365, 299, 1993.

III
Buffers and Their Effects on Cells

Chapter 7

New Buffer Systems for Cell Electrophoresis

Keith A. Knisley and L. Scott Rodkey

CONTENTS

I. INTRODUCTION

One of the perennial problems in cell biology involves isolation and characterization of subsets of cell populations of different types. Studies done over the last decade have demonstrated heterogeneity of many cell populations previously thought to be homogeneous. The fluorescence-activated cell sorter (FACS), although a powerful tool, has many drawbacks related to sorting efficiency and viability of the cells after long periods of sorting. Some problems associated with FACS cell fractionation might be overcome or minimized by using free flow electrophoresis. Bohn and Nebe[1] suggested free-flow cell electrophoresis as an alternative and/or complementary method to cell sorting for cell isolation. Bauer and Kachel[2] reported that electrophoresis enriched human peripheral blood cells and that coupling electrophoretic preseparation with FACS greatly enhanced the efficacy of flow cytometry. Hansen[3] and Hansen and Hannig[4] have reviewed pertinent literature on electrophoresis of lymphoid cells and cited numerous examples of its usefulness for separation of lymphoid cell subsets. Hannig[5] has summarized studies of electrophoretic characterization of immunocompetent cells as well as human blood cells, bone marrow cells, kidney cells, tumor cells, malarial parasites, chloroplasts, and a variety of cell organelles and membranes along with bacteria and viruses. Brown and

0-8493-8918-6/94/$0.00+$.50
© 1994 by CRC Press, Inc.

Holborow[6] described detectable alterations of electrophoretic mobility in B-cells infected with Epstein-Barr virus. Electrophoretically distinct cell subsets have been reported by Hymer et al.,[7] Morrison et al.,[8,9] Lewis et al.,[10] and Todd et al.[11] Todd[12] has summarized several of the studies of free-flow electrophoretic separation of heterogeneous cell subsets in microgravity. More recently, Bauer et al.[13] reported that the change in negative surface charge density which can be detected electrophoretically may be a marker for maturation of B-lymphocytes.

Antibodies have been used in free flow electrophoresis to specifically alter the electrophoretic mobilities of cell subsets for which the antibodies are specific. Hansen and Hannig[14] described the method of antigen-specific electrophoretic cell separation and showed that T- and B-cell subpopulations could be isolated readily with this approach. Hansen et al.[15] showed that this technique is applicable to a wide range of immunological investigations.

Thus, the accumulating evidence indicates that separation of viable cells based on electrophoretic differences has great potential. The purpose of the present studies was to explore the effects of alternative buffer formulations on resolution of separated cells in the McDonnell-Douglas continuous flow electrophoresis system. Band spreading during electrophoresis causes resolution problems in most preparative electrophoresis formats. The authors hypothesized that alternative buffers might partially alleviate band spreading. Thus, the possibility of using organic buffers for the separation of viable cell populations was explored. Specifically, organic zwitterionic buffers were selected for analysis of their utility for viable cell separations by preparative zonal electrophoresis. Initial studies focused on analysis of cell viability and cell growth in zwitterionic buffers formulated from existing zwitterion mixtures commonly used as ampholytes in isoelectric focusing. The availability of zwitterionic buffers formulated specifically for cell electrophoresis then prompted a comparison of these buffers with standard buffers used in past cell separation studies.

II. MATERIALS AND METHODS

A. INSTRUMENTATION

Free-flow preparative electrophoresis was done using the continuous flow electrophoresis system (CFES) which was developed by McDonnell-Douglas. Briefly, this instrument is a vertical flow liquid-phase zonal electrophoresis instrument equipped for continuous operation. A cross-sectional diagram of the separation chamber and photo of the laboratory instrument are shown in Figure 7-1. Samples are injected continuously into the separation chamber (1.5 mm × 6 cm), and sample residence time is varied by the carrier buffer flow rate. The 6-cm separation path width is partitioned into 197 discrete fractions at the output end so that each fraction can be individually sampled and analyzed. Cell distribution for the red blood cell (RBC) separation studies was estimated following electrophoresis by reading the absorbance of individual aliquots in a spectrophotometer at 280 nm. Preliminary studies in the authors' laboratory showed this to be an accurate, rapid, and reliable way of estimating relative RBC numbers in the different separated fractions. Rabbit and goose RBC were easily distinguished microscopically since goose RBC are nucleated. Hybridoma cell concentrations were counted microscopically using a hemocytometer following electrophoretic separation.

Electrophoresis runs were standardized on $E\tau$, a measure of field strength in volts per centimeter multiplied by cell residence time in the instrument[7] using either 150 $E\tau$ (42.9 V/cm, 3.5-min residence time) or 200 $E\tau$ (57.1 V/cm, 3.5-min residence time). Sample injection rates were 4.1 ml/h for RBC and for hybridoma cells. Cell suspensions at a cell

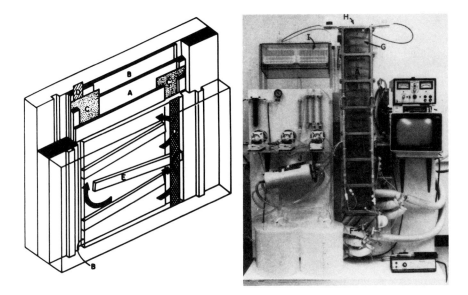

Figure 7-1 McDonnell-Douglas continuous flow electrophoresis system. The laboratory unit is shown. **Left:** diagram indicating how the separation chamber is surrounded by the coolant/electrode flow chambers. (A) Separation chamber, which is 1.5 mm thick and 6 cm wide; (B) front and back coolant/electrode chambers, through which electrode buffer (which also serves as coolant) is pumped in a serpentine path under pressure control; (C) two semipermeable membranes that separate coolant/electrode chambers from sample chamber fluid while permitting free passage of ions for the maintenance of a stable electric field across the chamber; (D) a platinum ribbon electrode placed parallel to the semipermeable membrane in each of the two coolant/electrode chambers; (E) baffles that force serpentine flow of coolant/buffer fluid between the electrodes and membranes. **Right:** laboratory unit showing (A) separation chamber, (F) sample inlet, (G) flared region of the chamber of the laboratory model which accommodates 197 outlet tubes, (H) 197 sample outlets that permit fractions to exit through Tygon® tubes at the top of the chamber, (I) sample collection tube rack, (J) buffer degassing cartridges.

density of 10^6 per milliliter were used for both RBC and hybridoma cells. Samples were always injected at the position on the CFES corresponding to fraction 48. pH Measurements were made using a Beckman pHI 70 pH meter with a combination electrode, and conductivity measurements were made with a Yellow Springs Instruments Model 31 conductivity bridge equipped with a 3403 dip cell.

B. RED BLOOD CELLS
Formalin-fixed RBC from rabbit and goose were gifts from Dr. W. Hymer, Department of Molecular and Cell Biology, Center for Separation Science, The Pennsylvania State University, University Park, PA 16802.

C. VACCINIA VIRUS INFECTION OF FIBROBLAST CELLS
Rabbit skin fibroblasts were seeded in 96-well plates at 1×10^4 cells per well in Dulbecco's modified Eagle's medium containing 10% fetal bovine serum, 2 m*M*

L-glutamine, 100 U/ml penicillin, and 100 μg/ml streptomycin. Serial dilutions of vaccinia virus ranging from 10^{-3} to 10^{-9} were added to each of three replicate plates containing fibroblasts. Zwitterion mixture was added so that each plate would contain a final concentration of 0% (control), 0.06%, or 0.5% zwitterion. Plates were monitored daily for the appearance of virus-induced cytopathic effects on the cells. The tissue culture infective dose ($TCID_{50}$) per 0.1 ml was calculated by standard means for each plate as an indicator of the effects of zwitterion on virus infectivity and replication.

D. HYBRIDOMA CELLS

Mouse hybridoma cell lines 5AF6 and 97D8, as well as the fusion partner SP2 cell line, were grown *in vitro* in RPMI 1640 medium supplemented with 10% fetal calf serum. 5AF6 and 97D8 were gifts from Dr. A. Brown at Kansas State University, Manhattan, and each synthesized an IgG1κ antibody specific for the arsonate hapten.

E. STANDARD BUFFERS

The standard buffers developed for use in the CFES are derivatives of triethanolamine cell separation buffers.[16] Briefly, the STS-8 buffer is a triethanolamine buffer which was modified for microgravity electrophoresis studies on Space Shuttle missions STS-3, STS-7, and STS-8.[12] The working STS-8 buffer consisted of 22.5 μM triethanolamine (Aldrich, 98%), 9.2 mM glycine (Aldrich, 99+%), 70 μM potassium acetate (Fisher, crystalline, 99+%), 73 mM glycerol (Aldrich, 99.5+%), 14.6 mM sucrose (Fisher, crystal, 99+%), 9 μM calcium chloride dihydrate (Fisher, 78%), and 10 μM magnesium chloride hexahydrate (Fisher, crystal, 99+%), with a final pH of 7.4 to 7.5. The G-1 buffer[7] is also a triethanolamine-based buffer which was developed at McDonnell-Douglas; it contained 0.675 mM triethanolamine (Aldrich, 98%), 75 mM glycine (Aldrich, 99+%), 165 μM potassium acetate (Fisher, crystalline, 99+%), 3.66 mM glucose (Dextrose, Fisher, 99+%), and 18.8 mM sorbitol (Aldrich, 99+%), with a final pH of 7.4 to 7.8.

F. ZWITTERIONIC BUFFERS

Zwitterion mixtures equivalent to wide-range ampholytes commonly used in isoelectric focusing protocols were synthesized from pentaethylenehexamine and acrylic acid according to established protocols (see Reference 17). Zwitterion concentrations were varied in tissue culture media by addition of the appropriate volume of a 40% solution of zwitterion reagent to media to attain the final concentration desired.

The Q buffer used in these studies is a proprietary zwitterionic buffer (Ampholife Technologies, Inc., 2828 North Crescent Ridge Drive, The Woodlands, TX 77381). The buffer is formulated from two available reagents. The first reagent consists of two proprietary zwitterion solutions, a basic solution and an acidic solution. Each are 1000× concentrated zwitterion solutions which can be diluted to any desired final concentration. The conductivity of the final buffer is adjusted by varying the dilution factor above or below the recommended factor of 1000. The desired pH is obtained by titrating the diluted acid solution with the basic solution. The second reagent is a dry reagent weighing approximately 30 g, sufficient for making 1 l of Q buffer, consisting of a mixture of glycine, glucose, sorbitol, Mg^{2+}, and Ca^{2+}. The final concentrations of each component in the second reagent of Q buffer are 229 mM glycine, 11 mM glucose, 56 mM sorbitol, 272 μM Ca^{2+}, and 295 μM Mg^{2+}. The recommended buffer as supplied by Ampholife Technologies, Inc. has a final pH of 6.9 to 7.1 and a conductivity of 100 to 110 μS when the zwitterion reagents are diluted 1:1000, which gives a final zwitterion concentration of 0.04%. The conductivity of Q buffer can be altered by varying the final concentration of the zwitterion reagent, and successful runs have been made in the CFES using Q buffer with conductivity as low as 80 μS and as high as 210 μS.

III. RESULTS

A. CELL GROWTH IN ZWITTERION-CONTAINING MEDIA

Studies were initiated to assess the effects of zwitterions on cell growth in culture over varying time periods. The effects on tritiated thymidine incorporation as a measure of cell growth over periods of 24 h and 72 h were tested by incubating identical cultures of each cell type in increasing concentrations of zwitterions ranging from 0.05 to 4% final concentration. The results are shown in Figure 7-2. The three continuous cell lines were all unaffected in their growth, as measured by tritiated thymidine incorporation, when cultured in zwitterion-containing media for 24 h. The HeLa and 3T3 cell lines were unaffected in any significant way when cultured for 72 h with less than 0.1% zwitterion concentration, but growth was affected by concentrations greater than 0.1%. The MA104 cells were affected at 72 h of growth by all concentrations of zwitterions. Primary rabbit skin fibroblasts displayed a titratable growth pattern over the 24 h tested. Growth of primary rabbit skin fibroblasts was enhanced by adding zwitterions to the culture medium, as compared to the control cultures lacking added zwitterion, up to concentrations in the range of 0.06 to 0.25%, and growth dropped significantly at 0.5% concentration.

B. VACCINIA VIRUS PRODUCTION IN CELLS CULTURED IN ZWITTERION-CONTAINING MEDIA

Further analysis of cells cultured in zwitterion-containing media was done by studying the ability of such cells to support virus infection and replication. Primary rabbit skin fibroblasts were seeded in wells containing culture medium with 0 (control), 0.06, or 0.5% zwitterion, and virus was seeded. Cytopathic effects were scored daily. The $TCID_{50}$ of virus produced per 0.1 ml of supernatant was then scored as a measure of the ability of cells to support physiological processes necessary for viral propagation while in the presence of zwitterions. The data in Table 7-1 show that no significant deleterious effects on virus production in primary fibroblasts were detected for zwitterion concentrations up to 0.5%.

C. PHYSICAL PROPERTIES OF Q BUFFER

Shortly following the above studies related to cell viability in mixed zwitterion-containing media, studies were concluded in which chemistry was developed for direct chemical synthesis of zwitterion mixtures which would distribute in narrow pH ranges of only two to three pH units when they were used for isoelectric focusing. Among these mixtures one formulation exhibited a pH range of only two pH units, and the unfractionated mixture had outstanding buffering capacity. Q buffer was derived from that ampholyte and was made available from Ampholife Technologies, Inc. for the studies described below.

1. Buffering Capacity

The superior buffering capacity of Q buffer was documented in a simple titration experiment. The STS-8 and G-1 buffers used routinely in earlier studies with the CFES instrument were also titrated for comparison. This comparison was prompted by the poor performance of the STS-8 and G-1 buffers in several preliminary studies. Equal quantities of each of the three buffers at the same concentration used for electrophoresis were stirred while adding aliquots of 1 M NaOH or 1 M HCl. The titration experiment showed that Q buffer had a substantially greater buffering capacity than either of the two triethanolamine-based buffers (Figure 7-3) across a wide range of acid and base addition.

2. pH Stability

The three buffers were compared for their pH stability in the CFES instrument during a standard electrophoresis run with no added samples. Buffer with no included cells was

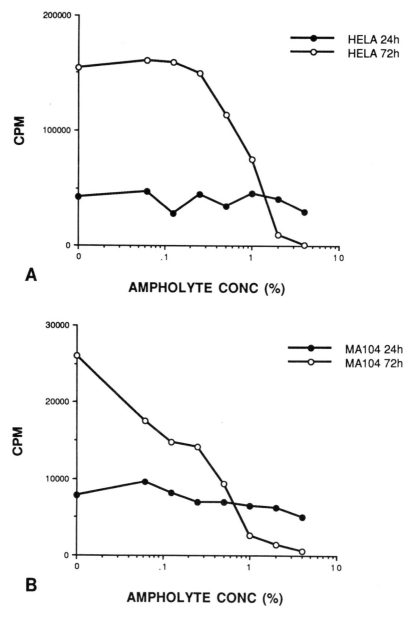

Figure 7-2 Effects of zwitterion mixtures on thymidine incorporation by cultured cells. (**A**) HeLa, (**B**) MA104, (**C**) 3T3, and (**D**) primary fibroblasts were cultured in complete medium containing the indicated concentrations of zwitterion mixture. HeLa, MA104, and 3T3 cells were cultured for 72 h, pulsed with tritiated thymidine, and harvested at 24 or 72 h to determine rates of thymidine incorporation. Fibroblasts were cultured for 24 h in complete medium containing the indicated concentrations of zwitterion mixture, pulsed with tritiated thymidine, and harvested 24 h later.

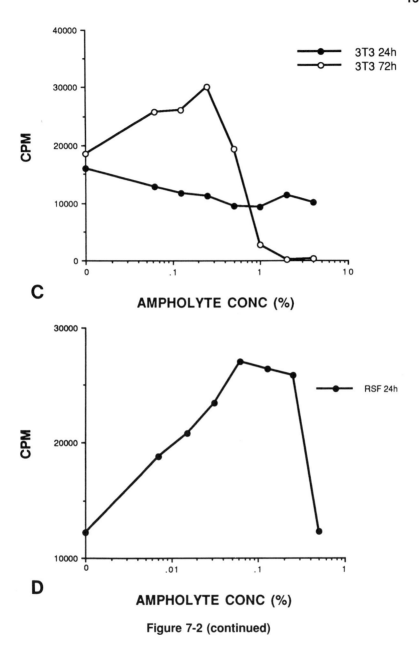

Figure 7-2 (continued)

injected during the run. Following each run the 197 collected buffer fractions were each checked for pH. Similar runs using each of the three buffers showed different pH profiles for each following electrophoresis. The results of these studies are shown in Figure 7-4. The STS-8 buffer was found to be unstable at both the anode and the cathode side of the buffer column, and the G-1 buffer was shown to be unstable on the anode side. Q buffer maintained excellent pH stability across the total width of the electrophoresis buffer column.

Table 7-1 **Propagation of vaccinia virus in fibroblasts grown in zwitterion-containing medium**

Ampholyte Concentration (%)	$TCID_{50}$[a]
0	$10^{-6.5}$
0.06	$10^{-6.5}$
0.5	$10^{-6.55}$

[a] $TCID_{50}$ = tissue culture infective dose.

3. Conductivity

The three buffers were then compared for stability of conductivity across the width of the electrophoresis path. Similar CFES runs of each of the three buffers showed, once again, significant differences in behavior of the three buffers. The results are shown in Figure 7-5. The conductivity of the STS-8 buffer varied across nearly the complete width of the electrophoresis buffer column, with the conductivity of samples toward the anode being threefold higher than that of the beginning buffer and the conductivity of samples toward the cathode dropping to only one third that of the beginning buffer, resulting in nearly a tenfold difference in conductivity across the width of the electrophoresis chamber. The conductivity of the G-1 buffer, although better than that of STS-8, was unstable in the nine samples on the cathode side, and the conductivity rose over twofold in the 50 tubes on the anode side. Q buffer was very stable throughout the width of the electrophoresis buffer column, with only five tubes on the cathode side showing conductivity differences, these were very minor.

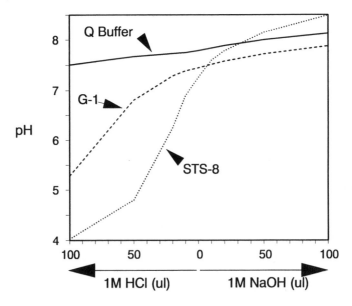

Figure 7-3 Relative buffering capacity of STS-8, G1, and Q buffers. Identical aliquots of each buffer were titrated with 1.0 *M* HCl or 1.0 *M* NaOH. Each buffer was titrated at the final concentration used for electrophoresis in the continuous flow electrophoresis system as described in Section II. (From Rodkey, L. S., *Appl. Theor. Electrophoresis*, 1, 243, 1990. With permission.)

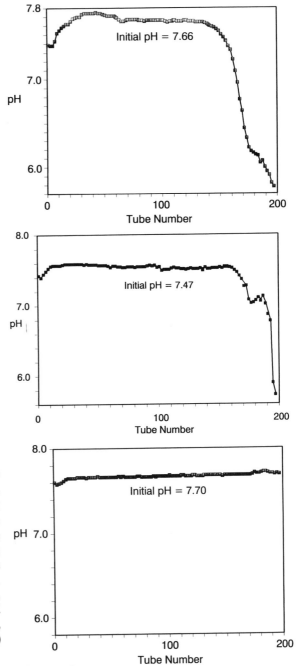

Figure 7-4 pH profiles of STS-8, G-1, and Q buffers after electrophoresis. The pH values of individual fractions were analyzed after 2-h runs of (**A**) STS-8, (**B**) G-1, or (**C**) Q buffer in the continuous flow electrophoresis system. The anode is to the right. (From Rodkey, L. S., *Appl. Theor. Electrophoresis,* 1, 243, 1990. With permission.)

D. RED BLOOD CELL SEPARATIONS

Mixtures of rabbit and goose RBC were used to compare the relative resolving powers of the three buffers. Formalin-fixed RBC were washed by centrifugation and resuspended in the buffer being tested. Electrophoresis in the CFES was then done using the buffer being tested. The results of these separations showed substantial differences in the resolving powers of the three buffers (Figure 7-6). Electrophoretic profiles of both cell

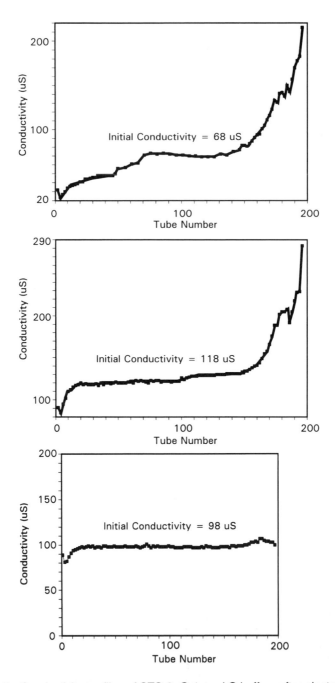

Figure 7-5 Conductivity profiles of STS-8, G-1, and Q buffers after electrophoresis. The conductivity of individual fractions were analyzed after 2-h runs of (**A**) STS-8, (**B**) G-1, or (**C**) Q buffer in the continuous flow electrophoresis system. The anode is to the right. (From Rodkey, L. S., *Appl. Theor. Electrophoresis,* 1, 243, 1990. With permission.)

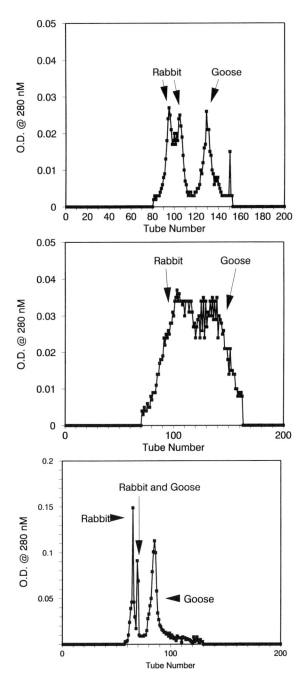

Figure 7-6 Electrophoretic separation of red blood cells. Mixtures of formalin-fixed rabbit and goose RBC were separated using (**A**) STS-8, (**B**) G-1, or (**C**) Q buffer in the continuous flow electrophoresis system. The anode is to the right. (From Rodkey, L. S., *Appl. Theor. Electrophoresis*, 1, 243, 1990. With permission.)

types were very broad with both STS-8 and G-1 buffers. The rabbit RBC were polydispersed in the STS-8 buffer, with two apparent peaks resolved over a spread of about 30 tubes. The goose cells were also polydispersed in STS-8 buffer, with a spread of about 30 tubes. The cell mixture emerged from the chamber in a 90-tube pattern using G-1 buffer with rabbit RBC and goose RBC not being well separated, but with the anodic fractions being mostly goose cells and the more cathodic fractions being mostly rabbit cells. Microscopic examinations of the intermediate fractions between tubes 100 and 145 showed that nearly all contained some mixture of both types of cells. However, the rabbit and goose RBC were well resolved with Q buffer. Indeed, a sharp cathodic peak of 8 fractions was found microscopically to be pure rabbit RBC, and an anodic fraction of approximately 15 fractions was found microscopically to be pure goose RBC. Furthermore, two fractions containing aggregates of rabbit and goose RBC (determined microscopically) were found with intermediate mobility between the pure rabbit and goose RBC major peaks. The apparent lower electrophoretic mobility of the cells in Q buffer (Figure 7-6C) was probably a function of the lower pH of Q buffer in this run (pH 7.1) as compared to the standard buffers.

E. EFFECTS OF Q BUFFER ON CELL VIABILITY

The RBC electrophoresis studies suggested that Q buffer might have some distinct advantages for preparative cell separations as compared with the triethanolamine buffers which are widely used for cell separations. It was postulated that Q buffer could have some advantages for viable cell separation studies if the buffer had no significant deleterious effects on cell viability when cells were suspended in the buffer for periods approximating the length of a typical electrophoretic separation run. To assess this, the hybridoma fusion partner cell line SP2 and two hybridomas derived from fusion of murine spleen cells with SP2 were studied for their viability after suspension for up to 4 h in Q buffer. Results of a representative study are shown in Figure 7-7. The data were normalized by plotting the results as percentage of the original time 0 viability of each

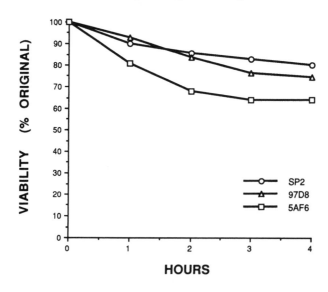

Figure 7-7 Effects of Q buffer on cell viability. Myeloma (SP2) or hybridoma cells (97D8 and 5AF6) were suspended in Q buffer for 4 h at room temperature. Samples were removed at 1-h intervals, and the percent viability was determined by trypan blue exclusion. The results were plotted as percentage of original viability.

cell line since the three varied slightly in the percentage of viable cells in the starting cell sample. It is evident that incubation for up to 4 h at room temperature in Q buffer had no significant effect on the viability of these three cell lines. Furthermore, similar viabilities were found for the same cell lines after incubation in Hank's balanced salt solution alone (data not shown; see below).

F. HYBRIDOMA CELL ELECTROPHORESIS

The success in enhancing the resolution of goose and rabbit RBC separations using Q buffer and the ability of this buffer to support cell viability successfully over short periods of time suggested that this buffer might have properties desirable for separation of viable cells. The earlier studies on cell viability and growth in zwitterion mixture buffers also suggested that this application was ideal for zwitterionic buffers. Prior to electrophoretic analysis of viable hybridoma cells the ability of Q buffer to support viability was again measured. Cells were harvested and washed twice by centrifugation in cold phosphate-buffered saline, pH 7.2, to wash out media components. Hybridoma 97D8 cells were split into two aliquots; one aliquot was incubated in Hanks' balanced salt solution, and the other aliquot was incubated in Q buffer. The cells were incubated for a period of 2 h, and the viability was then measured by trypan blue exclusion. The results showed that 88% of the cells were viable after incubation at room temperature for 2 h in either Hanks' or Q buffer. Preliminary studies with SP2 cells suggested that Q buffer offered an excellent separation medium for viable cells (data not shown). In order to verify and extend these studies, further studies using a hybridoma derived by fusion with SP2 cells were done. Stock cultures of 5AF6 hybridoma cells were split into four equal aliquots. Two aliquots were each grown up to mid-log phase on two different days, and two other aliquots were each grown up to stationary phase on two different days. Thus, four separate experiments were done on four separate days. The resulting log-phase or stationary-phase cells were then run in the CFES using the Q buffer. The cell concentrations of all 197 fractions were counted and plotted to determine the reproducibility of individual electrophoretic runs in the CFES using Q buffer. The results are shown on Figure 7-8. The results suggested that the separations of log-phase and stationary-phase cells were highly reproducible using Q buffer.

IV. DISCUSSION

These studies were initiated after reviewing the electrophoretic behavior of several kinds of particles and cells in the McDonnell-Douglas CFES preparative electrophoresis instrument. Numerous studies were done with a variety of cell types using the STS-8 and G-1 buffers developed for this instrument. The net result of several studies done in the authors' laboratory over a long period of time separating a variety of cells and charged particles suggested that the CFES instrument was giving results which were frequently unpredictable; run-to-run reproducibility ranged from marginal to nonexistent. Many of the operating parameters were changed in an attempt to make the separation runs more consistent and reproducible. Repeated attempts to improve the reproducibility of results obtained from the system convinced us that the problem was not in the hardware or in the design of the system. Since the buffers were the only variable outside of the hardware that could be manipulated, a series of studies was undertaken to find alternative buffers.

Early studies suggested that buffers based on the zwitterionic mixtures such as those which are widely used for isoelectric focusing might have some utility for cell separation. The early studies utilized zwitterion mixtures which functioned as wide-range ampholytes when used for isoelectric focusing experiments such as those described by Binion and Rodkey.[17] These studies showed that the crude zwitterion mixtures were not significantly toxic or otherwise deleterious to cells when added to cultures that were then incubated for

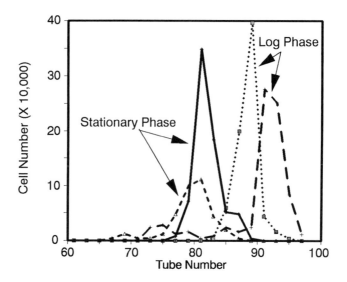

Figure 7-8 Viable hybridoma cell electrophoresis. Electrophoretic profiles of four separate continuous flow electrophoresis runs of 5AF6 hybridoma cells. The cells were suspended in Q buffer for sample injection, and electrophoresis was done in Q buffer. The anode is to the right. (From Rodkey, L. S., *Appl. Theor. Electrophoresis,* 1, 243, 1990. With permission.)

periods of 24 h or less (Figure 7-2). Some effects were seen after 72 h of incubation at higher concentrations of zwitterions, but no effects were noted after 24 h with one interesting exception. Primary rabbit skin fibroblasts were stimulated in their growth over a low range of concentrations, with optimal concentration being in the range of 0.06 to 0.25%. Cell viability dropped at a concentration of 0.5%. This finding of enhanced viability has not yet been pursued, but should be studied more closely. Additionally, the studies on vaccinia infectivity of cells grown in zwitterion-containing media show that such cells are capable of sustaining viral propagation even at 0.5% zwitterion concentration (Table 7-1). This is further evidence that essential cellular functions are not compromised by the presence of zwitterions in the media for relatively short periods of time. Since the presence of 0.25% zwitterions had no discernable effects on cell metabolism for short time periods, this suggested that concentrations lower than 0.25% should be acceptable for short-term use with viable cells. Thus, the use of Q buffer with zwitterion concentrations of 0.04 to 0.1% for short periods of electrophoresis time is not likely to cause any direct effects on the cells being separated.

Simple titration experiments subsequently showed that the standard triethanolamine buffers developed specifically for use in this instrument were actually quite poor buffers at the low concentrations at which they must be used (Figure 7-3) when compared to Q buffer. Low ionic strength buffers are necessary for most preparative electrophoresis systems to accommodate both the physiological demands of the cells and the physical constraints of the hardware for minimal heat generation during electrophoresis as well as to eliminate or minimize bubble formation in the separation chamber.

The subsequent studies with the standard triethanolamine buffers suggested that they were unstable in the electrical field and appeared to be chemically breaking down or were either gaining or losing ions to the anode or cathode chambers (Figures 7-4 and 7-5). It is clear that Q buffer has substantially better buffering properties than either the STS-8 or the G-1 buffer (Figure 7-3), and this should substantially reduce any effects of ions gained or lost at either electrode. Thus, the general conclusion that the standard triethanolamine buffers were chemically unstable in the electrical field seems warranted and could explain the marginal to nonexistent degree of reproducibility experienced in the earlier studies using them. Q buffer has given superior titration results, and both its pH and conductivity profiles after electrophoresis indicate that it has a high degree of chemical stability at low ionic strength during the electrophoresis process.

Additionally, the electroosmotic fluid flow caused by the zeta potential of the chamber walls may be affected by the chemical nature of the Q buffer. The zwitterionic nature of the buffer may tend to offset the wall charge of the separation chamber, leading to a decrease in the slip motion on particles close to the walls. Since these buffers can be shown by isoelectric focusing to contain individual species of molecules with individual isoelectric points (pI) ranging across a spectrum of charge densities, a subset of molecules with net cationic charge densities relative to the average pI of the mixture would likely be available to offset the negative charge on the chamber walls. The flowing buffer would have the effect of "coating" the chamber wall with a continuous supply of neutralizing cationic buffer subspecies, which would reduce fluid flow on the walls and, thus, would allow sharper resolution of particle subsets. Since the zwitterions are present in a constantly renewing supply due to buffer flow, any coating molecule which leached off would be immediately replaced and the excellent buffering capacity of the zwitterionic buffer could easily compensate for slight transient decreases in the most strongly cationic species. Albumin coating at 2 to 3% has been used in attempts to reduce wall effects (see Reference 5), but albumin will slowly leach away during the run. The behavior of the fixed RBC suspensions in Q buffer showed dramatically improved resolution as compared with the standard triethanolamine buffers. Band spreads of as little as 9 to 13 tubes using Q buffer, as compared to the 40- to 50-tube band spread (25% of the total path width) obtained with the STS-8 and G-1 buffers, suggest that resolution was enhanced by at least four- to fivefold with the use of Q buffer.

The hybridoma cell studies shown in Figure 7-8 were done with four separate aliquots of cells grown and separated in the CFES on four different days. The slight variation in the position of the log-phase cells on two different days might be the result of slightly different total growth times prior to electrophoresis. Our results show that hybridoma cells in log phase bear a slightly higher average net negative charge than do stationary-phase cells. Furthermore, Q buffer in the CFES provides a degree of resolution which can consistently and reproducibly distinguish the two populations. Sharnez and Sammons[18] showed that subsets of yeast cells with lower net negative charge were present in log-phase cells, but were not detectable in stationary-phase cells. Bauer et al.[13] showed that consistently higher net negative charges associated with B-lymphocytes were appearing to be in later developmental stages than B-cells with lower negative charge density. The results obtained in the hybridoma cell electrophoresis studies suggest that the electrophoretic differences found in diverse cell types in different phases of the growth cycle may be exploitable for most eukaryotic systems.

V. SUMMARY AND CONCLUSION

Studies of a zwitterionic buffer formulated for cell electrophoresis were done using the McDonnell-Douglas continuous flow electrophoresis system. Initial studies showed that several zwitterion mixtures and a proprietary zwitterionic buffer supported cell viability and growth during short-term exposures. Two standard triethanolamine-based buffers used previously for cell electrophoresis were analyzed for their stability in the electrical field, and the results showed that both triethanolamine-based buffers tested were chemically unstable. Furthermore, titration studies showed that the standard buffers buffered

160

poorly at the pH employed for electrophoresis. The zwitterionic Q buffer buffered well at its nominal pH and was shown to be stable in the electrical field. Studies comparing the zwitterionic Q buffer with the two triethanolamine cell separation buffers using formalin-fixed rabbit and goose red blood cells showed that the zwitterionic Q buffer gave dramatically improved resolution for separation of the cells when compared with the triethanolamine-based buffers. Studies with viable hybridoma cells showed that Q buffer supported cell viability and that hybridoma cells in different stages of the growth cycle demonstrated reproducible differences in their electrophoretic mobilities.

The organic zwitterionic buffers appear to offer significant advantages over conventional buffers for improvement of resolution of viable cells in zonal electrophoresis separation protocols. This increased resolution is gained with no significant loss of viability, although cell viability does vary according to the individual cell origin or line and the length of exposure to the zwitterion buffer. High resolution and excellent reproducibility from run to run are the principal attributes of these zwitterionic buffers when they are used for preparative cell electrophoresis applications.

REFERENCES

1. **Bohn, B. and Nebe, C. T.,** Analytical and preparative free-flow cell electrophoresis: an alternative and complementary method to flow cytometry and sorting, in *Cell Electrophoresis,* Schütt, W. and Klinkmann, H., Eds., Walter de Gruyter, Berlin, 1985, 305.
2. **Bauer, J. and Kachel, V.,** Separation accuracy of free flow electrophoresis as proved by flow cytometry, *Electrophoresis,* 9, 62, 1988.
3. **Hansen, E.,** Preparative free-flow electrophoresis of lymphoid cells: a review, in *Cell Electrophoresis,* Schütt, W. and Klinkmann, H., Eds., Walter de Gruyter, Berlin, 1985, 287.
4. **Hansen, E. and Hannig, K.,** Electrophoretic separation of lymphoid cells, *Methods Enzymol.,* 108, 180, 1984.
5. **Hannig, K.,** New aspects in preparative and analytical continuous free-flow cell electrophoresis, *Electrophoresis,* 3, 235, 1982.
6. **Brown, K. A. and Holborow, E. J.,** Epstein-Barr (EB) virus infected B lymphocytes possess an altered electrophoretic mobility, in *Cell Electrophoresis in Cancer and Other Clinical Research,* Preece, A. W. and Light, P. A., Eds., Elsevier/North-Holland, Amsterdam, 1981, 217.
7. **Hymer, W. C., Barlow, G. H., Blaisdell, S. J., Cleveland, C., Farrington, M. A., Feldmeier, M., Grindeland, R., Hatfield, J. M., Lanham, J. W., Lewis, M. L., Morrison, D. R., Olack, B. J., Richman, D. W., Rose, J., Scharp, D. W., Snyder, R. S., Swanson, C. A., Todd, P., and Wilfinger, W.,** Continuous flow electrophoretic separation of proteins and cells from mammalian tissues, *Cell Biophys.,* 10, 61, 1987.
8. **Morrison, D. R., Barlow, G. H., Cleveland, C., Farrington, M. A., Grindeland, R., Hatfield, J. M., Hymer, W. C., Lanham, J. W., Lewis, M. L., Nachtwey, D. S., Todd, P., and Wilfinger, W.,** Electrophoretic separation of kidney and pituitary cells on STS-8, *Adv. Space Res.,* 4, 67, 1984.
9. **Morrison, D. R., Lewis, M. L., Barlow, G. H., Todd, P., Kunze, M. E., Sarnoff, B. E., and Li, Z.,** Properties of electrophoretic fractions of human embryonic kidney cells separated on Space Shuttle flight STS-8, *Adv. Space Res.,* 4, 77, 1984.

10. **Lewis, M. L., Barlow, G. H., Morrison, D. R., Nachtwey, D. S., and Fessler, D. L.,** Plasminogen activator production by human kidney cells separated by continuous flow electrophoresis, in *Progress in Fibrinolysis,* Davidson, J. F., Bachmann, F., Bouvier, C. A., and Kruithof, E. K. O., Eds., Churchill Livingstone, Edinburgh, 1983, 143.

11. **Todd, P., Plank, L. D., Kunze, M. E., Lewis, M. L., Morrison, D. R., Barlow, G. H., Lanham, J. W., and Cleveland, C.,** Electrophoretic separation and analysis of living cells from solid tissues by several methods. Human embryonic kidney cells as a model, *J. Chromatogr.,* 364, 11, 1986.

12. **Todd, P.,** Microgravity cell electrophoresis experiments on the space shuttle: a 1984 overview, in *Cell Electrophoresis,* Schütt, W. and Klinkmann, H., Eds., Walter de Gruyter, Berlin, 1985, 3.

13. **Bauer, J., Kachel, V., and Hannig, K.,** The negative surface charge density is a maturation marker of human B lymphocytes, *Cell. Immunol.,* 111, 354, 1988.

14. **Hansen, E. and Hannig, K.,** Antigen-specific electrophoretic cell separation (AESECS): isolation of human T and B lymphocyte subpopulations by free-flow electrophoresis after reaction with antibodies, *J. Immunol. Methods,* 51, 197, 1982.

15. **Hansen, E., Wustrow, T. P. U., and Hannig, K.,** Antigen-specific electrophoretic cell separation for immunological investigations, *Electrophoresis,* 10, 645, 1989.

16. **Zeiller, K., Loser, R., Pascher, G., and Hannig, K.,** Free flow electrophoresis. II. Analysis of the method with respect to preparative cell separation, *Hoppe-Seyler's Z. Physiol. Chem.,* 356, 1225, 1975.

17. **Binion, S. and Rodkey, L. S.,** Simplified method for synthesizing ampholytes suitable for use in isoelectric focusing of immunoglobulins in agarose gels, *Anal. Biochem.,* 112, 362, 1981.

18. **Sharnez, R. and Sammons, D. W.,** Continuous flow electrophoresis: resolving real and apparent heterogeneity, *Biopharm,* 2, 54, 1989.

19. **Rodkey, L. S.,** Free flow electrophoresis using zwitterionic buffer, *Appl. Theor. Electrophoresis,* 1, 243, 1990.

Chapter 8

Alteration of Cellular Features after Exposure to Low Ionic Strength Medium

Ingolf Bernhardt

CONTENTS

I. INTRODUCTION

The application of external electromagnetic fields of relatively high field strength on cells (e.g., in electrorotation, electrofusion, or to some extent in dielectric breakdown experiments) requires the use of nonphysiological low ionic strength solutions to ensure favorable conditions for polarization and to avoid heating. In addition, low ionic strength solutions are often used to perform cell electrophoresis and to investigate the influence of parameters such as the transmembrane potential or the surface potential on various cellular characteristics. Furthermore, it has to be taken into account that some cells (e.g., plant cells) exist under "normal" conditions in media of low ionic strength. This chapter summarizes some of the effects of low ionic strength media on cells and concentrates on red blood cells since many investigations in this field have been carried out on this type of cell.

In general, the lowering of the ionic strength of an extracellular solution by reducing the extracellular salt concentration (mainly NaCl) results in a change of the ion concentration, and possibly of the osmolarity, of the medium. These changes affect the outer surface potential of the cell as well as the transmembrane potential directly. Each of the parameters mentioned has consequences for a large variety of cellular characteristics.

II. SURFACE POTENTIAL AND SURFACE STRUCTURE

Due to the Boltzmann distribution of ions in an electric field, the local ion concentration near the cell surface as well as the cell surface pH differ from their bulk values when there is a difference in electric potential.[1] Therefore, knowledge of the electric potential profile near a cell surface is crucial for the investigation of a large variety of biochemically related processes.[2,3] In low ionic strength solutions, in particular, significant changes in these processes have to be expected.

It is known that fixed charges are distributed in a layer of some nanometers thickness around the cell surface (the glycocalyx). They are generated by the adsorption and dissociation of ionogenic groups of the surface coat glycoproteins and glycolipids, as well

0-8493-8918-6/94/$0.00+$.50
© 1994 by CRC Press, Inc.

as membrane phospholipids. Under physiological conditions the glycocalyx of most animal cells carries a surface charge on the order of 0.02 C/m².[4,5] Data exist for the human red blood cell showing that about 1.5×10^7 elementary electric charges are distributed in a surface layer of approximately 6 nm thickness and a surface area of about 140 μm².[6] The classical description of the electric potential profile near the cell surface (Gouy-Chapman theory) is based on the assumption that the fixed charges are homogeneously distributed over the surface. The calculations of the electric potential profile perpendicular to the membrane surface, $\psi(x)$, are based on the Poisson-Boltzmann equation:

$$\psi''(x) = -\frac{\rho(x)}{\varepsilon\varepsilon_o} = -\frac{eN}{\varepsilon\varepsilon_o}\sum_i z_i c_i \exp\left(-z_i e\psi(x)/kT\right)$$

(8-1)

Here, $\rho(x)$ is the charge density of mobile ions, N Avogadro's number, ε_0 the dielectric constant, ε the permittivity, e the elementary electric charge, z_i the valence of the ith ionic species, c_i the bulk ionic concentrations, k the Boltzmann constant, and T the absolute temperature. In high ionic strength solutions the electric potential near the cell surface tends to be relatively low and the linearized solution of Equation 8-1 is sufficient:

$$\psi(x) = \psi_o(x)\exp(-\kappa x)$$

(8-2)

Here ψ_0 is the surface potential at $x = 0$; κ is the inverse of the Debye length and is given by the equation:

$$\kappa = \sqrt{\frac{e^2 N \sum_i z_i^2 c_i}{\varepsilon\varepsilon_o kT}}$$

(8-3)

However, in low ionic strength solutions the nonlinear solution of Equation 8-1 is more precise. In this case, the function of the electric potential, dependent on the distance from the cell surface, is given by the implicit equation:[7]

$$x = \frac{1}{\kappa}\ln\left\{\frac{\tanh(ze\psi_o/4kT)}{\tanh(ze\psi(x)/4kT)}\right\}$$

(8-4)

In the linear approximation $1/\kappa$ is the distance necessary for the surface potential ψ_0 to be reduced by $1/e$ (1/2.718). Although the Gouy-Chapman theory was extended by Stern also taking into account the adsorption processes at the membrane-solution interface, both classical approaches can be considered only as simple approximations. As stated above, the fixed electric charges are spatially distributed in the glycocalyx and not only at the membrane surface. Therefore, a fixed space charge density $\rho_f(x,y,z)$ has to be taken into account instead of the surface charge density. As pointed out by Voigt and Donath, the problem of an adequate description of the surface electrostatics can be solved relatively easily when the quotient of the mean fixed charge distance inside the glycocalyx, a, and the Debye length $1/\kappa$ fulfills the relation $a \times \kappa < 1$.[7] In this case, fixed charges are considered to be smoothly distributed in the cell surface layer and the space charge density depends only on the x coordinate (perpendicular to the membrane surface). If $a \times \kappa > 1$, which is quite common under certain experimental conditions, a discrete charge

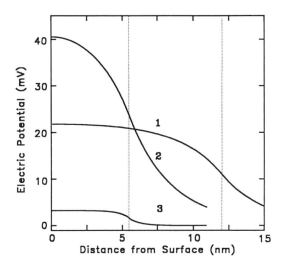

Figure 8-1 Dependence of the electric potential near the cell membrane surface on the distance from the surface at different ionic strengths of the extracellular solution. Calculations were made on the basis of the space charge model of Donath and Voigt.[11] Surface charge and dielectric constant were assumed to be 0.02 C/m² and 78, respectively; temperature = 25°C. Curve 1 — 10 mM NaCl solution; distribution of the surface charge in a glycocalyx of 12 nm. Curve 2 — 10 mM NaCl solution; distribution of the surface charge in a glycocalyx of 5.5 nm. Curve 3 — 150 mM NaCl solution; distribution of the surface charge in a glycocalyx of 5.5 nm.

distribution model has to be considered. The calculations of the surface electrostatics are more complicated in this case, the mathematical solution for the electric potential being expressed in terms of Fourier-Bessel integrals.[8]

When varying the ionic strength of the solution surrounding a cell one has to consider diverse effects caused by changes in the electric potential profile near the cell surface. For example, in a low ionic strength solution, the increase of the electric potential results in an enhancement of the intra- and intermolecular electrostatic repulsion or attraction of the charged sites. This, in turn, leads to the structural changes of the glycocalyx molecules (glycoproteins, glycolipids) as well as changes in the physical characteristics of the whole glycocalyx (density profile, thickness).[9] It was calculated from electrophoretic measurements that the glycocalyx of human erythrocytes suspended in a low ionic strength solution of 10 mM NaCl increased in thickness to about 12 nm.[10]

Figure 8-1 shows the electric potential profile near the outer surface of human erythrocytes suspended in solutions of 150 mM and 10 mM NaCl concentration assuming a fixed space charge density in the glycocalyx. The calculations are based on the concept of Donath and Voigt.[11]

In addition, depending on the electric potential of a microenvironment of a given site on the membrane surface or of the glycocalyx, local changes in the ion concentration as well as in the pH value have to be considered in accordance with the Boltzmann equation. The changes may occur not only in the direction of the x coordinate, but also in the y,z direction (plane of the membrane). These variations are of importance for changes in a large variety of cellular and membrane parameters, e.g., membrane lipid phase state,[12,13] local binding constants of ions and various substances,[2,14-17] membrane transport,[18-20] cell adhesion,[21-24] cell-cell interaction,[25-28] and photosynthetic processes.[29,30]

III. TRANSMEMBRANE POTENTIAL, CELL VOLUME, AND INTRACELLULAR PH

Whether or not there is a change in the transmembrane potential of cells after reduction of the ionic strength of the extracellular solution depends on the cell type as well as on the actual medium ion concentration. The transmembrane potential can be calculated according to the Goldman-Hodgkin-Katz equation:

$$\Delta\psi = \frac{RT}{F} \ln \frac{P_K C_K^o + P_{Na} C_{Na}^o + P_{Cl} C_{Cl}^i}{P_K C_K^i + P_{Na} C_{Na}^i + P_{Cl} C_{Cl}^o} \tag{8-5}$$

Here P_K, P_{Na}, P_{Cl}, C_K, C_{Na}, and C_{Cl} are the permeability coefficients and the ion concentrations of K^+, Na^+, and Cl^-, respectively. The superscripts i and o denote inside and outside, respectively, R is the gas constant, F the Faraday constant, and T the absolute temperature.

In most animal cells the K^+ permeability is much higher than the Na^+ and Cl^- permeabilities.[31] Therefore, a decrease in the NaCl concentration of the extracellular solution (at a constant osmolarity and at a constant K^+ concentration) will lead to a remarkable change in the transmembrane potential only in the case of dramatic variations of the NaCl concentration. Since the extracellular K^+ concentration of animal cells is relatively small, its further reduction will have only minor effects. The situation is different in cells with the highest membrane ion permeability for Na^+ or Cl^-, such as erythrocytes, where the Cl^- permeability is several orders of magnitude higher than the Na^+ and K^+ permeabilities.[32] A decrease in the extracellular NaCl concentration results in a dramatic change in the transmembrane potential of these cells, from about –8 mV in a physiological ionic strength solution (NaCl) to about +50 mV in an isotonic low ionic strength solution (sucrose addition).[33]

When studying changes in intracellular pH and in cell volume one has to take into account the time scale of the following processes: (1) water equilibration due to the osmotic gradient, (2) pH equilibration process (proton and hydroxyl ion permeability), and (3) exchange of inorganic anions (chloride, bicarbonate) and of inorganic cations. Decreasing the ionic strength of the extracellular solution without balancing its osmolarity usually results in an increase of the cell volume. However, in buffered isotonic low ionic strength solutions (when the osmolarity is kept constant by the addition of uncharged impermeable substances), changes in the intracellular pH and in the cell volume also can occur. These changes are of importance for cells with relatively fast pH equilibration processes (in the same time scale as the equilibration process of the ion determining the transmembrane potential). In this case not only the intracellular pH is changed in a relatively short time, but the cell volume also becomes different from the "normal" value. This happens because the isoelectric point and the buffer capacity of intracellular osmotically active substances are influenced by the intracellular pH and, therefore, the net charges of these substances are changed.[33]

The red blood cell is a classical example. The lowering of the ionic strength of the extracellular solution by decreasing the (NaCl + KCl) concentration to about 10 mM in a buffered solution (5 mM phosphate buffer, pH = 7.4) and the addition of sucrose (250 mM) to keep the osmolarity constant (300 mosmol/l) will effect changes in the transmembrane potential (from –8 mV to +45 mV), in the intracellular pH (from 7.3 to about 8.1), and also in the cell volume, which decreases by about 10%.

For human erythrocytes Glaser et al. have proposed a stationary state model for simultaneous calculations of the transmembrane potential, intracellular pH, and cell volume under different extracellular conditions.[33,34] Some calculations based on this model are presented in Figure 8-2.

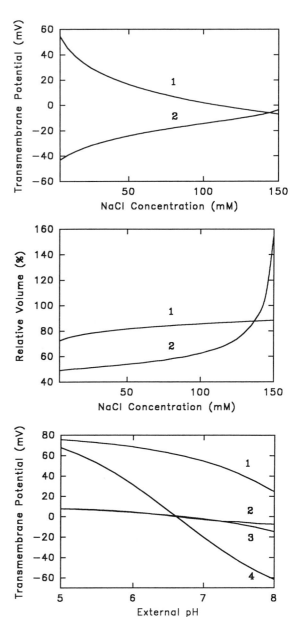

Figure 8-2 Dependence of the transmembrane potential and cell volume of human erythrocytes on the extracellular NaCl concentration at a constant pH of 7.4 as well as transmembrane potential dependence on the extracellular pH at different extracellular NaCl concentrations. Calculations were made on the basis of the equations given by Glaser et al.[33,34]Temperature = 37°C, osmolarity = 300 mosmol/l. **(a,b)** Curve 1: chloride equilibrium state (C state); curve 2: Donnan state (equilibrium state of Cl⁻ as well as Na⁺ and K⁺, D state) **(c)** Curve 1: 10 mM NaCl, C state; curve 2: 145 mM NaCl, D state; curve 3: 145 mM NaCl, C state; curve 4: 10 mM NaCl, D state.

After an initial swelling of vertebrate cells in hypertonic media by osmotic equilibration, they subsequently shrink due to a net loss of KCl and a concomitant loss of cell water. Most vertebrate cells regulate their volume by an activation of specific volume-sensitive ion transport mechanisms. However, there are also cells which lose KCl and cell water relatively slowly since a volume-sensitive KCl transport mechanism is lacking or latent in these cells. In mature human erythrocytes particularly, the volume-sensitive K⁺/Cl⁻ cotransport is only activated by definite maneuvers (high hydrostatic pressure, N-ethylmaleimide treatment).[35]

For the regulatory volume decrease (RVD) in different cell types, several transport mechanisms are of importance: (1) the activation of a K⁺/Cl⁻ cotransport (carrier) mechanism, (2) the activation of separate K^+ and Cl^- conductive pathways, or (3) the activation of a K^+/H^+ exchange mechanism functionally coupled to a Cl^-/HCO_3^- exchange mechanism (for cellular mechanisms of volume and pH regulation see, e.g., Reference 36).

IV. MEMBRANE ELECTRIC FIELD

Another important parameter which can influence membrane constituents and, therefore, modify a large variety of membrane processes is the membrane electric field. The electric field inside a biological cell membrane is determined by the gradient of the electric potential in the three dimensions x,y,z. Concerning the direction perpendicular to the membrane surface (x), in a constant field approximation assuming zero charge density inside the membrane, the electric field is given by the difference between the actual electric potentials on both sides of the cell membrane divided by the membrane thickness. Therefore, the transmembrane potential as well as both surface potentials are of importance for the electric field inside a cell membrane. The electric field strength in a biological cell membrane can be assumed to be of the order of up to 10^7 V/m.[37] It is easy to understand that a reduction of the ionic strength of the extracellular solution — e.g., by decreasing the NaCl concentration — results in a change of the electric field inside the cell membrane due to alterations of the transmembrane potential and the outer surface potential (see above). It has to be mentioned that the constant field approximation is useful only for a general discussion of electric field effects in cell membranes. In fact, the electric potential profile across the cell membrane is not linear since electrically charged and polarizable groups are present and, besides, are located in different positions relative to both sides of the membrane surface. This concerns the head groups of the phospholipids as well as parts of membrane proteins. The electric potential profiles proposed for ion channel proteins may serve as a classical example.[38-40] In addition, a change of the electric potential gradient leads to considerable alteration of the mechanical tension inside the membrane, given by the Maxwell stress, P_e:[41]

$$P_e = \frac{\varepsilon \varepsilon_o \left(d\psi / dx \right)^2}{2}$$

(8-6)

The electric field in the cell membrane in the y,z direction (plane of the membrane) depends on the present localization of the membrane proteins and lipids. Of special importance are the distribution of charged and polarized membrane constituents as well as their lateral movement.[42] Furthermore, the existence of electrogenic pumps and electrodiffusive pathways can contribute to the actual electric field in the membrane.[43]

Considering molecular mechanisms, one has to take into account electric field effects on membrane phospholipids and membrane proteins. It is known that the electric field influences the mobility and the position of the hydrocarbon chains of the phospholipids as well as the phase transition temperature.[12,44,45] The head groups of the phospholipids are also affected; an increase in the electrostatic repulsion (possibly due to a reduced ionic strength of the extracellular solution) contributes to the interfacial tension, which in turn can have mechanical consequences (e.g., morphological changes have been reported for erythrocytes).[46]

The influence of the membrane electric field on membrane proteins can be direct or indirect via the discussed changes in the phospholipid environment of the proteins leading to changes in lipid-protein interactions. Direct modifications of the protein conformation induced by changes in the electric field are based on one or more of the following mechanisms: charge displacement, dipole reorientation, and dipole induction.[3,47,48]

V. THE EFFECT OF LOW IONIC STRENGTH ON ION TRANSPORT

Under conditions of low ionic strength treatment of cells, the ion transport processes in biological membranes can be influenced greatly due to the reasons discussed above. Ion fluxes can be affected directly by the change in ion concentration of the medium, as well as indirectly via stoichiometrically coupled ion transport mechanisms.

Intracellular pH changes influence the activity of many membrane proteins and, therefore, ion transport processes too. The changes in cell volume modify the ion transport through the membranes, and it is known that there are some specific volume-sensitive ion transport pathways in membranes of a large variety of animal cells.[36] The ionic strength itself can influence the conformation of membrane proteins as well as the surface charge of the cell. A change in any of these parameters will have an effect on the ion transport (and the transport of other substances). It has to be considered that changes in the surface charge will give rise to changes in ion concentration near the cell surface and, therefore, also near the binding sites of specific membrane transport proteins (see above).

In addition, ion transport processes also can be influenced by changes in the transmembrane potential due to different reasons: (1) it will have a direct influence on electrodiffusive fluxes, and (2) together with changes of the surface charge (surface potential) it will modify the electric field strength in the membrane, affecting the conformation of membrane proteins.[3]

Thus, one must assume that changes in the ionic strength of the extracellular solution lead to significant variations in ion transport (and transport of other substances). It is often difficult to decide which of the altered parameters is responsible for the observed changes. As an example, the low ionic strength effect on monovalent cation transport through the red blood cell membrane will be reviewed in more detail.

The first observation that there is an increase in net K^+ efflux of human erythrocytes after reduction of the NaCl concentration of the isotonic extracellular solution was made by Davson in 1939 and Wilbrandt in 1940.[49,50] The replacement of NaCl by sucrose results in a change in the ionic strength of the solution and, hence, in the transmembrane potential (the transmembrane potential of erythrocytes is a Nernst potential of chloride). Attempts have been made to explain the increased K^+ efflux of human erythrocytes solely on the basis of electrodiffusion. However, Donlon and Rothstein had to assume a change in the permeability coefficient with respect to the transmembrane potential because the measured increase in K^+ efflux in a low NaCl solution was higher than expected from the Goldman flux equation.[51,52] It is important to emphasize that the Goldman flux equation is derived from the linear relationship:

$$\text{Flux } (J) = \text{Linear coefficient } (L) \times \text{Driving force } (X)$$

Thus, L is a constant independent of X. This means that if the permeability coefficient depends on the transmembrane potential (as proposed by Donlon and Rothstein), the Goldman flux equation does not hold under these conditions. Another attempt was made to solve this problem by extending the Goldman flux equation, taking the inner and outer surface potentials into account. In this case the increase in K^+ efflux observed in human erythrocytes suspended in solutions of low NaCl concentration could be explained without considering a change in the permeability coefficient.[53] However, it seems unlikely that this explanation is correct, since no enhanced K^+ flux was observed in bovine erythrocytes under the same experimental conditions.[54,55] This is despite the fact that the transmembrane potential (measured as the ^{36}Cl distribution ratio) of bovine erythrocytes changes by the same amount as that of human erythrocytes after reduction of the extracellular Cl⁻ concentration.[54,56] In addition, cell electrophoretic measurements dependent on the extracellular NaCl concentration have shown that there is a similar change in the external surface potential of cow erythrocytes compared with human erythrocytes.[54] Furthermore, it can be assumed that the internal surface potentials of cow and human erythrocytes do not differ significantly, since all of the negatively charged membrane phospholipid, phosphatidylserine is located in the inner leaflet of the cell membrane.[57]

Comparison of the rate constants of the K^+ efflux from erythrocytes of different mammalian species in physiological ionic strength solutions as well as those of low ionic strength showed that there was a significant increase in the K^+ efflux in low ionic strength solution in human, cat, rat, horse, and rabbit erythrocytes, whereas no significant change was observed in pig and cow erythrocytes.[56,58] It is noteworthy that there is no correspondence between the low ionic strength-induced enhancement of the K^+ efflux and the presence of the Ca^{2+}-induced K^+ channel in these cells.[58] The main differences between erythrocytes of various mammalian species are the relative content of the membrane phospholipids and the structure of their acyl chains.[59,60] Therefore, the attempt was made to correlate the measured rate constant of the K^+ efflux in solutions of physiological and low ionic strength with data for the membrane phospholipids of the different mammalian species. There is some evidence that the ionic strength-induced alteration of the passive cation movement correlates with the arachidonic acid content of the membrane phospholipids.[61]

The possible importance of the arachidonic acid content of the membrane phospholipids for the low ionic strength effect is based on three other observations:

1. Although cow erythrocytes do not show the low ionic strength effect, erythrocytes from newborn calves show the same increase in the rate constant of K^+ efflux in a solution of low ionic strength as human erythrocytes. The increase depends on the age of the calves. Investigations of the lipid composition of these erythrocytes demonstrate that the head group composition of the membrane phospholipids does not change with the age of the calves. However, there is a change in the relative content of arachidonic acid (20:4 [n-6]), as well as that of linoleic acid (18:2 [n-6]), in the membrane phospholipids. It was thus possible to correlate the rate constant of the K^+ efflux from calf erythrocytes in a solution of low ionic strength with the content of arachidonic acid depending on the age of the calves (correlation coefficient r = 0.951, $p = 0.05$).[61]
2. Using a phospholipid exchange protein, Kuypers et al.[62] showed that human erythrocytes were leaky for K^+ if the native phosphatidylcholine was partly replaced by a phosphatidylcholine containing arachidonic acid.
3. For the anion transport protein in the erythrocyte membrane (band 3) of different species, correlations were found between the phosphate influx as well as the bicarbonate-chloride exchange and the arachidonic acid content of the phospholipids of the erythrocyte membrane.[63,64]

From these results it is difficult to decide whether the low ionic strength effect on the K^+ efflux is influenced by the membrane fluidity or by the content of arachidonic acid in the membrane phospholipids. In this respect two problems have to be considered: (1) a change in the content of a highly unsaturated fatty acid of the membrane phospholipids can change the membrane fluidity, and (2) a specific effect of arachidonic acid could include other highly unsaturated fatty acids which are present in the erythrocyte membrane only in minor amounts.[65] In both cases it seems reasonable to assume that membrane proteins are involved in the low ionic strength effect whereby the surrounding lipids can influence these proteins by changing their conformation and/or by modifying the state of the lipid-protein interaction.

In the last few years a variety of transport pathways have been described in the red blood cell membrane (see, e.g., Reference 66). It has, therefore, been necessary to take these transport pathways into account in order to find out whether a specific cation transport or the residual K^+ flux is involved in the low ionic strength effect. At present, the best experimental measure of residual K^+ fluxes through erythrocyte membranes involves a unidirectional K^+ uptake or loss in the presence of ouabain, bumetanide or furosemide, and EGTA, with chloride replaced by nitrate or methylsulfate, to suppress the Na^+/K^+ pump, $Na^+/K^+/Cl^-$ cotransport, Ca^{2+}-activated K^+ transport, and any volume-sensitive K^+/Cl^- flux, respectively.[66]

With reference to the different transport pathways for monovalent cations it has been shown that there is no involvement of a partial flux mediated by the Na^+/K^+ pump or the Ca^{2+}-activated K^+ channel in the increase of K^+ efflux in low ionic strength media.[54] Also, 4-aminopyridine, which is used to block voltage-sensitive cation channels in membranes of different cell types, does not significantly change the rate constant of K^+ efflux in a physiological as well as a low ionic strength solution.[54,57,58] The addition of 4,4'-diisothiocyanato-2,2'-stilbenedisulfonic acid (DIDS, an inhibitor of anion transport via the band 3 protein), however, results in a significant decrease of the K^+ efflux in a low ionic strength medium as compared to the same solution without DIDS.[54,69] In a physiological ionic strength solution there is no significant influence of DIDS on K^+ efflux.[54,69] It still remains unclear whether the band 3 protein participates directly in the low ionic strength effect on K^+ efflux (see also below for a discussion of a specific protein involved in the low ionic strength effect).

The investigation of the possible involvement of $Na^+/K^+/Cl^-$ cotransport in the increase in the rate constant of the K^+ efflux in a low ionic strength medium in human erythrocytes has shown that in addition to the increase in the residual K^+ transport there is also an increase in the $Na^+/K^+/Cl^-$ cotransport.[54]

The investigations of the low ionic strength effect were extended to measurements of the residual K^+ influx. Marked enhancement of the residual (i.e., ouabain-, bumetanide-, and Ca^{2+}-insensitive) K^+ influx was observed when human red cells were suspended in a low ionic strength medium.[70] Figure 8-3 shows the dependence of both the residual K^+ efflux and the residual K^+ influx in human erythrocytes on the extracellular NaCl concentration.

As the K^+/Cl^- cotransport system in human erythrocytes is dependent on Cl^-, $CH_3 SO_4^-$ replacement suppresses the K^+ flux via this transport system. It was shown, however, that the low ionic strength-induced flux is independent of Cl^-.[70] Volume sensitivity (i.e., activation by cell swelling) is another important characteristic of K^+/Cl^- cotransport.[71] In contrast, the low ionic strength effect decreases slightly on cell swelling.[70] Furthermore, the K^+/Cl^- cotransport system excludes Na^+, whereas the low ionic strength effect is shown equally for Na^+ fluxes. The low ionic strength-induced fluxes cannot be identified with any of these criteria, and it is therefore concluded that K^+/Cl^- cotransport is not involved in the low ionic strength effect.[70]

It is known that human red blood cells suspended in sucrose solutions with the same osmolarity as the physiological NaCl solutions have a smaller volume (see above). This shrinkage is not responsible for the elevated K^+ influx observed in low ionic strength solutions, since a sucrose medium with an osmolarity of 200 mosmol (which leads to about the same erythrocyte cell volume as in a physiological ionic strength solution) only slightly influences the residual K^+ influx.[70]

When discussing the effect of the enhancement of the residual K^+ influx of human erythrocytes in a low ionic strength solution there are additional indications that the transmembrane potential is not responsible for the low ionic strength effect (see above). It is not possible to explain a dramatic increase in the residual K^+ influx of human

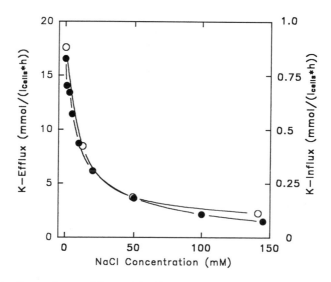

Figure 8-3 Dependence of the residual K⁺ efflux (filled circles) and residual K⁺ influx (open circles) of human erythrocytes on the extracellular NaCl concentration of isotonic solutions (sucrose addition, osmolarity = 300 mosmol/l). Measurements of the K⁺ fluxes were made in the presence of ouabain (0.1 mM), bumetanide (0.1 mM), and EGTA (0.1 mM). The extracellular KCl concentration was 7.5 mM and the concentration of Na_2HPO_4/NaH_2PO_4 was 5.8 mM; pH = 7.4. Experimental values were taken from Reference 88.

erythrocytes by reducing the extracellular NaCl concentration on the basis of the Goldman flux equation.[70] Under these conditions one would predict a threefold decrease in the unidirectional residual K⁺ influx. This is because of a depolarization of the membrane from the physiological value of about –10 mV to +50 mV in low NaCl solutions. This point was explored further by altering the anion concentration and adding the impermeant anions gluconate or glucuronate. Under these conditions there was only a small increase in the K⁺ influx, in marked contrast to sucrose, although the transmembrane potential and pH_i values were equivalent in both cases.[70] However, there are some indications that the induced flux is sensitive to the internal pH, since the residual K⁺ uptake is higher at pH_i of either 6.2 or 8.3 than at 7.3, but this effect is relatively small in comparison to the tenfold effect of low ionic strength at $pH_0 = 7.4$.[70] Chipperfield and Shennan[72] and Zade-Oppen et al.[73] have also demonstrated a small pH dependence for the residual K⁺ transport in human red blood cells.

The elevated K⁺ flux observed in low ionic strength media was reversible when the cells were resuspended in a solution of physiological ionic strength.[53,70] The selectivity of the low ionic strength-induced flux was investigated by comparing the K⁺, Na⁺, Ca²⁺, and lysine uptake in physiological and low ionic strength media. It was shown that Ca²⁺ and lysine were excluded from the low ionic strength-induced flux pathway, with only K⁺ and Na⁺ showing a large stimulation (about 10×) of transport.[74] In addition, these results show that the characteristics for the enhancement of the residual K⁺ and Na⁺ fluxes in a low ionic strength solution differ from those reported for a voltage-activated cation channel recently presented by Halperin et al.[75] This channel, activated by increasing the transmembrane potential of human red blood cells, is characterized by an enhancement of K⁺, Na⁺, and Ca²⁺ transport. This effect is not reversible and occurs in high-potassium-type

(HK) but not low-potassium-type (LK) sheep red blood cells.[75,76] In contrast, the effect of the increase of residual K^+ and Na^+ fluxes in a low ionic strength solution was observed in LK but not in HK sheep red blood cells.[77]

The investigation of the concentration dependence of the residual K^+ and Na^+ influx over the range 0.25 to 10 mM revealed that the low ionic strength media elevated K^+ and Na^+ residual transport by about the same amount. In both cases the flux was linear with the concentration, showing no signs of saturation over the range studied.[70] However, if high K_m values are a property of the flux, saturation will be apparent only at higher substrate concentrations. In fact, Canessa et al.[78] have shown that residual K^+, as well as Na^+ effluxes into K^+- and Na^+-free solutions, saturate when the intracellular K^+ or Na^+ concentration is increased.

Since it has been demonstrated that the residual K^+ flux in human red cells suspended in a physiological ionic strength medium increases with a decrease in temperature from 12 to 0°C,[79,80] the flux in low ionic strength medium over the temperature range 0 to 37°C was measured to determine if there are any similarities between the two stimulatory maneuvers. Since it has been shown that Li^+ is a potent inhibitor of the paradoxical temperature effect, the experiments were also carried out in the presence of this cation.[79] In a sucrose medium the residual K^+ influx was much higher and showed a more pronounced paradoxical effect of lowering the temperature, with a clear minimum at about 8°C. In this case, the addition of 1 mM LiCl led to a substantial but not complete inhibition of this low-temperature-induced K^+ flux. At 0°C, when 1 mM NaCl was added to assess the contribution of ionic strength per se, the inhibition of the residual K^+ influx in a low ionic strength solution was only 15%.[70] Therefore, addition of Li^+ reduces the residual K^+ influx by slightly raising the ionic strength, but principally by a more specific inhibitory action. At 37°C LiCl slightly reduced the K^+ influx in a sucrose medium, whereas the addition of 1 mM NaCl inhibited the low ionic strength effect to the same degree.[70] This indicates that at the higher temperature the reduction of the residual K^+ influx by LiCl is due solely to the change in the ionic strength of the solution. These results led to the conclusion that the residual K^+ influx in a solution of low ionic strength at a low temperature could be resolved into two independent components: (1) a low-temperature-induced component which is Li^+ sensitive and (2) a low ionic strength-induced component which is not Li^+ sensitive.[70,79]

If the present phenomenon resembles free diffusion — i.e., a weak interaction with a water-filled "pore" — a low temperature dependence would be expected. In fact, this was not the case since the K^+ uptake was significantly temperature dependent. The estimated activation energy of 50 to 60 kJ/mol in a low ionic strength solution containing 1 mM LiCl is much higher than the expected value for electrodiffusion through a water-filled "pore" (about 20 kJ/mol).[81]

High hydrostatic pressure has two major effects on cation transport in human red cells: it activates the latent K^+/Cl^- cotransport pathway, and it stimulates an anion-independent monovalent cation (e.g., K^+ or Na^+) pathway in these cells.[82] Experiments were carried out changing both the ionic strength and the hydrostatic pressure simultaneously. In these experiments the high hydrostatic pressure-induced K^+/Cl^- cotransport was eliminated by measuring the transport in red blood cells suspended in solutions containing $CH_3SO_4^-$. Hydrostatic pressure enhanced the residual K^+ influxes measured in a low ionic strength medium synergistically so that a very large (about 10 mmol/[$l_{cells} \times$ h]) flux was obtained at 40 MPa in the low ionic strength solution. Under these conditions the residual K^+ influx was about two orders of magnitude higher than the residual flux in physiological saline at normal pressure. No hemolysis occurred under these conditions.[70] Since it is known that pressure inhibits the K^+ efflux in liposomes, it also must be assumed from these experiments that proteins play an important role in the low ionic strength effect on ion transport. This proposal is supported by comparative physiology (see above for discussion on LK vs. HK sheep red blood cells).

There are several possible explanations for the low ionic strength effect on residual cation transport. Lowering the ionic strength of the solution might result in a conformational change of proteins involved in residual cation fluxes and/or in a changed lipid-protein interaction, leading to regions able to allow transmembrane ion movements at instabilities between protein and lipid molecules. The facilitation of the low ionic strength effect by high hydrostatic pressure suggests that a change in the lipid-protein interaction, rather than a direct protein conformational change, is involved in this process. This is because in general lipids are more compressible than proteins.[83] Therefore, the "defects" in membrane structure at the lipid-protein interface, caused by low ionic strength, could be exacerbated by hydrostatic pressure.

Another possibility is that there is a specific membrane protein involved in the low ionic strength-stimulated residual K^+ and Na^+ fluxes. At the present stage of investigation it is not possible to answer this question. The inhibition of the low ionic strength effect by DIDS (see above) suggests a role of the band 3 protein. Also, Solomon et al.[84] have already speculated that the anion transport protein could be involved in cation transport. It is possible, of course, that the DIDS effect could be due to a more nonspecific DIDS-membrane protein interaction. An exclusive role for band 3 (possibly via a conformational change) seems unlikely since it is known that hydrostatic pressure in the range used in the described experiments decreases anion equilibrium exchange via band 3 in human red blood cells, as well as zero-trans exchange via other transporters,[85,86] although it could be possible that the "slippage" process via band 3 is increased under pressure. In addition, if the low ionic strength effect is due only to band 3 then the phenomenon should be present in red blood cells from all mammalian species (band 3 is known to be absent only in lamprey erythrocytes).[87] Furthermore, it cannot be ruled out that a still unidentified protein plays a role in the low ionic strength effect, possibly as a further carrier molecule responsible for stoichiometrically coupled monovalent cation transport. Recently, the increase in the residual K^+ and Na^+ efflux as well as influx of human red blood cells in low ionic strength solutions was explained on the basis of a carrier model.[88]

VI. CONCLUSIONS

In the present chapter an attempt was made to discuss the possible mechanisms of the effect of low ionic strength solutions on the cells. As was shown in Sections II to IV, various effects caused by the reduction of ionic strength of the external solution are due to changes in several parameters, such as surface potential, transmembrane potential, electric field strength inside the cell membrane, cell volume, local ion concentration, and local pH. Modifications of these parameters result in a large variety of direct or indirect effects on the molecular as well as cellular level. Since it is impossible to emphasize all aspects of that complex phenomenon in details, attention has been given to one example, i.e., to the effect of low ionic strength on the residual monovalent cation transport through red blood cell membranes (Section V). Analysis of the data available reveals that it is rather difficult to determine which of the changed parameters in a low ionic strength solution is responsible for the observed increase in the residual monovalent cation transport of human red blood cells. Further investigations must be done to clarify the molecular mechanisms.

ACKNOWLEDGMENT

The author would like to thank Dr. E. Donath and Mrs. A. Yu. Bogdanova for stimulating discussions and Mrs. K. Kießling for technical assistance. The work was supported by Deutsche Forschungsgemeinschaft (DFG-project Be1655).

REFERENCES

1. **McLaughlin, S.,** Electrostatic potentials at membrane-solution interfaces, *Curr. Top. Membr. Transp.,* 9, 71, 1977.
2. **Wojtczak, L. and Nalecz, M. J.,** Surface charge of biological membranes as a possible regulator of membrane-bound enzymes, *Eur. J. Biochem.,* 94, 99, 1979.
3. **Ostroumov, S. A. and Vorobiev, L. N.,** Membrane potential and surface charge densities as possible generalized regulators of membrane protein activities, *J. Theor. Biol.,* 75, 289, 1978.
4. **Parsegian, V. A. and Gingell, D.,** Some features of physical forces between biological cell membranes, *J. Adhesion,* 4, 283, 1972.
5. **Parsegian, V. A. and Gingell, D.,** A physical force model of biological membrane interactions, in *Recent Advances in Adhesion,* Lee, L. H., Ed., Gordon & Breach, London, 1991, 153.
6. **Donath, E. and Pastushenko, V.,** Electrophoretical study of cell surface properties. The influence of the surface coat on the electric potential distribution and on general electrokinetic properties of animal cells, *Bioelectrochem. Bioenerg.,* 6, 543, 1979.
7. **Voigt, A. and Donath, E.,** Cell surface electrostatics and electrokinetics, in *Biophysics of the Cell Surface,* Glaser, R. and Gingell, D., Eds., Springer-Verlag, Berlin, 1990, 75.
8. **Arakelian, V. B., Walther, D., and Donath, E.,** Electric potential distribution around discrete charges in a dielectric membrane — electrolyte solution system, *Colloid Polym. Sci.,* 270, 268, 1993.
9. **Donath, E. and Voigt, A.,** Electrophoretic mobility of human erythrocytes: on the applicability of the charged layer model, *Biophys. J.,* 49, 493, 1986.
10. **Donath, E. and Lerche, D.,** Electrostatic and structural properties of the surface of human erythrocytes. I. Cell-electrophoretic studies following neuraminidase treatment, *Bioelectrochem. Bioenerg.,* 7, 41, 1980.
11. **Donath, E. and Voigt, A.,** Streaming current and streaming potential of structured surfaces, *J. Colloid Interface Sci.,* 109, 122, 1986.
12. **Träuble, H. and Eibl, H.,** Electrostatic effects on lipid phase transitions: membrane structure and ionic environment, *Proc. Natl. Acad. Sci. U.S.A.,* 71, 214, 1974.
13. **Belaya, M. L. and Feigelman, M. V.,** Melting temperature of charged lipids lattice on the membrane surface, *Biol. Membr.,* 1, 438, 1984.
14. **Sauve, R. and Ohki, S.,** Interactions of divalent cations with negatively charged membrane surfaces. I. Discrete charge potential, *J. Theor. Biol.,* 81, 157, 1979.
15. **Loosley-Millman, M. E., Rand, R. P., and Parsegian, V. A.,** Effects of monovalent ion binding and screening on measured electrostatic forces between charged phospholipid bilayers, *Biophys. J.,* 40, 221, 1982.
16. **Norde, W. and Lyklema, J.,** The adsorption of human plasma albumin and bovine pancreas ribonuclease at negatively charged polystyrene surfaces. III. Electrophoresis, *J. Colloid Interface Sci.,* 66, 277, 1978.
17. **Marra, J.,** Direct measurement of the interaction between phosphatidylglycerol bilayers in aqueous electrolyte solutions, *Biophys. J.,* 50, 815, 1986.

18. **Neumcke, B. and Läuger, P.,** Space charge-limited conductance in lipid bilayer membranes, *J. Membr. Biol.,* 3, 54, 1970.
19. **Aytyan, S. Kh., Dukhin, S. S., and Chizmadshev, Yu. A.,** Effekty sil izobrazenija pri peremescanii zarjada v membranach, *Electrochimia,* 13, 779, 1977.
20. **Pastushenko, V. F. and Chizmadshev, Yu. A.,** Energetic profile of dipole molecules in membranes, *Biofizika,* 26, 458, 1981.
21. **Maroudas, N.,** Adhesion and spreading of cells on charged surfaces, *J. Theor. Biol.,* 49, 417, 1975.
22. **Donath, E. and Gingell, D.,** A sharp cell surface conformational transition at low ionic strength changes the nature of the adhesion of enzyme treated red blood cells to hydrocarbon, *J. Cell Sci.,* 63, 113, 1983.
23. **Wolf, H. and Gingell, D.,** Conformational response of the glycocalyx to ionic strength and interaction with modified glass surfaces: study of live red cells by interferometry, *J. Cell Sci.,* 63, 101, 1983.
24. **Belintsev, B. V.,** Cell adhesion physical factors governing the cell-substrate interaction, *Biol. Membr.,* 5, 1100, 1988.
25. **Ohsawa, K., Ohshima, H., and Ohki, S.,** Surface potential and surface charge density at the cerebral cortex synaptic vesicle and stability of vesicle suspensios, *Biochim. Biophys. Acta,* 648, 206, 1981.
26. **Lerche, D. and Glaser, R.,** Investigation of artificial aggregation of washed human erythrocytes caused by decreased pH and reduced ionic strength, *Acta Biol. Med. Ger.,* 30, 973, 1980.
27. **Lerche, D.,** Spontaneous aggregation of washed human erythrocytes in isotonic media of reduced ionic strength. Conclusions about the spatial arrangement of the N-terminal part of the glycophorins, *Biorheology,* 19, 587, 1982.
28. **Bäumler, H., Lerche, D., Paulitschke, M., and Meier, W.,** Aggregation of red blood cells in hypotonic and hypertonic salt solutions determined by light back scattering technique, *Stud. Biophys.,* 115, 5, 1986.
29. **Barber, J.,** Membrane surface charges and potentials in relation to photosynthesis, *Biochim. Biophys. Acta,* 594, 253, 1980.
30. **Rubin, B. T. and Barber, J.,** The role of membrane surface charge in the control of photosynthetic processes and the involvement of electrostatic screening, *Biochim. Biophys. Acta,* 592, 87, 1980.
31. **Lev, A. A.,** *Ionic Selectivity of Cell Membranes,* Part II, Academy of Sciences of the USSR, Nauka, Leningrad, 1975, chap. 2.
32. **Passow, H.,** Molecular aspects of band 3 protein-mediated anion transport across the red blood cell membrane, *Rev. Physiol. Biochem. Pharmacol.,* 103, 61, 1986.
33. **Glaser, R. and Donath, J.,** Stationary ionic states in human red blood cells, *Bioelectrochem. Bioenerg.,* 13, 71, 1984.
34. **Glaser, R., Brumen, M., and Svetina, S.,** Stationäre Ionenzustände menschlicher Erythrozyten, *Biol. Zentralbl.,* 99, 429, 1980.
35. **Ellory, J. C., Hall, A. C., and Stewart, G. W.,** Volume-sensitive cation fluxes in mammalian red cells, *Mol. Physiol.,* 8, 235, 1985.
36. **Hoffmann, E. K. and Simonsen, L. O.,** Membrane mechanisms in volume and pH regulation in vertebrate cells, *Physiol. Rev.,* 69, 315, 1989.
37. **Tverdislov, V. A., Tichonov, A. N., and Jakovenko, L. V.,** *Fisicheskije Mechanismi Funktionirovanija Biologicheskich Membran,* MGU Press, Moscow, 1987, 15.
38. **Hille, B. and Schwarz, W.,** Potassium channels as multi-ion single-file pores, *J. Gen. Physiol.,* 72, 409, 1978.
39. **Läuger, P.,** Microscopic calculation of ion-transport rates in membrane channels, *Biophys. Chem.,* 15, 89, 1982.

40. **Jordan, P. C.,** Effect of pore structure on energy barriers and applied voltage profiles. I. Symmetric channels, *Biophys. J.,* 45, 1091, 1984.

41. **Landau, L. D. and Lifschitz, E. M.,** *Lehrbuch der theoretischen Physik,* Vol. 8, Elektrodynamik der Kontinua, Akademie-Verlag, Berlin, 1974, chap. 2.

42. **McLaughlin, S. and Poo, M. M.,** The role of electro-osmosis in the electric-field-induced movement of charged macromolecules on the surface of cells, *Biophys. J.,* 34, 85, 1981.

43. **Fromherz, P.,** Spatio-temporal patterns in the fluid-mosaic model of membranes, *Biochim. Biophys. Acta,* 944, 108, 1988.

44. **Jähnig, F.,** Electrostatic free energy and shift of the phase transition for charged lipid membranes, *Biophys. Chem.,* 4, 309, 1976.

45. **Forsyth, P. A., Marcelja, S., and Mitchell, D. J.,** Phase transition in charged lipid membranes, *Biochim. Biophys. Acta,* 469, 335, 1977.

46. **Coakley, W. T. and Deeley, J. O. T.,** Effects of ionic strength, serum protein and surface charge on membrane movements and vesicle production in heated erythrocytes, *Biochim. Biophys. Acta,* 602, 355, 1980.

47. **Urry, D. W., Long, M. M., Jacobs, M., and Harris, R. D.,** Conformation and molecular mechanisms of carriers and channels, *Ann. N.Y. Acad. Sci.,* 264, 203, 1975.

48. **Glaser, R.,** The influence of membrane electric field on cellular functions, in *Biophysics of the Cell Surface,* Glaser, R. and Gingell, D., Eds., Springer-Verlag, Berlin, 1990, 173.

49. **Davson, H.,** Studies on the permeability of erythrocytes. VI. The effect of reducing the salt of the medium surrounding the cell, *Biochem. J.,* 33, 384, 1939.

50. **Wilbrandt, W.,** Die Ionenpermeabilität der Erythrozyten in Nichtleiterlösungen, *Pflügers Arch.,* 242, 537, 1940.

51. **Goldman, D. E.,** Potential, impedance and rectification in membranes, *J. Gen. Physiol.,* 27, 37, 1943.

52. **Donlon, J. A. and Rothstein, A.,** The cation permeability of erythrocytes in low ionic strength media of various tonicities, *J. Membr. Biol.,* 1, 37, 1969.

53. **Bernhardt, I., Donath, E., and Glaser, R.,** Influence of surface charge and trans-membrane potential on rubidium-86 efflux of human red blood cells, *J. Membr. Biol.,* 78, 249, 1984.

54. **Bernhardt, I.,** Untersuchungen zur Regulation des Ouabain-insensitiven Membrantransports monovalenter Kationen an Erythrozyten, D.Sc. thesis, Humboldt University, Berlin, 1986.

55. **Bernhardt, I., Erdmann, A., Vogel, R., and Glaser, R.,** Factors involved in the increase of K^+ efflux of erythrocytes in low chloride media, *Biomed. Biochim. Acta,* 46, 36, 1987.

56. **Bernhardt, I., Erdmann, A., Glaser, R., Reichmann, G., and Bleiber, R.,** Influence of lipid composition on passive ion transport of erythrocytes, in *Topics in Lipid Research. From Structural Elucidation to Biological Function,* Klein, R. A. and Schmitz, B., Eds., The Royal Society of Chemistry, London, 1986, 243.

57. **Van Dijck, P. W. M., Van Zoelen, E. J. J., Seldenrijk, R., Van Deenen, L. L. M., and De Gier, J.,** Calorimetric behaviour of individual phospholipid classes from human and bovine erythrocyte membranes, *Chem. Phys. Lipids,* 17, 336, 1976.

58. **Erdmann, A., Bernhardt, I., Herrmann, A., and Glaser, R.,** Species-dependent differences in the influence of ionic strength on potassium transport of erythrocytes. The role of membrane fluidity and Ca^{2+}, *Gen. Physiol. Biophys.,* 9, 577, 1990.

59. **Nelson, G. J.,** Lipid composition of erythrocytes in various mammalian species, *Biochim. Biophys. Acta,* 144, 221, 1967.

60. **Wessels, J. M. C. and Veerkamp, J. H.,** Some aspects of the osmotic lysis of erythrocytes, *Biochim. Biophys. Acta,* 291, 190, 1973.

61. **Bernhardt, I., Seidler, G., Ihrig, I., and Erdmann, A.,** Species-dependent differences in the influence of ionic strength on potassium transport of erythrocytes — the role of lipid composition, *Gen. Physiol. Biophys.,* 11, 287, 1992.

62. **Kuypers, F. A., Roelofsen, B., Op den Kamp, J. A. F., and Van Deenen, L. L. M.,** The membrane of intact human erythrocytes tolerates only limited changes in the fatty acid composition of its phosphatidylcholine, *Biochim. Biophys. Acta,* 769, 337, 1984.

63. **Gruber, W. and Deuticke, B.,** Comparative aspects of phosphate transfer across mammalian erythrocyte membranes, *J. Membr. Biol.,* 13, 19, 1973.

64. **Lu, Y. and Chow, E. I.,** Bicarbonate/chloride transport kinetics at 37°C and its relationship to membrane lipid in mammalian erythrocytes, *Biochim. Biophys. Acta,* 689, 485, 1982.

65. **Boggs, J. M.,** Intermolecular hydrogen bonding between lipids: influence on organization and function of lipids in membranes, *Can. J. Biochem.,* 58, 755, 1980.

66. **Bernhardt, I., Hall, A. C., and Ellory, J. C.,** Transport pathways for monovalent cations through erythrocyte membranes, *Stud. Biophys.,* 126, 5, 1988.

67. **De Coursey, T. E., Chandy, K. G., Gupta, S., and Cahalan, M. D.,** Voltage-gated K$^+$ channels in human T lymphocytes: a role in mitogenesis, *Nature,* 307, 465, 1984.

68. **Hermann, A. and Gorman, A. L. F.,** The effect of 4-aminopyridine on potassium currents in a moluscan neuron, *J. Gen. Physiol.,* 78, 63, 1981.

69. **Jones, G. S. and Knauf, P. A.,** Mechanism of the increase in cation permeability of human erythrocytes in low-chloride media. Involvement of the anion transport protein capnophorin, *J. Gen. Physiol.,* 86, 721, 1985.

70. **Bernhardt, I., Hall, A. C., and Ellory, J. C.,** Effects of low ionic strength media on passive human red cell monovalent cation transport, *J. Physiol.,* 434, 489, 1991.

71. **Ellory, J. C. and Hall, A. C.,** Human red cell volume regulation in hypotonic media, *Comp. Biochem. Physiol.,* 90A, 533, 1988.

72. **Chipperfield, A. R. and Shennan, D. B.,** The influence of pH and membrane potential on passive Na$^+$ and K$^+$ fluxes in human red blood cells, *Biochim. Biophys. Acta,* 886, 373, 1986.

73. **Zade-Oppen, A. M. M., Tosteson, D. C., and Adragna, N. C.,** Effects of pH, potential, chloride and furosemide on passive Na$^+$ and K$^+$ effluxes from human red blood cells, *J. Membr. Biol.,* 103, 217, 1988.

74. **Ellory, J. C., Hall, A. C., and Bernhardt, I.,** Physical and biochemical factors modulating red cell cation transport, *Stud. Biophys.,* 127, 215, 1988.

75. **Halperin, J. A., Brugnara, C., Tosteson, M. T., Van Ha, T., and Tosteson, D. C.,** Voltage-activated cation transport in human erythrocytes, *Am. J. Physiol.,* 257, C986, 1989.

76. **Halperin, J. A., Brugnara, C., Van Ha, T., and Tosteson, D. C.,** Voltage-activated cation permeability in high-potassium but not low-potassium red blood cells, *Am. J. Physiol.,* 258, C1169, 1990.

77. **Erdmann, A., Bernhardt, I., Pittman, S. J., and Ellory, J. C.,** Low potassium-type but not high potassium-type sheep red blood cells show passive K$^+$ transport induced by low ionic strength, *Biochim. Biophys. Acta,* 1061, 85, 1991.

78. **Canessa, M., Brugnara, C., Cusi, D., and Tosteson, D. C.,** Modes of operation and variable stoichiometry of the furosemide sensitive Na and K fluxes in human red cells, *J. Gen. Physiol.,* 87, 113, 1986.

79. **Blackstock, E. J. and Stewart, G. W.,** The dependence on external cation of sodium and potassium fluxes across the human red cell membrane at low temperatures, *J. Physiol.,* 375, 403, 1986.

80. **Stewart, G. W., Ellory, J. C., and Klein, R. A.,** Increased human red cell cation passive permeability below 12°C, *Nature,* 286, 403, 1980.

81. **Hall, A. C. and Willis, J. C.,** The temperature dependence of passive potassium permeability in mammalian erythrocytes, *Cryobiology,* 23, 395, 1986.

82. **Hall, A. C. and Ellory, J. C.,** Effect of high hydrostatic pressure on passive monovalent cation transport in human red cells, *J. Membr. Biol.,* 94, 1, 1986.

83. **Macdonald, A. G.,** The effects of pressure on the molecular structure and physiological functions of cell membranes, *Philos. Trans. R. Soc. London,* B304, 47, 1984.

84. **Solomon, A. K., Chasan, B., Dix, J. A., Lukacovic, M. F., Toon, M. R., and Verkman, A. S.,** The aqueous pore in the red cell membrane: band 3 as a channel for anions, cations, nonelectrolytes, and water, *Ann. N.Y. Acad. Sci.,* 414, 97, 1983.

85. **Canfield, V. A. and Macey, R. I.,** Anion exchange in human erythrocytes has a large activation volume, *Biochim. Biophys. Acta,* 778, 379, 1984.

86. **Hall, A. C. and Ellory, J. C.,** Hydrostatic pressure effect on transport in liposomes and red cells, in *Current Perspectives in High Pressure Biology,* Jannasch, H. W., Marquis, R. E., and Zimmerman, A. M., Eds., Academic Press, London, 1987, 191.

87. **Nikinmaa, M. and Railo, E.,** Anion movements across lamprey *(Lampetra fluviatilis)* red cell membrane, *Biochim. Biophys. Acta,* 899, 134, 1987.

88. **Denner, K., Heinrich, R., and Bernhardt, I.,** Carrier-mediated residual K^+ and Na^+ transport of human red blood cells, *J. Membr. Biol.,* 132, 137, 1993.

Chapter 9

Electrophoresis of Cells at a Physiological Ionic Strength

M. V. Golovanov

CONTENTS

I. INTRODUCTION

The present chapter concerns the analysis of the physiological state of cells both in the human environment (habitat) and in the medium of electrophoretic measurement with a large ionic force. The physiological state of cells comprises such a notion as the preservation of the vital activity of cells until irreversible metabolic processes set in, including the extreme (e.g., boundary, high, low, and optimal) values of the ionic force of the extracellular medium.

The vital activity of cells during preparation, and in the process of measurement of their electrophoretic mobility, is affected by many factors influencing the electrical state of the cell surface. Such factors are full replacement of the natural incubation medium, the addition of anticoagulants, and others.

Various preparatory operations may detrimentally affect the structure and the function of the surface components of a membrane and its electric double layer. This has led to development of a methodological analysis of the reaction of living cells to the effect of an electric field (both an external field and that induced by the cell surface), which has in turn provided the possibility of correctly evaluating the diagnostic significance of the cell electrophoresis method.

II. ELECTROPHORESIS OF CELLS IN A MEDIUM WITH A PHYSIOLOGICAL IONIC FORCE

The behavior of cells in an electric field is the main indicator which guarantees that the use of electrophoretic mobility enables one to characterize the surface properties. It is determined, in particular, by the quantitative and qualitative features of the electrolyte in which a cell is living and is characterized by changes in the shape, the volume, and the composition of the surface of a plasma membrane and its external, near-membrane layer, the glycocalyx.

0-8493-8918-6/94/$0.00+$.50
© 1994 by CRC Press, Inc.

A. THE SURFACE OF CELLS (ERYTHROCYTES, LEUKOCYTES)

It is known that a cell as a living entity is separated from its ambient medium by a plasma membrane. The thickness of this membrane is approximately 10 nm. Many types of cells have an external, near-membrane mucopolysaccharide layer, or glycocalyx, up to 4 nm thick on the surface of their plasma membrane.[1,2]

The plasma membrane contains proteins, lipids, and carbohydrates. The glycocalyx also includes special-purpose polyanions such as sialuronic acid which determine the blood groups for erythrocytes and the antigens that are characteristic for different types of leukocytes. In addition, the plasma membrane possesses pores through which a cell surface effects the water-electrolyte exchange with the surrounding medium.

During their vital activity cells absorb nutrients from the ambient medium and release metabolic products into it via the plasma membrane. They may further change the form of the entire cellular surface, or part of it, during motion, during proliferation, during the synthesis of external vacuoles, and so on. Thereby, the structural elements shift relative to one another.

It would be natural to suppose that the ionic groups on the surface of the cells also change their position and that their electric charges disappear, reappear, decrease, and increase due to the redistribution of anions and cations, not only on the cell surface, but also in the adjacent layer of the external medium (the so-called diffuse layer).

It can be assumed that the structure of the electric double layer of a plasma membrane (its adsorption and diffuse portions) of a viable cell differs substantially from the surface of a plane-parallel condensator, which is the basic concept in the theory of electrophoresis for nonliving, physical particles. Consequently, the development of the theory of the electric double layer of living cells requires taking into account the dynamic state of electric charges on the cell surface.

The cell surface is determined by the functions the cells perform in an organism. For example, there are many types of erythrocytes: normal erythrocytes, megalocytes, microspherocytes, poikilocytes, anisocytes, macrocytes, microcytes, and many others. Some hereditary hemolytic anemias have ovalocytes (elliptocytes), target-like erythrocytes, sickle-shaped erythrocytes (drepanocytes), and/or acanthocytes. Some pathological and regenerative forms of erythrocytes are stomatocytes, erythrocytes with Heinz bodies, polychromatophils, and reticulocytes. To test the physical characteristics of the surface of such cells, it is important to know the functional peculiarities of each type of erythrocyte, in addition to the shape.

Among leukocytes there are also varieties with regard to their shape, size, and functional properties: these are lymphocytes (T- and B-cells, killer cells, helper cells, and others), monocytes, granulocytes (neutrophils, basophils), as well as many pathological and blast forms.[3] It may be supposed that cells with the aforesaid forms might possess very different electrophoretic mobility (EPM) values (Figure 9-1).

B. THE EXTERNAL NEAR-MEMBRANE LAYER — THE GLYCOCALYX

Until recently, the reaction of cells to the action of a high-concentration electrolyte was quite unknown. It is expressed as the development of a gel-like formation near the leukocyte surface. Within 4 to 10 s after the addition of a mixture of a 15% (2.55 M) NaCl solution and some India ink to a blood smear or to a cellular suspension, a gel is formed around some leukocytes due to the swelling of polymers found on the surface of cells. The gel around the cells is likely to attain 30 μm in diameter. It must be noted that the gel is arranged unevenly around some leukemic cells. This might be evidence of violations of the functional and the structural character of the cell surface.

Near the cell surface the gel represents a ball-shaped formation called a halo. It forms near the surface of leukocytes of blood, of leukocytes of immunocompetent tissues and organs, as well as of leukemic and tumor cells (Figure 9-2).

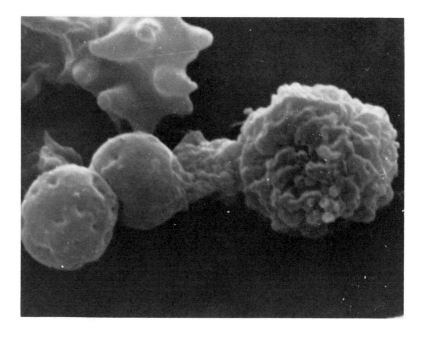

Figure 9-1 Scanning electron micrograph of surface of blood cells in isotonic NaCl solution. (Magnification × 300.)

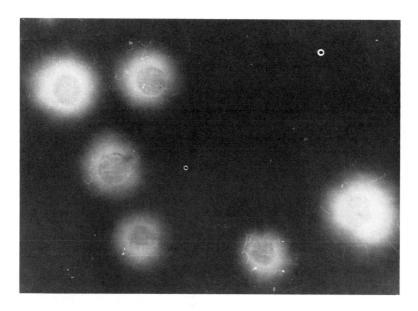

Figure 9-2 Gel-glycocalyx around leukocytes of peripheral blood. (India ink; magnification × 100.)

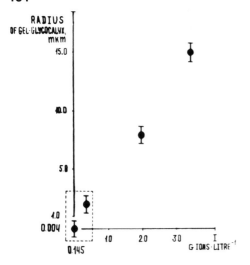

RADIUS
OF GEL-GLYCOCALYX,
mkm

15.0

10.0

5.0

1.0
0.004

0.145

10   20   30   I
G IONS·LITRE⁻¹

Figure 9-3 Dependence of the size of gel-glycocalyx on the ionic force of NaCl solution (the gel formation time is about 4 to 10 s.). An especially important region of the investigation of the adsorption and the diffuse layers of the cell surface is marked by a square.

Haloes are formed in the medium consisting of India ink particles over a period of several seconds. They appear at a rate of one halo every 2 to 3 min (Figure 9-3). This is due to an increase in the volume of the components of the external, near-membrane cell layer on contact with a high-concentration electrolyte. The aforesaid components have been partly isolated and identified as glycoproteins. Further biochemical, immunological, and other analyses of the isolated components of the gel-glycocalyx will be of practical interest.

The method for detecting the halo-forming cells (HFC) of leukocytes is as follows: ten drops of a 15% NaCl solution are added to one drop of blood. One drop of the mixture is spread on a slide and covered to provide a preparation of the "drop between two glasses" variety. Then the reactions occurring between the cells are examined with the aid of an optical light microscope.[4,5]

Underlying the halo formation reaction is a gel which is characterized as a rounded space free of visible objects. It takes 10 to 15 min for this free space to form in the medium of erythrocytes (blood), where it may attain up to 300 μm in diameter (Figure 9-4). Outside of the particle-free center of the halo there are visible objects, such as India ink particles, erythrocytes, leukocytes, yeast cells, polymer balls, etc.[5,6]

Haloes arising in the erythrocytes-containing medium are an order of magnitude larger than those arising in the medium containing India ink particles. The size of the halo also depends on the type of added or acting halogen (fluoride, chloride, bromide, iodide in increasing order of the lyophilic series; see Figure 9-5) and on the density of the cellular suspension. The lower the cell density, the larger the diameter of the halo. It depends, furthermore, on the activity of the HFC itself. The scatter of the halo diameters in the medium containing erythrocytes is in the range of 20 to 300 μm. It should be noted that the number of HFC in the medium with erythrocytes is smaller by a factor of 10 to 12 than that in the medium with India ink particles. In the author's opinion, this is due to the different weights of the India ink particles and of the erythrocytes.

The possibility of the participation of electrostatic forces in the formation of a halo cannot be excluded, since the halo-forming leukocytes possess a negative zeta potential of −7.5 mV while that of the background erythrocytes is about −4.0 mV. A mutual repulsion of equally charged cells occurs.[5,6] The method of halo formation in the medium of erythrocytes is put to practical use in the oncological and pathophysiological clinic and in experimental biology.[7-10]

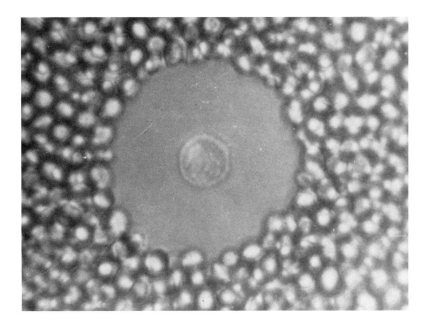

Figure 9-4 Halo around leukocyte in peripheral blood. (India ink; magnification × 150.)

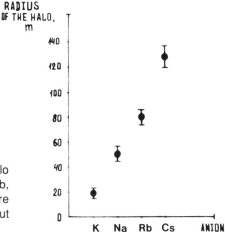

Figure 9-5 Dependence of the halo size on the kind of anion (for K, Na, Rb, and Cs cations the differences are slight). The halo formation time is about 10 to 15 min.

C. THE PHYSIOLOGICAL MEANING OF IONIC FORCE

The physiological meaning of physicochemical parameters (e.g., the ionic force and pH value) is commonly associated with regularities in the functioning of living cells. These functions concern the nature of basic life processes and general manifestations of vital activity such as metabolism, the properties of biological membranes, and the reactions of the cells to the effects of the surrounding medium. These are the most important phenomena that distinguish living beings from nonliving matter.

Irritability and, in certain periods of life, growth and proliferation are peculiar to living cells. Irritability as a physiological reaction of cells is the capability of responding actively to external effects in several ways. These are irritation, motion, compression, the enhancement of metabolism and growth, acceleration of division, release of metabolites, electrical pulses, and others.

The physiological values of the ionic force and pH comprise the extreme (critical, boundary) and optimal values of the concentration and composition of the salt in artificial or natural media, which ensure the vital activity of the cells during incubation and measurement. Media are subdivided into hypotonic, isotonic, and hypertonic categories. All these correspond to the preservation of vital activity of a cell or to reversible processes when life functions have been temporarily disturbed (stopped) and have then recovered.

It must be noted that the cells of animals and plants, as well as the cells of microorganisms (e.g., bacteria, fungi, water plants), live in a fairly wide range of salt concentrations. Living media (habitats) range from freshwater to halophiles, practically from low concentrations to 30% (5.2 M) and even higher proportions of salts in water. Yet the electrolyte composition (the physiological ionic force) has its specific limits of physiological survival for each kind of cell. For example, extremely halophilic bacteria of the genera *Halobacterium* and *Halococcus* are able to grow in a saturated NaCl solution. They are characterized by the fact that the lowest critical limit of NaCl content for their growth is about 12 to 15%. Many marine bacteria need the presence of NaCl in a concentration of about 3% (about 0.5 M) for their growth, but they are able to survive considerably higher concentrations of NaCl, such as 20, 25, and 30%.[11]

For leukocytes and erythrocytes of human blood such limits have not yet been studied. The extreme physiological concentrations of electrolytes occur on the verge of cell death, shortly before irreversible processes of cell destruction are induced.

The pH value of the medium plays an immense role in the physiological state of an organism. The concentration of hydrogen ions affects the ionic state and, hence, the accessibility of cells to many metabolites and inorganic ions. It is impossible to overestimate its influence on the stability and functions of biological processes. Like bacteria and other microorganisms, the cells of mammals and in particular those of human blood, can normally function only at certain pH values (both the boundary and the optimal ones). It is known that the blood of mammals and human beings is weakly alkaline: the pH of aerated blood is about 7.35 to 7.47, while that of dark-red blood is lower by 0.02 unit. The extreme pH limits of blood, which are still compatible with the life of an organism (but not of the blood cells), are 7.0 to 7.8. The pH value of aerated blood averages 7.4, while that of dark-red blood is somewhat lower.[12] In the case of an uncompensated acidosis the pH value might be reduced to 7.1 to 6.9.

Yet variations in the pH value of blood within the range of optimum values also might be characteristic of the pathological state of blood cells (erythrocytes and leukocytes). Consequently, changes in the conformational states of the macromolecules of cell membranes will also cause changes in the electrical properties and the electrophoretic mobility of the cells. Unfortunately, the pH values that are extreme but compatible with the physiological parameters of the viability of blood cells (the boundary ones) have been insufficiently studied until now. This circumstance hinders the determination of the diagnostic significance of the electrophoretic mobility of erythrocytes and leukocytes.

The author's investigations are associated with the study of the reactions of normal and leukemic cells to the effect of electrolytes having a high ionic force, such as NaCl, bromide, fluoride, and other compounds at a concentration from 0.9 to 20% or 0.154 to 3.4 M. The reactions include

1. Compression and swelling of cells (changes in the hypertonic stability of erythrocytes and leukocytes)
2. Formation of vacuoles near leukocytes
3. Formation of gel-glycocalyx near the surface of normal and leukemic cells

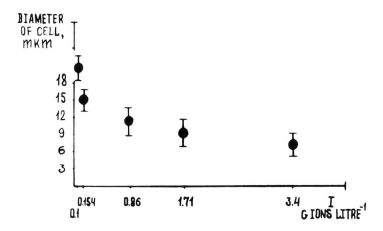

Figure 9-6 Dependence of the diameter of cells (ovarian carcinoma cells) on the ionic force of the NaCl solution.

4. Formation of haloes
5. Change in the electrokinetics of surface potential
6. Spontaneous formation of periodic hexagonal structures, colonies, and other reactions of cells

One of the first reactions of cells to the action of a high-concentration electrolyte consists of the compression and swelling of the cell content (which is characteristic of both blood leukocytes and erythrocytes). Such a reaction is determined by alterations in the structure of the plasma membrane, the content of the cytoplasm, and the amount of free water inside the cell. When examining the surface of the cells it is possible to note the compression of the contraction of the surface membrane and a 40 to 50% diminution of the cell volume. If the high-concentration electrolyte is now removed, then the cell recovers its original size (Figure 9-6).

In addition to the visual examinations, an electron microscope has been used to analyze the wholeness of leukocytes and erythrocytes and the ability of plasma membranes to release water and the substances of a different nature dissolved in it. Photographic data confirm the wholeness of plasma membranes when erythrocytes are exposed to electrolytes of different concentrations. An investigation of the quantitative content of the cells in the incubation salt medium has shown that the number of cells decreases during 1- and 20-h incubations. However, a NaCl concentration of 6.5% (1.11 M) is, in the author's view, the optimum electrolyte concentration for the maintenance of a constant number of erythrocytes in the solution.

In connection with these facts, a method has been developed for the determination of the stability of the plasma membranes of erythrocytes during their incubation in the salt medium. In the author's opinion, this method indicates the stability of the membranes to the osmotic effect, and this characteristic may be used for diagnostic purposes both clinically and experimentally.[13]

In addition to the aforesaid compression-swelling property, the effect of the appearance of vacuoles of different values and diverse quantities near the cell surface has been observed. A vacuole represents a formation which is free of India ink particles and is confined by a thin shell or membrane connected to a cell. The character of the connection of a vacuole to a cell is still unknown to us. The content of a vacuole is optically transparent. The vacuole is colored by trypan blue, as are the membrane and the content

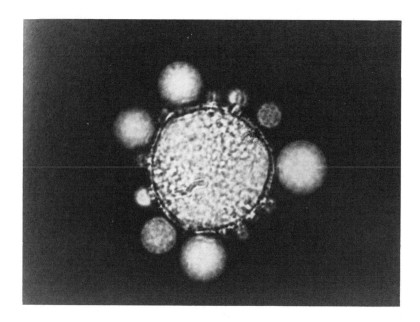

Figure 9-7 Formation of vacuoles in the 3.4 *M* NaCl solution. (India ink; magnification × 300.)

of the cell. When the cell is shifted to the right or to the left by pushing on the preparation glass, the vacuole is elongated or compressed, following the cell, while retaining its wholeness (Figure 9-7).[14]

The size of a vacuole can exceed that of a cell by a factor of 1.5 to 2; it can attain a diameter up to 50 μm. In this case, the India ink particles constantly collide with the membrane surface of the vacuole and the cell. The vacuoles usually can form during a 5-min time period, whereupon their volume further increases. Then the vacuoles separate from a cell and can be found near the cell or moving chaotically at a distance of 20 to 40 μm from the cell. Most frequently, about 1 to 20 vacuoles are usually formed near the cell. A gigantic vacuole represents essentially a bag or a bubble which is confined by an elastic membrane (as colored by methylene blue or trypan blue) and is able to deform when the liquid preparation is pressed.

The author has not succeeded in determining the content of vacuoles by the use of dyestuffs such as neutral red, trypan blue, or methylene blue. The vacuole shape is most frequently oval, round, or rounded. A vacuole is attached to or stems from one spot of the membrane of leukocytes. After separation from the cell the vacuole assumes the shape of a round ball. It is known from cytology that vacuoles inside a cell are storage vesicles for water and mineral salts (the adapting reaction of cells). It can be assumed that in the present case, too, the vacuole outside a cell is characteristic of an adapting reaction to the action of a high-concentration electrolyte.

D. VITAL ACTIVITY OF CELLS IN AN ELECTROLYTE WITH A HIGH IONIC FORCE

Answering this basic question (vital activity of cells) would enable one not only to approach the intimate processes of halo formation, but also to assess the practical significance of this phenomenon, particularly in oncology. The differences of tumor cells in the ability to form haloes, as observed by the author, might be associated with the peculiarities of their proliferation, invasion, and metastasis. Furthermore, the halo-forming activity of the protective cells of an organism, i.e., of the leukocytes, may serve as an indicator of their functional state, as shown in a series of papers.[5,7] The formation of haloes around the cells occurs when they are affected by electrolytes of different concentrations and indicates the state of the electric charges of the ionic groups of molecules on the surface of the cell membranes.

From numerous parameters that are characteristic of the vital activity of the cells forming haloes, two have been chosen by the author for study: the ability of cells to proliferate after being subjected to the effect of a hypertonic electrolyte solution, and the penetrability of the cell membranes to trypan blue.

The ability of tumor cells to grow in mice of the DBA/2 stock was studied using several varieties of tumors, in particular the ascites form of L-1210-type leukosis. First, 0.5 ml of a 5% (0.85 M) NaCl solution was introduced into the abdomen of each control animal. In each of four test animals 0.5 ml of a 5% (0.85 M) NaCl solution containing 7 million living tumor cells was introduced. Prior to introduction the tumor cells had been incubated in the 5% (0.85 M) sodium chloride solution for 10 to 20 min at room temperature (+19°C). By day 7 all the test animals had died from the development of tumors, whereas the control animals survived. In the following experiment tumor cells (7 million per 0.5 ml) were preincubated either in 5% or in 0.9% NaCl solution for 30 min, 60 min, 17 h. The tumor (leukemic) cells in 0.9% NaCl solutions were introduced into the control mice, while the tumor cells suspended in 5% NaCl solutions were introduced into the test mice of the DBA/2 stock. Each group consisted of five mice. Both test and control animals died in a period of 7 to 9 days after introduction if the cells had been preincubated for 10 to 20 min, 30 to 35 min, or 60 to 65 min. However, when leukemic cells had been preincubated for 17 h, the animals survived.

Mice also died if the leukemic cells had been preincubated in a 10% (1.7 M) NaCl solution for 15 to 20 min, 30 min, or 60 min. The experiments were carried out on mice of the DBA/2 stock, six animals being used for each incubation time.

If only 3.2 million living leukemic cells were introduced, the test animals died from the development of the tumor process by the 11th to 12th day. In this case, the period of development of the tumor process increased on the average by 3 days.

From these results on the growth of tumor cells in DBA/2 mice one can conclude that the tumor cells have retained their vitality during pretreatment with a physiological and a high-concentration electrolyte solution.

The fact that the formation of haloes is a result of a vital activity of HFC is substantiated by the observation that the HFC are not colored by trypan blue after 5 to 10 min in the hypertonic solution, when the formation of the haloes begins. The experiments were carried out on the leukemic cells of the L1210 stock of mouse lympholeukosis cells. It has been established by numerous research workers that trypan blue exclusion is a marker of cell vitality. Consequently, the haloes are formed by vital cells.

It has been established that the HFC whose vital activity has been suppressed by treating them with such substances as potassium permanganate or potassium bichromate do not form haloes or a gel around them.[5] It is believed that when a cell is affected by a hypertonic electrolyte solution numerous biological processes are mobilized in it, although the cell cannot resist the effect of a harmful factor for long. It then either passes over into anabiosis or perishes after a period of time. It may be supposed that the degree of stability of cells to the acting factor (in the present case, to the effect of a concentrated electrolyte solution) is determined by the potentialities of cells and, in particular, of HFC.

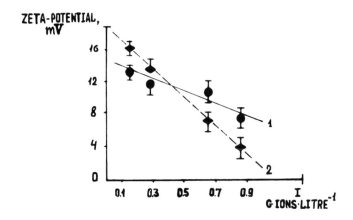

Figure 9-8 Dependence of zeta potential on the ionic force of an NaCl solution. (1) Leukocytes; (2) erythrocytes.

E. ELECTROPHORESIS OF CELLS IN AN ELECTROLYTE WITH A HIGH IONIC FORCE

The motion of cells under the effect of an external electric field is used extensively in practice. The analytical electrophoresis and preparative methods are based on the presence of an electric charge on the cell surface, the distribution conditions of which are determined by the presence of an adsorption layer and a diffuse layer around the cell surface. The theory of an electrical layer (EDL) of colloid particles and their behavior in an external electric field when particles are affected by various factors (electrolyte and others) is well known.

In 1979 Golovanov and Derjaguin showed that living, halo-forming leukocyte cells in an electrolyte with a high ionic force possessed a negative zeta potential of -7.5 mV, whereas the zeta potential of background erythrocyte cells was -4.0 mV. The peculiarities of the electrophoresis chamber and the incubation medium are set forth in detail in References 5 and 6 (Figure 9-8).

In connection with the fact that a redistribution of the electrostatic potential on the surface of cells (both leukocytes and erythrocytes) occurs in a high-concentration electrolyte, an experimental investigation method was developed: electrophoresis of the blood cells in an electrolyte with a high ionic force — in particular, in a 5% (0.85 M) NaCl solution. The method consists of the following: one drop of blood is diluted in 60 ml of 0.855 M NaCl solution (pH 7.2) and incubated for 30 min at a temperature of 20°C. A cell suspension with a concentration of 2 million cells per milliliter is transferred into an electrophoretic chamber of a "Parmoquant" instrument (of German manufacture; see Chapter 13 of this book). Then the electrophoretic mobility measurements of the cells are performed. The measurement results are recorded in a histogram. Control measurements are performed with blood diluted in a similar way in an isotonic 0.154 M NaCl solution (Table 9-1).

Measurements revealed that an electrostatic surface charge is retained on a cell under conditions of high electrolyte concentrations in connection with the conservation of the vitality of cells and the presence of a gel-glycocalyx. The membrane components that are responsible for the electric charge of a cell are gradually supplied onto its surface, and the electric charge is thus retained in the electrolyte.

The essential novelty and importance of the information obtained from such electrophoretic measurements is revealed by the following:

Table 9-1 **Electrophoretic mobility (EPM) of blood cells in 0.9% (isotonic) and in 5% (high-concentration) NaCl**

EPM (µm-s/V-cm)	Cell Number	Histogram
Isotonic (0.9%)		
0.50		
0.55	1	X
0.60		
0.65	1	X
0.70		
0.75		
0.80	4	X
0.85	27	XXXXXXXXXXX
0.90	85	XXXXXXXXXXXXXXXXXXXXXXXXXXXXXXXXXXXXXXXXXX
0.95	77	XXXXXXXXXXXXXXXXXXXXXXXXXXXXXXXXXXXXXX
1.00	16	XXXXXX
1.05	4	X
High NaCL (5%)		
0.50		
0.55		
0.60	10	XXXX
0.65	30	XXXXXXXXXXXXX
0.70	11	XXXXX
0.75	79	XXXXXXXXXXXXXXXXXXXXXXXXXXXXXXXXXXXXXXXXXXXX
0.80		
0.85	46	XXXXXXXXXXXXXXXXXXXXX
0.90		
0.95	13	XXXXXX
1.00		
1.05		

1. The surface of cells differs in its physical characteristics from the model of a flat condenser underlying the modern theory of an electric double layer. Thus, certain corrections are required in the theory and practice of the electrophoresis of cells, particularly with regard to the structure and function of an adsorption layer.
2. A cell conserves its vitality in a high-concentration electrolyte.
3. A gel-glycocalyx appears which functions in a cell as a cation exchange agent and changes the structure of a diffuse layer.
4. On the surface of some leukemic cells the gel-glycocalyx is arranged unevenly (according to the data obtained with the aid of the light microscope).
5. During electrophoresis not only a decrease in the zeta potential but also a qualitative subdivision of the cells according to their electrophoretic mobilities can be noted (Table 9-1).[15,16]

The data on the influence of the physiological electrolyte solutions with a high ionic force and different pH values on the electrophoretic mobility of the erythrocytes and leukocytes are at present undergoing mathematical treatment.

It is believed that the physicochemical substantiation and calculation of all the processes taking place both near and on the surface of the plasma membrane will open new possibilities in the study of metabolism and in research on the electrokinetic properties of normal and leukemic cells. Here, however, it should be noted that the dilution of blood by a factor of 1000 to 2 million cells per milliliter (the technical condition of electrophoresis) destroys all the physiological parameters peculiar to the native state of blood. This relates to the osmotic and oncotic pressures as well as to the destruction of the buffer system of blood (e.g., carbonate, phosphate, protein-plasma, hemoglobin). Due to this the pH values also become indefinite, although they are maintained by artificial media. Such changes undoubtedly must affect the electrical state of the cell surface and electrophoretic mobility, too.

F. PERIODIC STRUCTURES OF LEUKEMIC CELLS

The spontaneous formation of the periodic structure of halo-forming cells (PS HFC) and colonies or groups consisting of leukemic cells is considered to be one of the last reactions of cells to extreme physiological salt solutions (1 to 20% = 0.17 to 3.4 M NaCl). The periodic structure is formed during a period of time of up to 1 month after all blood erythrocytes have been destroyed. The distance between the halo-forming cells in the model of chronic human leukosis can reach a maximum value of about 60 to 120 µm (Figure 9-9).[17,18]

The following aspects are characteristic of the periodic suspension of halo-forming cells: (1) the physicochemical and biological preconditions for the investigation of PS HFC; (2) the conditions of the formation of PS in blood and in the leukocytory mass; (3) the morphology of cells forming the PS; (4) the boundaries of PS with regard to their formation conditions, their stabilities, and their dynamics; (5) the structure and periodicity of suspensions; (6) the deformation of the PS HFC lattice; (7) the formation of bonds or links; (8) the destruction of gels; and (9) the disintegration of PS.

The basic regularities of formation and the behavior of different types of periodic colloidal structures were established by application of the so-called DLVO theory (Derjaguin-Landau-Verway-Overbeek), which concerns the processes of gelatination of sols, suspensions, and emulsions. It has been shown that different regular structures arise over long distances as a result of the interaction of microobjects, depending on the prevailing conditions.[17] A description of the conditions of formation and stability of the periodic suspensions of halo-forming leukemic cells follows.

The experiments were carried out as follows: a blood droplet (about 30 µl) from a patient suffering from one form of leukosis was diluted 1:10 into a mixture of 15% (2.55 M) NaCl solution and India ink and mixed. Then one droplet thereof was placed on a slide and a cover glass was applied; the edges of the cover glass were smeared with petrolatum to protect the preparation from drying. The spacing between the objective and the cover glass was on average 100 µm. First, an arbitrary arrangement of separated halo-surrounded cells was sighted in the visual field of the microscope. During 2 to 4 days of incubation of the preparation at room temperature, the beginning of the formation of separate groups or colonies was observed. The latter consisted of different numbers of cells (5 to >500) surrounded by haloes. After the formation of the periodic structures a boundary was established between the latter and the space without any HFC. The periodic suspension of HFC is a dynamic system because not only the progressive motion of halo-forming cells in the direction of their lower concentration occurs, but also the rotational motion of separate cells as well as whole groups and HFC colonies. The periodic suspension of halo-forming cells can be preserved for a period of more than 1 to 3 months. The relative arrangement of separate HFC is continuously changing, but the mean spacing between the centers of the cells is maintained. It is the author's belief that the main role in the formation and stability of PS HFC is played by the long-range repulsion and attraction forces.

The presence of a sharp boundary between the periodic structures consisting of HFC indicates repulsion forces being balanced at large distances by the attraction forces. In the

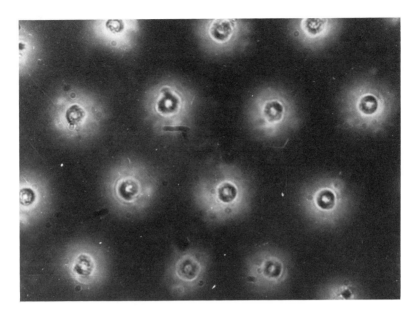

Figure 9-9 The periodic structure of halo-forming leukemic cells. (India ink; magnification × 300.)

absence of attraction forces the HFC would tend to fill up all the free space while propagating without any hindrance.

A substantial distinction of the structures detected by the author as compared to the known periodic, biocolloidal structures consists of a considerably (by an order of magnitude) larger distance between the HFC.[19] In the author's experiments, periodic regular structures might have been formed by normal donor lymphocytes and leukemic blood cells of a patient with a diagnosis of chronic lympholeukosis. The difference between these two morphological groups of cells is revealed by the stability, duration, and character state of the periodic structure.[20] Thus, the form of a PS may provide evidence that physically homogeneous cells are present.

One can suppose that the cells included in colonies and in periodic structures are morphologically identical in spite of the fact that the long-range repulsion forces are predominant in these structures, whereas the long-range attraction forces are predominant in the colonies.[17,18] The character of the physical interaction between the colonies of HFC and the cells of periodic structures is revealed by the continuous shifting of all the cells in different directions and at different distances. A periodic suspension composed of HFC may be dynamic, developing, or decaying with time and at the same time be stable and equilibrated.

The term "preservation of structure" can be used to describe the conservation of the distances between the elements or cells of a periodic cellular suspension, i.e., of the average distances between the centers of HFC. In spite of the Brownian motion of the India ink particles, and in spite of the processes associated with the appearance of new haloes, separate fragments of the cellular suspension retain their periodicity as the colonies are formed. In this case, the mean distance between HFC amounts to about 30 to 120 μm depending on the density of the cellular suspension and other factors.

In the author's experiments the PS retained their stability for 1 year and even longer — up to 101 weeks. It is believed that the long-term stability of PS is guaranteed by such processes as the life activity of the HFC under conditions of hypoxia and the

194

physicochemical reactions that are characteristic of the interaction of leukemic cells. During the process of the dynamic formation of PS HFC, deformations of the PS lattice also occur — that is, the form of suspension changes. Simple types of deformation are an impairment of the periodicity of a regular hexagonal structure or changes in the distances and angles between the HFC. It should be noted that after the mechanical destruction or some shift of the PS HFC a tendency of a suspension to return to its original state can be observed, obviously owing to the elastic effect of intercellular forces and the elasticity of the gel.

The mechanism of the formation of PS and colonies is mainly determined by the same processes. In the author's opinion, the process of the formation of PS and colonies occurs according to the following flow sheet:

1. All the cells — both leukocytes and erythrocytes — are at first located chaotically in a suspension.
2. Then haloes are chaotically formed around separate cells in the medium of background cells.
3. There is then a mass destruction of the background erythrocyte cells.
4. Thereafter, the formation of the linear periodic structure —the linear arrangement of HFC — begins.
5. Then the formation of the periodic structure in the plane sets in.
6. Arrangement of HFC into hexagonal PS takes place, having a mean intercellular distance of about 60 μm.
7. Finally, a high-periodicity structure forms having a mean intercellular distance of 60 to 120 μm.

A method for the determination of the degree of periodicity cellular suspensions has been developed for diagnostic purposes.[21] The long-range forces acting between the HFC might be either molecules of complexes found in their external, near-membrane layer or the whole cytoplasmic membrane of a cell. Mechanical hindrances of a diverse nature, such as glass, polymers, metal, and bioorganic compounds, did not distort the intercellular interactions occurring in the PS HFC, colonies, and haloes.

From the point of view of the classical electrophoresis theory for inorganic colloid particles, the thickness of an electric double layer diminishes approaching the thickness of the adsorption layer as the ionic force of the electrolyte increases. Thereby the zeta potential of a particle may even decrease to zero so that no electrostatic field remains near the particle.

The biophysical theory of cell electrophoresis states that if cells are affected by an electrolyte with a physiological ionic force, mutual repulsion of equally charged (negatively charged) cells at corresponding distances occurs when the zeta potential on the surface of a vital cell is not equal to zero but is permanently maintained by the cell metabolism, when the electrically neutral gel-glycocalyx surrounds a cell, and when the plasma is probably also neutral due to the adsorption of ions from the medium by the plasma molecules. At present, this hypothesis is actively supplemented by experimental facts from studies on electrical characteristics of cells and their habitat medium. Thus,

1. A physiological medium with a high ionic force (about 0.15 to 3.4 M) conserves the vitality of cells, depending on the treatment time.
2. The zeta potential of blood cells in an electrolyte with a high ionic force decreases and is redistributed both qualitatively and quantitatively.
3. When the physiological ionic force changes, the volume of cells and the shape of their surface influencing the adsorption layer also change.
4. The volume of the glycocalyx increases with the physiological ionic force. Thereby both the structure and the function of the adsorption and diffuse layers are changed.

5. The kind of anion determines the size of the gel-glycocalyx in accordance with its position in a lyotropic series.

III. CONCLUSIONS

Cell electrophoresis, as one of the effective modern methods, will undoubtedly help scientists develop exact diagnoses of diseases. In the author's opinion, the following parameters are required for more successful development and application of the biophysical theory of electrophoresis of cells; they will guarantee standard reproducible results of investigations:

1. Standard cells
2. The physical values of the ionic force and pH of the incubation medium and measurements of their limits and compositions
3. Morphological, functional uniformity of the cells being tested
4. Known reactions of cells to the incubation medium artificially provided for electrophoresis, and other parameters

The essential point of the theoretical investigation of the electrophoresis of cells consists of the physicomathematical analysis of the new concept of the structure of an electric double layer because the former theory of the electrophoresis of physical inorganic particles has been determined to be incorrect for the description of the behavior of living cells in an electric field.[16,22,23]

REFERENCES

1. **Policard, A.,** *La Surface Cellulaire et son Microenvironment,* Masson, Paris, 1972, 24.
2. **Leninger, A.,** *Biochemistry,* Mir Publishers, Moscow, 1974, 279.
3. **Abramov, M. G.,** *The Atlas of Hematology,* Meditsina, Moscow, 1985, 121 and 218.
4. **Golovanov, M. V.,** The method for revealing halo-forming leukocyte cells, The USSR Inventor's Certificate 930049, *Bulletin of Discovery,* 19, 1982.
5. **Golovanov, M. V.,** The Phenomena of the Formation of Haloes around Tumour and Normal Cells and Its Biological Significance, Ph.D. Thesis, Academy of Sciences, Kiev, 1984.
6. **Golovanov, M. V. and Derjaguin, B. V.,** On the repulsion forces resulting from some biological cells placed in a concentrated electrolyte solution, *Kolloidnyi Zh.,* 4, 649, 1979 (in Russian).
7. **Golovanov, M. V. and Tereshchenko, I. P.,** Dynamics of the charge in the halo-forming cells in pathological processes, *J. Microbiol. Epidemiol. Immunobiol.,* 2, 87, 1991 (in Russian).
8. **Kashulina, A. P., Aleksandrova, L. M., Tereshchenko, I. P., Sokolova, I. I., Trakhtenberg, A. K., and Popov, I.,** Functional state of neutrophil granulocytes of the peripheral blood; determinable in accordance with the halo-formation characteristic, tumor growth process, *Pathol. Physiol. Exp. Ther.,* 4, 37, 1990 (in Russian).
9. **Tereshchenko, I. P., Kashulina, A. P., Sobieva, Z. I., and Aleksandrova, L. M.,** On the new informative characteristic of the functional state of segmentonuclear neutrophiles, *Bull. Exp. Biol. Med.,* 10, 505, 1984.
10. **Tereshchenko, I. P., Kashulina, A. P., Sobieva, Z. I., Aleksandrova, L. M., and Karpova, M. N.,** The sympathic nervous system, hemocytes and tumour growth, *Pathol. Physiol. Exp. Ther.,* 4, 21, 1984 (in Russian).

11. **Kushner, D. J.,** *Microbial Life in Extreme Environments,* Academic Press, London, 1978, 367.

12. **Nozdrachev, A. D.,** *The Common Course of Physiology of Men and Animals,* Vysshaya Shkola, Moscow, 1991, chap. 2.

13. **Golovanov, M. V.,** Method for determination of the hypertonic stability of erythrocytes, *Hematol. Transfusiol.,* 5, 345, 1991 (in Russian).

14. **Golovanov, M. V. and Derjaguin, B. V.,** Behaviour of the plasmatic membrane and the near-to-membrane layer (glycocalyx) in a high-concentration electrolyte, *Bull. Exp. Biol. Med.,* 8, 210, 1992 (in Russian).

15. **Golovanov, M. V.,** Electrophoresis of cells in high-concentration electrolyte, *Kolloidnyi Zh.,* 53, 449, 1991 (in Russian).

16. **Golovanov, M. V.,** New concept on the structure of an electric double layer on the cell surface under the conditions of action of a high concentration electrolyte, *Biophysics,* 36, 463, 1991.

17. **Golovanov, M. V. and Derjaguin, B. V.,** On the formation and stability of periodic suspensions of halo-forming cells, *Dokl. Akad. Nauk SSSR,* 272, 479, 1983.

18. **Derjaguin, B. V. and Golovanov, M. V.,** On long-range forces of repulsion between biological cells, *Colloid Surf.,* 10, 77, 1984.

19. **Ephremov, I.,** The Periodicicity of Colloidal Structures, Ph.D. Thesis, Khimiya, Leningrad, 1971, 114.

20. **Golovanov, M. V., Basov, B. N., and Derjaguin, B. V.,** Leukemia and periodic structures of halo-forming cells, *Bull. Exp. Biol. Med.,* 8, 185, 1990 (in Russian).

21. **Basov, B. N. and Golovanov, M. V.,** Method of calculating the degree of periodicity of the arrangement of leukemic cells in suspension, *Biophysics,* 36, 114, 1991 (in Russian).

22. **Dukhin, S. S. and Derjaguin, B. V.,** *Electrophoresis,* Science, Moscow, 1976, 328.

23. **Miroshkinov, A. I., Phomchenkov, V. M., and Ivanov, A. Y.,** *Electrophysical Analysis and Division of the Cells,* Nauka, Moscow, 1986, 184.

IV

The Importance of Determination of the Negative Surface Charge Density of Cells

Chapter 10

Simultaneous Two-Parameter Measurements of the Electrophoretic Features of Cell Subpopulations and Their Different Sedimentation Characteristics

George G. Slivinsky

CONTENTS

I. INTRODUCTION

The cellular suspensions used for the investigation of the electrophoretic properties of cells are, as a rule, heterogeneous. This heterogeneity may be related to the presence of cells of several morphological types or to the presence of cells with diverse functional properties, such as the level of differentiation, the proliferative activity, the stage of the cell cycle, or the index of ploidy. The study of the electrophoretic heterogeneity of cell suspensions is of interest for the investigation of the electrokinetic properties of cells belonging to determined histological types and for the elucidation of the relationship between the phenotypic properties of the cell surface and the function of cells. At present, different methods are applied for the estimation of cell electrophoretic heterogeneity: (1) the analysis of the form of cell distribution according to electrophoretic mobility (EPM), (2) electrophoretic fractionation of cell suspensions with free-flow electrophoresis methods[1] or isoelectric focusing,[2] and (3) measurement of the EPM of cells from synchronized cell cultures.[3,4]

Another method of studying cell electrophoretic heterogeneity consists of a preliminary fractionation of cells according to some physical parameter and a subsequent study of the electrophoretic properties of the resultant fractions. The most suitable methods for the preliminary fractionation are sedimentation techniques. With their help cells can be fractionated either in concordance with their size (velocity sedimentation) or according to cell density (equilibrium density sedimentation).

It is known that the velocity sedimentation of a particle suspended in a liquid in a gravitation field is equal to

0-8493-8918-6/94/$0.00+$.50
© 1994 by CRC Press, Inc.

$$S = 2/9 \cdot \left[\left(\rho_c - \rho_o\right) \cdot g \cdot r^2\right]/D \qquad \textbf{(10-1)}$$

where S = particle sedimentation velocity, ρ_c and ρ_0 = cell and surrounding liquid densities, respectively, r = radius of particles, and D = viscosity of liquid.

In conditions where the density of the medium is lower than cell density $(\rho_c - \rho_0)$, the sedimentation velocity (SV) depends mainly on the cell size. Cell sizes and their corresponding SV values represent an information index, since cell fractionation by SV permits the study of fractions containing either cells of different morphological types or cells varying in important functional properties. In the case of sedimentation fractioning of bone marrow cells, cells with diverting directions of differentiation were isolated. Peters and Evans separated erythroid cells, lymphocytes, and cells of the granulocyte series.[5] When dispersed liver cells were fractionated, fractions of erythrocytes, lymphocytes, and endothelial cells were educed, in addition to fractions of diploid and tetraploid parenchymatous liver cells.[6] Cell differences in SV also correlated with the position in the cell cycle.[6]

Despite considerable efforts on cell fractionation by SV, the electrophoretic properties of cells with varying sedimentation characteristics have been described in very few papers so far. Josefowicz et al.[7] showed that faster sedimenting thymocytes from CBA/J mice tend to have a higher EPM. A combined method of preparative electrophoresis and sedimentation fractionation was used for studying lymphocyte heterogeneity.[8] A method to study cell heterogeneity according to two parameters simultaneously, SV and EPM, has been introduced.[9] In the present chapter the results obtained during utilization of this method for the investigation of electrophoretic heterogeneity of cell populations will be described. Data on the electrophoretic properties of cell fractions isolated by the sedimentation method will be adduced.

II. TECHNIQUES FOR SIMULTANEOUS MEASUREMENTS OF SEDIMENTATION VELOCITY AND ELECTROPHORETIC MOBILITY

The trajectories of cell movement in the chamber for microelectrophoresis are resulting trajectories of movement in the earth's gravitational field and in the electrical field. If the chamber is orientated laterally, the movement of every cell takes place in a stationary plane perpendicular to the axis of observation. If these trajectories are fixed — for example, with the help of a photograph — and their components measured, the EPM and SV can be judged.

For the two-parameter measurements an instrument for microelectrophoresis (Figure 10-1) was utilized consisting of a flat quartz chamber measuring $0.5 \times 15 \times 40$ mm. The electrical parts include a stabilizing source of direct current and chlorine-silver electrodes, isolated from the chamber with porous membranes and agar-agar bridges. In order to observe and photograph the cell movement trajectories, a photomicroscope with a "dark-field" condenser was used. The trajectories were registered for two directions.

Two types of buffers were used. The high ionic strength buffer was a modified Hanks' solution consisting of 144.42 mM NaCl, 5.37 mM KCl, 4.14 mM NaHCO$_3$, 0.22 mM Na$_2$HPO$_4$, and 0.44 mM KH$_2$PO$_4$. The low ionic strength buffer was similarly composed, but NaCl was substituted with 285 mM sorbitol. The cellule, microscope objective, and condenser were placed into a thermostatic chamber in which the temperature was $25 \pm 0.2°C$. Before the measuring the cellule was calibrated visually. Erythrocytes of healthy donors served as the standard (U = $1.04 \pm 0.03 \times 10^{-8}$ m^2 V^{-1} s^{-1}).

The resulting trajectories of cell movements appeared as shown in Figure 10-2. They were photographed by means of a long exposure time. After the electrophoretic chamber

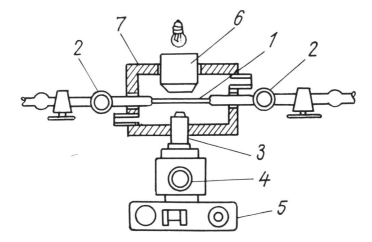

Figure 10-1 Diagram of the instrument used for the measurement of cell movement trajectories during electrophoresis. (1) Camera for electrophoresis; (2) electrodes; (3) objective of the microscope; (4) photomicroscope; (5) camera; (6) "dark-field" condenser; (7) thermostatic chamber. (From Slivinsky, G. G., *Tsitologiya,* 20, 839, 1978. With permission.)

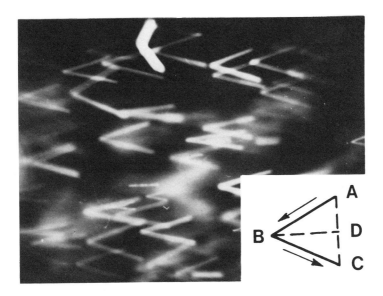

Figure 10-2 Cell movement trajectories of rat ovary ascites carcinoma (OT) as seen through the microelectrophoresis camera. The inset represents the trajectories and their components. (From Slivinsky, G. G., *Tsitologiya,* 20, 839, 1978. With permission.)

was filled and temperature equilibrium was reached, tension was given to the electrodes and the shutter of the camera opened simultaneously. During the time T, a period of 10 to 15 s, the cells moved toward the anode. Then the polarity of the electrodes was changed and the cells moved in the opposite direction for an equal stretch of time while the shutter of the photocamera remained open.

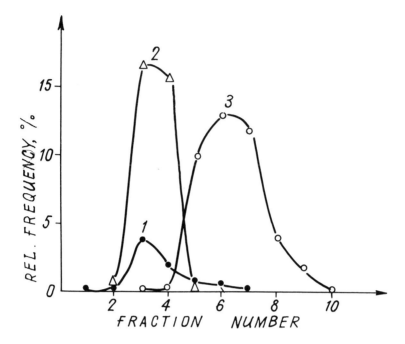

Figure 10-3 Cytological composition of cell fractions of rat bone marrow. (1) Lymphocytes; (2) erythroid cells; (3) myeloid cells. (From Slivinsky, G. G., Malkovsky, A. D., and Kirienko, S. I., *Tsitologiya,* 25, 976, 1983. With permission.)

In the case of the summary exposition term 2T, the horizontal component BD corresponded to the traces made by a cell in the electric field during time T, while the vertical component AC corresponded to the traces made by a cell in the gravitational field during time 2T. The elements of the trajectories were measured on magnified projections. In each experiment the trajectories of no less than 100 cells were evaluated. The values of SV and EPM determined for each cell from the traces AC and BD, respectively, were subgrouped into classes in concordance with an SV interval of 10^{-6} m s^{-1}. Then the average values of EPM for each class were calculated. Later, examples of utilization of this method for the study of electrophoretic heterogeneity in cell suspensions shall be adduced.

III. BONE MARROW CELLS

Bone marrow is an example of a very heterogeneous cell system. In concordance with their type and also their functions, the bone marrow cells can be identified according to their SV.[5,10] Studies of bone marrow cell heterogeneity in white rats were conducted using two different methods.[11] After fractionation at unit gravity[5] nine different fractions were educed. EPM and SV of the cells in the diverse fractions were registered and the cytological composition of each fraction was determined. In a second case EPM and SV of nonfractionated marrow cells were registered.

The distribution of cells according to fractions took the appearance of a two-apex curve. The first peak reflected the erythrocytes and lymphocytes and the second one the granulocyte series of cells. Data on the cytological composition of the fractions are shown in Figure 10-3. The erythroid series of cells and lymphocytes were present in fractions 3 and 4 and the myeloid cells in fractions 5-9. Erythroid cells were represented by normocytes.

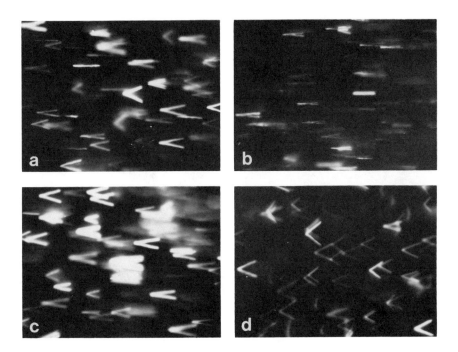

Figure 10-4 Cell movement trajectories of rat bone marrow as seen through a microelectrophoresis camera. (**a**) Nonfractionated marrow, (**b,c,d**) fractions 3, 6 and 8, respectively. (From Slivinsky, G. G., Malkovsky, A. D., and Kirienko, S. I., *Tsitologiya*, 25, 976, 1983. With permission.)

In fractions 5 to 7, 89 to 98% of cells were of the myeloid series, among which 65 to 72% were segmented neutrophils while the others were their precursor myelocytes and metamyelocytes. For the measurement of EPM, fractions 3, 6, and 7 were used. The resulting movement trajectories from a nonfractionated suspension (Figure 10-4a) and three different fractions are shown on Figure 10-4. It can be seen that the vertical components (AC) of the cell trajectories representing SV values increase from fraction 3 with increasing fraction number (Figure 10-4b,c,d). Figure 10-5 shows a comparison of simultaneous two-parameter measurements of nonfractionated (1) and fractionated (2) bone marrow cells. The values of EPM, plotted against the corresponding middle values of SV, revealed a sufficiently good coincidence between the results obtained by both methods (Figure 10-5).

Erythroid cells and lymphocytes were both collected in fraction 3. They did not differ according to their SV value, which was close to 10^{-6} m s^{-1} for both types of cells. The mean value of EPM of the fraction 3 cells was higher than in the rest of the marrow cells at the expense of the high percentage of erythrocytes. The analysis of the distribution of the EPM values in fraction 3 showed that it differs from the norm and can be represented in terms of two distributions with mean values of surface electric charge density equal to 5.62 and 8.56×10^4 electrons per mcm^2.[11] The last magnitude corresponds to the electric charge density of erythrocyte surfaces — 8.06×10^4 electrons per mcm^2.

The cells of the granulocyte series possessed a higher SV and were collected in fractions 5 to 9. With increasing SV (i.e., with increasing fraction number) the mean EPM of the cells rose as a consequence of increased content of immature myeloid cells in the

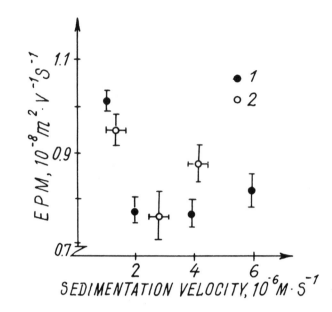

Figure 10-5 Electrophoretic mobility (EPM) of rat bone marrow cells with different sedimentation velocities. (●) Nonfractionated marrow; (○) 3rd, 6th, and 8th fractions of marrow. Ionic potential of medium = 0.153 *M*. (From Slivinsky, G. G., Malkovsky, A. D., and Kirienko, S. I., *Tsitologiya*, 25, 976, 1983. With permission.)

fractions. This observation is in accordance with results obtained from studies on electrophoretically fractionated marrow cells, which also show that promyelocytes and myelocytes have a higher electric charge than metamyelocytes and segmented neutrophils.[12]

IV. CELLS OF THE THYMUS AND SPLEEN

The thymus cell population consists of prethymocytes (subpopulations of dividing and maturing cortical forms), medular thymocytes, and mature lymphocytes. More than 60% of all thymocytes from white rats were cells with a low SV of 10^{-6} m s^{-1} (Figure 10-6a) and with a relatively low EPM. The rest of the thymocytes are rapidly sedimenting cells ($p < 0.001$) with increased EPM. For thymus cells of CBA/J mice analogous results were obtained by Josefowicz et al.[7] Cells with low SV and EPM are apparently cortical cells since the last mentioned are small in size and represent the dominant population, which amounts to 60 to 70% of all thymocytes. With the help of preparative electrophoresis it is possible to divide the thymus cells into two populations: cortisone-sensitive cortical cells with low EPM and cortisone-resistant, phytohemagglutinin (PHA)-sensitive medular cells with a higher EPM.[13,14] The electrophoretic heterogeneity of thymus cells apparently does not depend on the cell cycle and proliferative activity because the dividing populations of thymocytes reveal themselves in both the zone of high EPM and that of low EPM.[14] Therefore, the electrophoretic heterogeneity of thymus cells with different SV is related to the presence of cortical and medular cells.

The electrophoretic heterogeneity of nucleated spleen cells depends on the content of highly mobile T-lymphocytes and less mobile B-cells. As follows from data in Figure 10-6b, more than 80% of the nucleated spleen lymphocytes had an SV equivalent to 10^{-6} m s^{-1} or 2×10^{-6} m s^{-1}. These two cell groups did not differ according to their EPM values

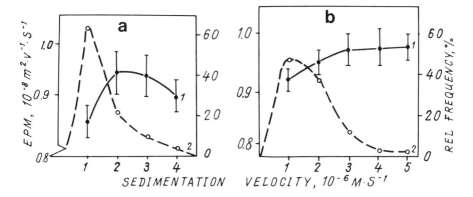

Figure 10-6 Electrophoretic mobility (EPM) of (**a**) thymus cells and (**b**) nucleated rat spleen cells with different SV. (1) EPM; (2) sedimentation profile.

($p > 0.05$). Hence, the difference in size of T-cells and B-cells and the corresponding difference in SV are not sufficient to divide them in concordance with these indices. Spleen lymphocyte fractionation by counterflow elutriation also showed that T- and B-lymphocytes were present in all educed fractions.[15]

V. LIVER CELLS

Polyploidy, a multiplication of the cell genome to the diploid one, is frequently found in the somatic tissue of mammals — for example, in the liver. That is why the investigation of electrophoretic properties of hepatocytes with a different ploidy index is of importance for the elucidation of the relation between the quantity of chromosome material (number of gene copies) and the properties of plasma membranes in these cells. It is known that the increase of ploidy leads to a multiple increase of the hepatocyte volume.[6] Consistent with Equation 10-1, the discrepancy in SV for di-, tetra-, and octaploid cells is characterized by the following proportion: 1:1.59:2.52. Consequently, the electrophoretic properties of hepatocytes with different ploidy indices were investigated by simultaneously measuring hepatocyte EPM and SV, the cells having been isolated from intact and regenerating liver.

White rats with a mass of 170 to 200 g were used for the tests. The partial hepatectomy operation was conducted in compliance with the Higgins and Anderson method.[16] Before isolating the cells, the liver was perfused with an oxygenated Hanks' solution without Mg^{2+} and Ca^{2+}, but containing 2 mM EDTA. The slices of liver tissue were dispersed carefully with a magnetic mixer. Then the suspension was pipetted and filtered.

Traces obtained by simultaneous two-parameter measurements of hepatocytes in the electrophoretic chamber are shown in Figure 10-7. The distribution of hepatocytes, in conformity to SV values, and the mean EPM values for hepatocytes with different SV are given in Figure 10-8. In the interval 3 to 16×10^{-6} m s^{-1} parenchymatous liver cells were predominantly obtained because nonparenchymal cells (Kupffer cells, endothelial cells) had lower SV.[17] The mean value of intact liver hepatocyte EPM was $1.02 \pm 0.06 \times 10^{-8}$ m^{2}V^{-1}s^{-1} ($n = 642$). EPM of hepatocytes with different SV could not be positively distinguished, except for the fraction of cells having SV of 9×10^{-6} m s^{-1}. EPM of these cells was lower than that of the remaining cells. The relative content of these cells in the liver was on average 25.6%. Sweeney et al.[18] had shown that a content of 3% diploid, 75.5% tetraploid, 19% octaploid, and 2.5% 16-ploid cells could be observed in rat liver.

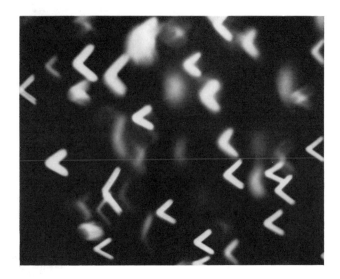

Figure 10-7 Cell movement trajectories of intact rat liver cells as seen through a microelectrophoresis camera.

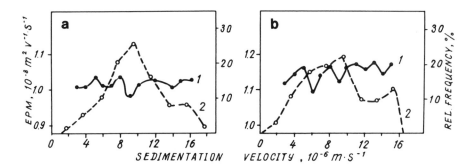

Figure 10-8 Dependence of electrophoretic mobility (EPM) on sedimentation velocity of hepatocytes from (**a**) an intact liver and (**b**) a rat liver 24 h after partial hepatectomy. (1) EPM; (2) sedimentation profile.

Hepatocytes with SV of 9×10^{-6} m s^{-1} enter into the interval corresponding to tetraploid cells. Therefore, this fraction does not represent the whole population of tetraploid cells.

Within 24 h after partial hepatectomy the hepatocyte EPM increased to $1.15 \pm 0.07 \times 10^{-8}$ m^2 V^{-1} s^{-1} ($n = 595$). The discrepancies in EPM between intact and regenerating liver cells were statistically significant ($p < 0.001$). The increase of hepatocyte EPM in the early stages after partial hepatectomy were described previously by other authors.[19,20] Other results were obtained with isoelectric equilibrium analysis.[2] In the last instance it was shown that the hepatocytes of a regenerating liver had a lower ionized group density in comparison with normal ones. This contradiction may be explained by the fact that in the case of isoelectric focusing the ionized group density is measured in a deeper layer of the perimembranous zone than it is in the microelectrophoresis method.

In the regenerating liver most hepatocytes with different SV had similar EPM, with the exception of the fraction of cells having SV equal to 6×10^{-6} m s^{-1}. The EPM of these cells was found to be $1.09 \pm 0.07 \times 10^{-8}$ m^2 V^{-1} s^{-1}.

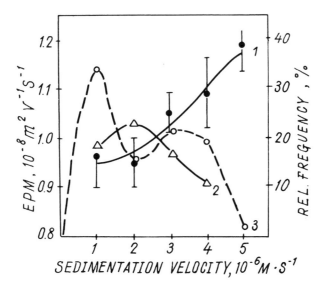

Figure 10-9 Electrophoretic mobility (EPM) of (1) Pliss ascites lymphosarcoma cells and (2) rat peritoneal cells with different sedimentation velocities; (3) sedimentation profile of tumorous cells, 4th day of tumor growth.

The distribution of EPM values of this fraction, which contains 11.9% of all cells, was narrower than in the rest of the hepatocytes ($p < 0.001$). The genesis of cells with lower EPM from intact and regenerating liver is not clear. The SV of cells with the lower surface electric charge corresponded to the SV of tetraploid cells, although their relative content was several times lower than the total content of tetraploid cells. The EPM of the remaining 75 to 90% of hepatocytes was equal; consequently, the ionized group density on the hepatocyte surface does not depend on the ploidy index. Thus, the induction of proliferation gave rise to a similar increase of ionized group density in the electrokinetic layer of hepatocytes with all ploidy indices.

VI. CELLS OF TRANSPLANTED TUMORS

Insofar as many functional properties of tumorous cells correlate with sedimentation characteristics, diverse properties of transplanted tumor cells fractionated by an isovelocity sedimentation method were investigated and compared with the results obtained by measuring the surface electric charge of these cells. In the early growth stage of transplanted tumors, the cells of which had sedimentation and electrokinetic characteristics similar to those of peritoneal cells, the possibility of contamination of different fractions with low-SV peritoneal cells was real (Figure 10-9). With this in mind, and in order to exclude the given factors, the distribution of cells regarding their EPM values was analyzed in each case, and those tumors in which the cells differed in their SV and EPM from peritoneal cells were studied.

A. MOUSE ASCITES TUMOR CELLS (NK/LY)

At the 10th day of tumor growth, cells were isolated from mouse ascites NK/Ly tumors and separated into five fractions by velocity sedimentation. Then the cells of each fraction were investigated.[21] The cytological characteristics of cells with different SV are given in Figure 10-10. The first two fractions contained diploid cells which were in the G_1 and S phases of the cell cycle. Fraction 3 consisted of diploid cells which were in the $G_2 + M$

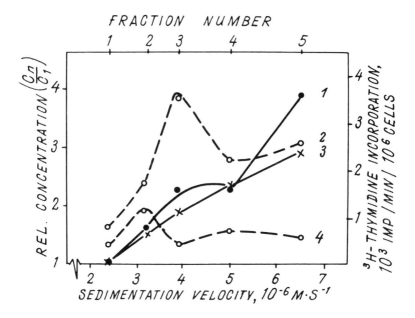

Figure 10-10 Characteristics of NK/Ly lymphoma cell fractions. (1,3) relative content of DNA and proteins in cells of different fractions in comparison with the same in cells of the first fraction. (2,4) intensity of ³H-thymidine incorporation into cells in the case of single and six-fold *in vivo* introduction, respectively.

phase and tetraploid cells which were in the early (G_1) stage of the cell cycle. ³H-thymidine incorporation data measured in prolonged labeling conditions revealed that fraction 1 contained cells with the lowest proliferation activity and fraction 3 contained those with the highest proliferation activity. Polyploid (tetraploid) cells, which were in a resting stage of the cell cycle, were enriched in fractions 4 and 5. These cells composed 6 to 9% of a tumor. Clonogenic capacity, survival after X-ray irradiation, and recovery from potential lethal damage of different NK/Ly tumor cells were also determined. For this purpose intact cells and cells irradiated with a dose of 5 Gy were fractionated in sterile conditions and cloned *in vivo* in diffusion chambers which contained soft agar and which had been implanted into the peritoneal cavity of recipient mice.[22] The data obtained are shown in Table 10-1. The cloning efficiencies of intact cells from fractions 1, 3, and 5 were 3.53%, 17.06%, and 1.57%, respectively, while 0.42%, 0.34%, and 0.86% of irradiated cells of the same fractions formed clones. The most significant loss of cells, which took place during cloning immediately after irradiation, was observed for fraction 3 (G_2 + M cells), where survival decreased 70- to 80-fold. Cells from fraction 1, as well as polyploid cells from fraction 5, were more resistant. When the cloning efficiencies were determined 24 h after irradiation, after the cells had had time to repair potentially lethal injuries, 3.39%, 0.34%, and 1.08% of the cells from fractions 1, 3, and 5, respectively, were able to form clones. This showed that the recovery from potentially lethal damage was highest of all in fraction 1 cells and lower in cells in the G_2 + M phase of the cell cycle and in polyploid cells.

The sensitivity of NK/Ly cells to the cytotoxic influence of natural killer cells (NK), dependent on the phenotypic properties of the surface, was also investigated. It is known that the NK — a subpopulation of effector lymphocytes — possess a spontaneous cytotoxic activity not requiring a preliminary sensitization by the tumorous cells. During

Table 10-1   **Cloning efficiency during control and after exposure to irradiation (5 Gy) and cell sensitivity to membrane-toxic activity of natural killer (NK) cells on different fractions of NK/Ly lymphoma cells**

	Fraction Number				
Indices	**1**	**2**	**3**	**4**	**5**
Cloning efficiency					
% of controls	3.53 ± 1.17	–	17.06 ± 2.57	–	1.57 ± 0.50
% 2 h after irradiation	0.42 ± 0.12	–	0.34 ± 0.08	–	0.86 ± 0.24
% 24 h after irradiation	3.39 ± 0.45	–	0.34 ± 0.11	–	1.08 ± 0.35
% NK sensitive cells	30.50 ± 7.5	43.00 ± 4.9	36.60 ± 2.3	41.90 ± 4.2	42.10 ± 5.00

Note: –, not determined.

activation of the NK cells the surface structure of the target cells plays an important role.[23,24] The sensitivity of NK/Ly cells to allogenic NK from mouse spleen was evaluated by performing the membrane toxicity test after supplying ^{3}H-uridine to the cells in the presence of exogenous ribonuclease.[25] According to the data obtained (Table 10-1) the cells of fractions 1 and 3 possessed a lower sensitivity to NK cells.

For the study of the relation between the characteristics described above and the electric charge of NK/Ly cells, EPM measurements were performed. In addition, distribution coefficients of the cells in a charged two-phase "dextran-polyethylene glycol" polymer system were determined according to the Albertson method,[26] and capacity of the cell surface to adsorb the cationic phthalocyanin dye Alcian blue was evaluated.[21]

With the increasing cell SV, EPM decreased monotonically (Figure 10-11). The differences in mobilities of cells from fractions 1, 3, and 5 were statistically significant ($p < 0.001$). The dependence of the distribution coefficient on the cell SV had another character. This observation (Figure 10-11b) is apparently related to the fact that cell distribution in two-phase systems depends not only on surface electric charge density, but also on hydrophobicity properties.[27] Cells from fractions 1 to 3 did not differ positively according to their AB adsorption capacity and adsorbed on average 51.6% more dye than polyploid cells. No correlation has been found between the Alcian blue adsorption capacity (Figure 10-11a) and the two other indices ($p > 0.05$); (Figure 10-11b,c).

In the late stages of NK/Ly tumor growth (10 days), diploid cells with a low cloning efficiency (fraction 1) had a higher EPM and also a higher distribution coefficient in comparison with proliferating diploid cells (Figure 10-11, fraction 3). The existence of a correlation between growth potential and EPM value in tumorous cells at a late growth stage also has been shown by Bischoff et al.[28] In cells being cultivated *in vitro*, changes of electric charge during the process of cell culture aging are of a different nature. In the latter case a decrease of EPM in aging fibroblasts has been observed[29] as well as an increasing percentage of cells with a low distribution coefficient.[30] The different nature of the changes in surface electric charge of tumorous cells during tumor growth and when these cells are cultivated *in vitro*, was demonstrated by Bischoff et al.[31]

From the comparison of data given in Table 10-1 and Figure 10-11 it also can be inferred that there is no correlation between different characteristics of superficial electric charge and the sensitivity of NK/Ly cells to membrane-toxic activity of NK. Consequently, in this case the electric charge density of target cells did not play a decisive role in the realization of the first stage of cytolysis, for which the establishment of a contact

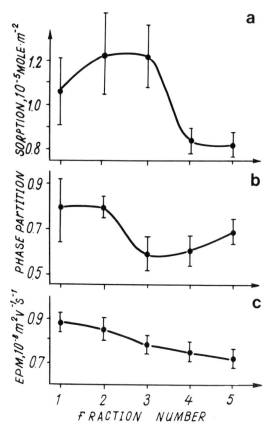

Figure 10-11 Surface charge of cells from different fractions of NK/Ly lymphoma. (**a**) Intensity of Alcian blue adsorption; (**b**) coefficient of cell distribution in two-phase polymer system; (**c**) electrophoretic mobility (EPM). The mean values and intervals for 0.1% level of significance are given. (From Slivinsky, G. G. and Magda, I. N., *Tsitologiya*, 32, 61, 1990. With permission.)

between the tumorous cell and the NK is essential. An unequivocal relationship between the surface electric charge characteristics, the loss of cells after exposure to irradiation, and the capacity to recover from potentially lethal damages was not detected either.

B. RAT OVARIAN CARCINOMA CELLS (OT)

As an example of another tumor, the cells of an ascites carcinoma of rat ovaries (OT) were investigated. For these cells a different relationship between EPM and SV was found. The cytological characteristics of the tumor cells with diverse SV that we investigated[32] are represented in Figure 10-12. Fractions 1 and 2 contained cells with low proliferation activity and cells in the G_1 phase of the cell cycle. Fractions 3 and 4 were represented by diploid S-phase cells. Fractions 5 and 6 contained diploid cells which were in the G_2 + M phase. Fractions 7 and 8 were enriched with polyploid cells with DNA content above 4c (c = species specific haploid DNA control).

Slowly sedimenting cells with low proliferative activity and G_1 cells had low mobility (Figure 10-13). Intermediate EPM values were measured for cells with high proliferative velocity, and polyploid cells with DNA content higher than 4c were characterized by maximum EPM.

OT clonal lines differing in two properties, the modal number of chromosomes and the capacity for metastatic activity, were also investigated.[33] Nonmetastatic clones were represented by two hyperdiploid lines with a modal chromosome number of 43 (OT-43) and 44 (OT-44), as well as by a polyploid clone with modal chromosome number of 80 to 90 (OT-P). Metastatic OT lines were an OT-af variant with metastases in the lungs, a

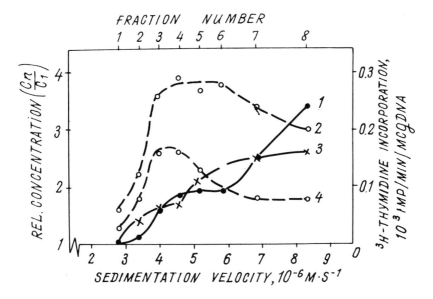

Figure 10-12 Characteristics for cell fractions of rat ovary ascites carcinoma (OT). (1,3) Relative contents of DNA and proteins in cells of different fractions in comparison with cells of the first fraction; (2,4) intensity of ^{3}H-thymidine incorporation into cells in the case of single and six-fold *in vivo* introduction, respectively. (From Slivinsky, G.G., Tsvetkova, T.V., and Tsokur, E.V., *Tsitologiya,* 34, 82, 1992. With permission.)

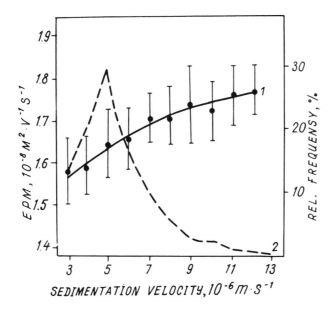

Figure 10-13 Dependence of OT cell electrophoretic mobility (EPM) on (1) their sedimentation velocity and (2) the sedimentation profiles of these cells; 7th day of tumor growth. The mean values and confidence intervals for 0.1% level of significance are given.

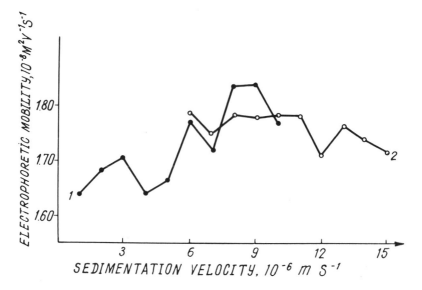

Figure 10-14 Dependence of cell electrophoretic mobility of (1) OT-44 and (2) OT-P on their sedimentation velocity. (From Slivinsky, G.G., Malkovsky, A.D., and Krasnoshtanov, V.K., *Eksp. Onkol.*, 6, 61, 1984. With permission.)

polyploid subvariant of this line called OT-P-af, as well as a clonal line, OT-L, metastasizing into the lymphatic glands and thymus. The method of obtaining the lines and their characteristics have been described previously.[34,35] The results of EPM and SV measurements of two nonmetastatic clones, OT-44 and OT-P, are given in Figure 10-14. The SV of OT-44 cells was situated in the interval 1 to 10×10^{-6} m s^{-1}, and the SV of OT-P cells was in the interval 6 to 15×10^{-6} m s^{-1}. In conformity with their SV values the cells of these two clones could be subdivided into three groups. Cells of the first group had low SV 2 to 5×10^{-6} m s^{-1} and were present only in the OT-44 clone. Their relative content in tumors was 49%. The EPM of these cells was minimal. Cells of the second group had SV of 6 to 10×10^{-6} m s^{-1}. They were present in both the first and the second clone. The mean EPM of these cells in both clones was the same. Cells of the third group, with SV of 11 to 15×10^{-6} m s^{-1}, could be found only in the polyploid clone OT-P. The range of EPM values for cells of this group did not differ from that for cells of group 2. The average EPM of cells from the first group was lower than the mean EPM of cells from the second and third groups ($p < 0.01$). The karyotype in cells of clone OT-44 differed insignificantly from that of the initial OT tumor. The electrokinetic properties of cells from this clone and the initial tumor were also similar. The existence of a SV interval of 6 to 10×10^{-6} m s^{-1}, common for cells of both clones, corresponded to the changes in SV during the cell cycle. It follows from the obtained data that the increase in EPM of OT cells takes place alongside a chromosome set increasing from hyperdiploid to tetraploid. A further increase in chromosome number to octaploid does not lead to a complementary increase in EPM; even a tendency to decrease can be observed.

The mean EPM of cells of metastatic clones OT-af and OT-L was higher than in nonmetastatic clones OT-43 and OT-P (Figure 10-15). At the same time, reliable divergences in the EPM of cells of the clones OT-43 and OT-P have not been discovered either in cells of the polyploid metastatic clone OT-P-af or in cells of the nonmetastatic lines OT-43 and OT-P ($p > 0.1$). Consequently, the superficial electric charge value in OT cells did not always correlate with the metastatic potential and the modal number of chromosomes.

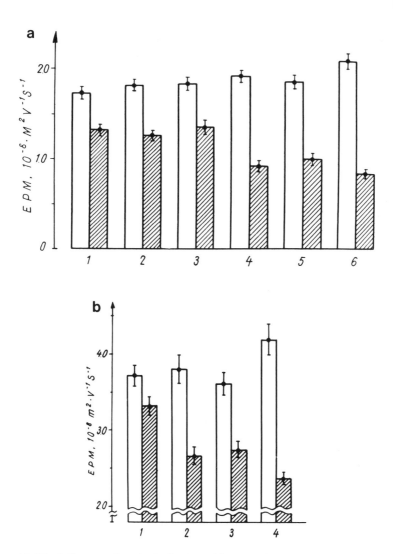

Figure 10-15 Influence of testicular hyaluronidase on electrophoretic mobility (EPM) of the cells from different OT lines. (1) OT; (2) OT-43; (3) OT-P; (4) OT-af; (5) OT-P-af; (6) OT-L. Open bars = control, hatched bars = hyaluronidase treatment. Ionic forces of medium during measurements were (**a**) 0.155 M and (**b**) 0.01 M. The mean values and confidence intervals for 0.1% level of significance are given. (From Slivinsky, G. G., Tsvetkova, T. V., Krasnoshtanov, V. K., and Fedoseenko, V. M., *Eksp. Onkol.,* 15, 28, 1993. With permission.)

The main contribution to the electric charge of the OT cell surface is made by carboxyl groups of sialic acid, removable by neuraminidase from *Vibrio cholerae* (61.7%), and by carboxyl and sulfate groups of hyposulfated glycosaminoglycans (GAG), sensitive to testicular hyaluronidase (23.7%).[35] Cells of metastatic and nonmetastatic OT lines did not differ in their sialic acid content, the latter being associated with proteins and lipids.[34] However, after removing hyposulfated GAG from the cell surface with hyaluronidase, the EPM of metastatic cells decreased much more intensely (Figure 10-15) than the EPM of either the cells of the initial tumor or the cells of nonmetastatic lines.[35] The density of the surface electric charge connected with the ionized groups of hyposulfated GAG in metastatic cells was equal to 7.7 to 12.0×10^{-3} C $\times$ m^{-2}, and that in nonmetastatic cells was 3.6 to 5.0×10^{-3} C $\times$ m^{-2}. The difference in concentration of ionized groups sensitive to testicular hyaluronidase also remained while measuring cell EPM in a medium with low ionic strength (0.01 M ions, Figure 10-15b).

VII. SUMMARY AND CONCLUSIONS

From the results summarized above and obtained from cells of two different tumors — NK/Ly lymphoma and OT carcinoma — it follows that for each of these tumors the dependence of EPM on cell SV was different. Proliferating cells, which comprised the bulk of the tumor mass, had less variation of mean EPM and SV values. At the same time, EPM varied significantly in cells with low SV and in rapidly sedimentating polyploid cells.

Cells with a low proliferative activity were localized in fractions with low and high SV. Fractions of OT cells with low SV also had low EPM in comparison with other cells. This corresponds with data demonstrating a higher surface charge of proliferating cells in comparison with cells at rest.[19,20,36,37] However, the EPM of slowly sedimentating NK/Ly cells was higher than in proliferating cells. Since NK/Ly cells with low SV were also characterized by a low cloning efficiency, it can be supposed that in this case we are not dealing with relative differences in EPM between proliferating cells and cells at rest (G$_0$ cells), but with differences between proliferating cells and cells not capable of proliferating.

Fractions containing cells at the beginning of the cell cycle (G$_1$ cells) and fractions containing G$_2$ + M cells positively differed in their EPM. However, this difference is apparently not related to the cell cycle. An increase of EPM in the mitosis phase has already been shown for cells from monolayer cultures.[36,38] At the same time, the dependence of EPM on the stage of the cell cycle has not been discovered in other studies.[14,39,40] In the nonsynchronized populations dealt with here, the percentage of cells in the mitosis phase usually does not exceed several percent and, hence, the electrophoretic properties of mitotic cells cannot be measured with a reasonable degree of reliability.

In contrast to liver cells, polyploid tumorous ascitic cells with a DNA content above 4c and cells with a chromosome set close to diploid differed in their EPM. Nevertheless, an unambiguous and single-valued interrelation between DNA content and the electric charge of the surface cannot be established. Consequently, the difference in the electrokinetic properties of tumorous cells with a differing modal chromosome number is not related to the quantity of chromosome material (the number of gene copies), but rather to the phenotypic properties of these cells, which are dependent on the gene expression level. The differences in the electric charge of the cell surface in cells of tumor lines with varying chromosome sets were also shown in other papers.[41,42] Polyploidy is one of the signs of tumorous progression. Tumors with an elevated content of polyploid cells are, as a rule, more malignant.[43] That is why the further study of the elecktrokinetic properties of tumorous cells with a varying ploidy index may contribute to the elucidation of the link between the molecular organization of plasma membranes in polyploid tumorous cells and their malignancy.

Cells from two of the three metastatic OT lines investigated had a higher EPM in comparison with nonmetastatic cells, which confirmed the data showing a higher electric charge for metastatic cells.[44-46] At present, a direct correlation between the metastatic potential and the degree of sialization plasma membranes has not been established.[44] The above results have shown that a high content of ionized groups of hyposulfated glycosaminoglycans (GAG), sensitive to testicular hyaluronidase, is concentrated on the surface of metastatic OT cells. In order to ascertain the role of the GAG localized in the perimembranous layers of tumorous cells in the shaping of their metastatic potential, further investigation is needed.

To sum up the results of the present chapter, it can be concluded that the simultaneous measurement of cell EPM and SV in heterogeneous cell systems may be helpful for studying the connections between cell electrokinetic properties and those properties correlating with their SV. These are, in particular, the functional properties, the level of differentiation, and the karyotype structure.

REFERENCES

1. **Hannig, K.,** Continuous free-flow electrophoresis as an analytical and preparative method in biology, *J. Chromatogr.,* 159, 183, 1978.
2. **Cajone, F., Gaja, G., and Sherbet, G. V.,** Plasma membrane changes in hepatocytes from normal and regenerating rat liver, *J. Cell Biol.,* 38, 174, 1983.
3. **Mayhew, E. and O'Grady, E. A.,** Electrophoretic mobilities of tissue culture cells in exponential and parasynchronous growth, *Nature,* 207, 86, 1965.
4. **Brent, T. P. and Forrester, J. A.,** Changes in surface charge of HeLa cells during the cell cycle, *Nature,* 215, 92, 1967.
5. **Peters, E. A. and Evans, W. H.,** Separation of bone marrow cells by sedimentation at unit gravity, *Nature,* 214, 824, 1967.
6. **Tulp, A., Welagen, J. J. M. N., and Emmelot, P.,** The versatility of the unit gravity sedimentation method, in *Cell Separation Methods,* Bloemendal, H., Ed., Elsevier, Amsterdam, 1977, 17.
7. **Josefowicz, J. Y., Ware, B. R., Griffith, A. L., and Catsimpoolas, N.,** Physical heterogeneity of mouse thymus lymphocytes, *Life Sci.,* 21, 1483, 1977.
8. **Petrov, R. V., Dozmorov, I. M., Lutsenko, T. V., Nikolaeva, I. S., and Sapozhnikov, A. M.,** Characterization of electrophoretically separated lymphocyte subpopulations in human peripheral blood, in *Cell Electrophoresis,* Schütt, W. and Klinkmann, H., Eds., Walter de Gruyter, Berlin, 1985, 421.
9. **Slivinsky, G. G.,** A method of sedimental and electrophoretical analysis of cell populations, *Tsitologiya,* 20, 839, 1978 (in Russian).
10. **Guerci, O., Huot-Marchand, F., and Schneider, O.,** Application de la sedimentation a 1 g sur gradient de Ficoll a la separation des cellules medullaires, *C.R. Soc. Biol.,* 175, 144, 1981.
11. **Slivinsky, G. G., Malkovsky, A. D., and Kirienko, S. I.,** Investigation of the electrophoretical properties of the bone marrow cells by methods of sedimental and electrophoretical analysis, *Tsitologiya,* 25, 976, 1983 (in Russian).
12. **Schubert, J. C. F., Walther, F., Holzberg, E., Pascher, G., and Zeiller, K.,** Preparative electrophoretic separation of normal and neoplastic human bone marrow cells, *Klin. Wochenschr.,* 51, 327, 1973.
13. **Sabolovic, D. and Dumont, F.,** Separation and characterization of cell subpopulations in thymus, *Immunology,* 24, 601, 1973.
14. **Hansen, E.,** Preparative free-flow electrophoresis of lymphoid cells: a review, in *Cell Electrophoresis,* Schütt, W. and Klinkmann, H., Eds., Walter de Gruyter, Berlin, 1985, 287.
15. **Thompson, C. B., Ryan, G. G., Siechman, D. I., Finkelman, F. W., Mond, G. G., and Scher, G.,** A method for size separation of murine spleen cells using counter-flow centrifugation, *J. Immunol. Methods,* 63, 299, 1983.

16. **Higgins, G. and Anderson, R.,** Experimental pathology of the liver. I. Restoration of the liver of the white rat following partial surgical removal, *Arch. Pathol.,* 12, 186, 1976.
17. **Deschenes, J., Lafleur, L., and Marcean, N.,** Sedimentation velocity distributions of cells from adult and fetal rat liver and ascites hepatoma, *Exp. Cell Res.,* 103, 183, 1976.
18. **Sweeney, G. D., Cole, F. M., Freeman, K. B., and Patel, H.,** Heterogeneity of rat liver parenchymal cells. Cell volume as a function of DNA content, *J. Lab. Clin. Med.,* 94, 718, 1979.
19. **Chaudhuri, S. and Lieberman, I.,** The surface of liver cells after partial hepatectomy, *Biochem. Biophys. Res. Commun.,* 20, 303, 1965.
20. **Zaret, B. L., West, L. R., and Becker, F. F.,** Regenerating of the mammalian liver. III. Electrokinetics of replicating cells treated with antimitotic antibiotics, *Proc. Soc. Exp. Biol. Med.,* 121, 1155, 1966.
21. **Slivinsky, G. G. and Magda, I. N.,** The surface charge of the NK/Ly ascites tumor cells differing in proliferative capacity, stage of cell cycle and ploidy, *Tsitologiya,* 32, 61, 1990 (in Russian).
22. **Magda, I. N., Gulverdashvili, N. A., Slivinsky, G. G., Afnasiev, G. G., and Pelevina, I. I.,** Clonogenic capacity, radiosensitivity and recovery from potentially lethal damages of some NK/Ly tumor subpopulations in mice, *Radiobiologiya,* 27, 303, 1987 (in Russian).
23. **Kiessling, R. and Grönberg, A.,** Cell-surface properties influencing target-cell sensitivity for NK-lysis, in *Mechanisms of Cell Mediated Cytotoxic Processes,* Clark, W. R. and Goldstein, P., Eds., Plenum, New York, 1982, 367.
24. **Kiessling, R., Hanson, M., and Grönberg, A.,** Natural killer cells as regulators of malignant and normal cell growth, *Prog. Immunol.,* 5, 1181, 1984.
25. **Slivinsky, G. G., Krifuks, O. I., and Magda, I. N.,** The influence of ionizing radiation on sensitivity of cells from some NK/Ly tumor subpopulations to a membrane-toxic action of natural killers, *Radiobiologiya,* 29, 564, 1989 (in Russian).
26. **Albertson, P. A.,** Phase partition of cells and cellular particles, in *Separation of Cell and Subcellular Elements,* Academic Press, Oxford, 1979, 17.
27. **Zaslavsky, B. Y., Mestechkina, N. M., Miheeva, L. M., and Rogozhin, S. V.,** Physico-chemical factors governing partition behaviour of solutes and particles in aqueous polymeric biphase system, *J. Chromatogr.,* 256, 49, 1983.
28. **Bischoff, P., Robert, F., and Donner, M.,** A comparative study of some growth characteristics and cell-surface properties of neoplastic cells, *Br. J. Cancer,* 44, 545, 1981.
29. **Yamamoto, K., Yamamoto, M., and Ooka, H.,** Changes in negative surface charge of human diploid fibroblasts TIG-1 during *in vitro* aging, *Mech. Ageing Dev.,* 42, 183, 1988.
30. **Pinaev, C., Hoorn, B., and Albertson, P. A.,** Counter-current distribution of HeLa and mouse mast cells in different stages of life cycle, *Exp. Cell Res.,* 98, 127, 1976.
31. **Bischoff, P., Robert, F., and Donner, M.,** Biophysical changes of cell membrane during the growth of an ascitic tumour, *Br. J. Cancer,* 39, 479, 1979.
32. **Slivinsky G. G., Tsvetkova T. V. and Tsokur E. V.,** Antioxidative properties of tumor cell subpopulations, *Tsitologiya,* 34, 82, 1992 (in Russian).
33. **Slivinsky G. G., Malkovsky A. D., Krasnoshtanov V. K.,** The electrical charge of tumorous cells with different caryological structures, *Exp. Onkol.,* 6, 61, 1984 (in Russian).

34. **Slivinsky, G. G., Tsokur, E. V., Tsvetkova, T. V., and Kranoshtanov, V. K.,** Concentration of sialic acids in the blood serum, ascitic fluid and ovary carcinoma cells with different ploidy and metastatic capacity in rats, *Eksp. Onkol.,* 14, 39, 1992 (in Russian).

35. **Slivinsky, G. G., Tsvetkova, T. V., Krasnoshtanov, V. K., and Fedoseenko, V. M.,** Electrophoretic and electron-microscopic study of the surface structure of cells from rat ovary carcinoma clonal lines, *Eksp. Onkol.,* 15, 28, 1993 (in Russian).

36. **Lloid, C. W.,** Sialic acid and social behaviour of cells, *Biol. Rev.,* 50, 325, 1975.

37. **Dolowy, K.,** Bioelectrochemistry of cell surface, *Prog. Surf. Sci.,* 215, 245, 1985.

38. **Mironov, S. L. and Dolgaya, E. A.,** Microelectrophoresis and fluorescence studies of the glycoprotein sheath covering the outer membrane surface in murine neuroblastoma cells, *Bioelectrochem. Bioenerg.,* 17, 429, 1987.

39. **Shank, B. B. and Burki, H. J.,** Surface charge and cell volume of synchronized mouse lymphoblasts (L5178-Y), *J. Cell. Physiol.,* 78, 243, 1971.

40. **Todd, P., Plank, L. D., Kunze, E., Lewis, M. L., Morrison, D. R., Barlow, G. H., Landham, J. W., and Cleveland, C.,** Electrophoretic separation and analysis of living cells from solid tissues by several methods. Human embryonic kidney cell cultures as a model, *J. Chromatogr.,* 364, 11, 1986.

41. **Jenforde, T. S., Adesida, P. O., Kelly, G. S., and Todd, P.,** Differing electrical surface charge and transplantation properties of genetically variant sublines of the TA3 murine adenocarcinoma tumor, *Eur. J. Cancer Clin. Oncol.,* 19, 227, 1983.

42. **Petrov, Y. P., Andreeva, E. V., and Pershina, V. P.,** Analysis of subfractant properties in two sublines of HeLa cells using a two-phase polymer system, *Tsitologiya,* 30, 1263, 1980 (in Russian).

43. **Friedlander, M. L., Hedley, D. W., and Taylor, J. W.,** Clinical and biological significance of aneuploidy in human tumors, *J. Clin. Pathol.,* 37, 961, 1984.

44. **Nikolson, G. I.,** Cancer metastasis. Organ colonization and cell surface properties of malignant cells, *Biochim. Biophys. Acta,* 695, 113, 1982.

45. **Carter, H. B. and Coffey, D. S.,** Cell surface change in predicting metastatic potential of aspirated cells from Dunning rat prostatic adenocarcinoma model, *J. Urol.,* 140, 173, 1988.

46. **Park, B. H., Fike, R. M., and Kim, U.,** Increased electrophoretic mobility of rat mammary tumor cells: a positive correlation with metastatic potential, *IRICS Med. Sci.,* 10, 96, 1982.

Chapter 11

Cell Interactions and Surface Hydrophilicity: Influence of Lewis Acid-Base and Electrostatic Forces

Carel J. van Oss

CONTENTS

I. INTRODUCTION

Nonelectrostatic repulsion between neutral hydrophilic particles suspended in water was first observed by Joffe and Mudd in 1935 with particle suspensions of a smooth strain of

0-8493-8918-6/94/$0.00+$.50
© 1994 by CRC Press, Inc.

219

Salmonella typhi.[1] A decade or so later a similar phenomenon was observed in conjunction with the stabilization of hydrophobic particles suspended in water by coating them with neutral hydrophilic polymers or nonionic surfactants; this was lucidly reviewed by Ottewill.[2] With a view to explaining this phenomenon a school of thought subsequently emerged which attempted to impute the mechanism of the repulsion caused by neutral polymers to "steric stabilization", a hazy but to this day still fairly popular concept which was extensively elaborated upon by Napper[3] (see also Reference 4 for a more recent analysis of this approach).

Meanwhile, through the efforts of Parsegian and his colleagues,[5-7] long-range nonelectrostatic repulsion forces, in the guise of hydration forces, were measured in various aqueous, mainly biological systems. These long-range forces and their (exponential) rate of decay could be determined with some accuracy (e.g., by osmotic pressure measurements).[5,7] However, given that "the surface determines the coefficient of an exponentially decaying force, while the medium determines the decay",[7] the hydration forces approach elucidated the long-range aspects of these forces, but did not deal with the actual surface properties of the cells, biopolymers, and other water-soluble polymers, which contain the primary underlying cause that gives rise to the long-range effects which are measurable by osmotic pressure[5,7] or force-balance determinations.[8,9]

The underlying cause of these nonelectrostatic repulsion forces originates in the interplay between the hydrogen-bonding forces of liquid water and the polar or electron-accepting and electron-donating (Lewis acid-base) capacities of the surfaces of cells as well as bio- and other polymers. The crucial role of (Lewis) acid-base interactions in interfacial interactions was first clearly identified by Fowkes.[10] However, a confusion between polar interactions (as in, e.g., van der Waals-Keesom dipole-dipole interactions) and polar (Lewis) acid-base interactions continued to persist and, in the minds of some workers, persists to this very day. Nonetheless, these two types of "polar" interactions are entirely different phenomena. Chaudhury[11] showed, by applying the Lifshitz approach to interfacial interactions, that on a macroscopic level the three different types of electrodynamic forces (i.e., the van der Waals-Keesom, the van der Waals-Debye, and the van der Waals-London forces) decay in the same manner as a function of distance; they can and should be grouped together and are most conveniently alluded to as Lifshitz-van der Waals (LW) or apolar interactions. After Chaudhury's treatment[11] it became unambiguously clear that the polar hydrogen bonding — or more generally, Lewis acid-base (AB) interactions — should be treated entirely separately from the quite unrelated electrodynamic dipole forces which pertain to LW interactions. The elucidation and measurement of AB interaction forces followed soon thereafter.[12,13]

From many (contact angle) measurements on cells, biopolymers, and related materials it soon became clear that all such highly hydrophilic substances manifest a very strong electron donor capacity ("electron-donicity") and have zero (or a negligible) electron acceptance capacity ("electron-accepticity"): all of these could be said to have pronounced monopolar surface properties.[12] Now, strongly monopolar surfaces immersed in water repel each other via nonelectrostatic, net hydrogen bonding (AB) repulsion forces. More recently it was concluded from a variety of observations that a qualitative proportionality exists between the ζ-potential of a hydrophilic surface and its electron donicity.[14]

Thus, the following statements can be made about electrical double layer (EL) and polar (AB) repulsions between similar hydrophilic cells as well as bio- or other polymers or particles immersed in water:

1. EL interactions are always accompanied by AB interactions, but
2. AB interactions can also occur alone, in the absence of EL forces

II. THEORY AND MEASUREMENT*

A. LIFSHITZ-VAN DER WAALS SURFACE TENSION COMPONENT

It should be reiterated that, on a macroscopic level, in interfacial interactions the three electrodynamic (van der Waals) forces — i.e., van der Waals-London or dispersion forces, van der Waals-Debye or induction forces, and van der Waals-Keesom or orientation forces — obey the same rules and may be grouped together as Lifshitz-van der Waals (LW) interactions.[11]

LW interactions between atoms and/or molecules, occurring in large molecules or particles, are to a significant degree additive, so their free energy of interaction, ΔG^{LW}, between two semi-infinite parallel flat slabs may be expressed as

$$\Delta G^{LW} = \frac{-A}{12\pi\ell^2} \tag{11-1}$$

where A is the Hamaker coefficient (which is linked to the physical properties of all interacting materials, including the liquid) and ℓ is the distance between the two parallel bodies or molecules. Hamaker coefficients have been calculated for a number of materials;[16] van der Waals interactions also can be measured via the Lifshitz approach[16] or via surface tension determinations.[13] The free energy ΔG^{LW}_{132} of the Lifshitz-van der Waals part of the interaction between materials 1 and 2 (in liquid medium 3), for $\ell = \ell_0$ (ℓ_0 is the minimum equilibrium distance between two parallel bodies or molecules: $\ell_0 \approx 0.157$ nm),[13] is obtained by using the Dupré equation for condensed media:

$$\Delta G^{LW}_{132} = \gamma^{LW}_{12} - \gamma^{LW}_{13} - \gamma^{LW}_{23} \tag{11-2}$$

where γ^{LW} stands for the interfacial tensions between the materials indicated by the subscripts and ΔG^{LW} is the same as in Equation 11-1. Once the LW components of the surface tensions of materials 1 and 2 and of the liquid medium (3) are known, the apolar interfacial tensions used in Equation 11-2 can be obtained according to the equation

$$\gamma^{LW}_{ij} = \left(\sqrt{\gamma^{LW}_i} - \sqrt{\gamma^{LW}_j}\right)^2 \tag{11-3}$$

ΔG^{LW}_{132} can be negative (attractive) or positive (repulsive), depending on the value of the Hamaker constant A_{33} of the liquid medium relative to the values of the Hamaker constants A_{11} and A_{22} of Ag and Ab. Thus, manipulation of some of the properties of the liquid medium may turn an attractive reaction (association) into a repulsive reaction (dissociation).[13] The Hamaker constant of water ($A_{ww} \approx 4.05 \times 10^{-20}$ J) is, however, so much lower than that of almost all biological compounds, that a net LW repulsion practically never occurs in biological systems in aqueous media. Thus, in such systems LW interactions are virtually always attractive; they are never the only interaction forces, usually representing less than 10% of the total interaction. The LW interaction energy is additive to the other components.

---

* The essence of the theoretical background given in this section also has been treated in van Oss, C. J., in *Structure of Antigens*, Vol. 1, Van Regenmortel, M. H. V., Ed., CRC Press, Boca Raton, FL, 1992, 99.

B. POLAR (LEWIS ACID-BASE) SURFACE TENSION COMPONENT AND ITS PARAMETERS

Polar forces are defined here as the forces operating in electron acceptor/electron donor or Lewis acid-base interactions. Polar forces are designated by the superscript AB, for acid-base interactions.[10,12,13] Since hydrogen bonds (Brønsted acid-base interactions) are a subcategory of Lewis acid-base interactions, all interactions that take place in a strongly hydrogen-bonding liquid, such as water, always involve polar (AB) interactions.

Even "hydrophobic" attractions ($\Delta G^{AB} < 0$) between totally apolar (macro) molecules or particles, when immersed in water, are preponderantly (i.e., $\approx 99\%$) polar; only about 1% are due to LW forces. Among cell suspensions and dissolved proteins one usually encounters the opposite phenomenon, i.e., "hydrophilic" repulsion ($\Delta G^{AB} > 0$).

The polar surface tension component, γ^{AB}, is additive to γ^{LW} to form the total surface (or interfacial) tension:

$$\gamma^{LW} + \gamma^{AB} = \gamma \tag{11-4}$$

However, γ^{AB} is decomposed into its electron-acceptor parameter, $\gamma^{\oplus}$, and its electron-donor parameter, $\gamma^{\ominus}$, since[12,13]

$$\gamma^{AB} = 2\sqrt{\gamma^{\oplus}\gamma^{\ominus}} \tag{11-5}$$

Polar free energies obey an equation analogous to Equation 11-2, but the γ_{12}^{AB} term is expressed by an equation which is very different from the γ_{12}^{LW} term (see Equation 11-3):

$$\gamma_{12}^{AB} = 2\left(\sqrt{\gamma_1^{\oplus}\gamma_1^{\ominus}} + \sqrt{\gamma_2^{\oplus}\gamma_2^{\ominus}} - \sqrt{\gamma_1^{\oplus}\gamma_2^{\ominus}} - \sqrt{\gamma_1^{\ominus}\gamma_2^{\oplus}}\right) \tag{11-6}$$

The polar component of the interaction energy between entities 1 and 2 when they are immersed in a liquid medium, 3, can also be expressed by the Dupré equation for condensed media:

$$\Delta G_{132}^{AB} = \gamma_{12}^{AB} - \gamma_{13}^{AB} - \gamma_{23}^{AB} \tag{11-2A}$$

Inserting γ_{ij}^{AB} values found via Equation 11-2A, this becomes

$$\Delta G_{132}^{AB} = 2\left[\sqrt{\gamma_3^{\oplus}}\left(\sqrt{\gamma_1^{\ominus}} + \sqrt{\gamma_2^{\ominus}} - \sqrt{\gamma_3^{\ominus}}\right) + \sqrt{\gamma_3^{\ominus}}\left(\sqrt{\gamma_1^{\oplus}} + \sqrt{\gamma_2^{\oplus}} - \sqrt{\gamma_3^{\oplus}}\right) \right. \\ \left. - \sqrt{\gamma_1^{\oplus}\gamma_2^{\ominus}} - \sqrt{\gamma_1^{\ominus}\gamma_2^{\oplus}} \right] \tag{11-7}$$

ΔG_{132}^{LW} and ΔG_{132}^{AB} are also additive, yielding the total interfacial (IF) free energy, ΔG_{132}^{IF}. If an electrostatic interaction also exists, then the total free energy of interaction is expressed as

$$\Delta G^{TOT} = \Delta G^{LW} + \Delta G^{AB} + \Delta G^{EL} \tag{11-8}$$

C. DETERMINATION OF γ^{LW}, $\gamma^{\oplus}$, AND $\gamma^{\ominus}$

For homogeneous solids as well as for pure liquids, γ_i^{LW}, $\gamma_i^{\oplus}$, and $\gamma_i^{\ominus}$ can be determined, e.g., via contact angle measurements (liquid/solid) or interfacial (IF) tension measurements (liquid/liquid).[12,13] By contact angle (θ) determinations on dry (or on hydrated) layers of proteins or cells, with three completely characterized liquids L (two of which liquids must be polar), the Young equation can be solved for the three unknown parameters γ_s^{LW}, $\gamma_s^{\oplus}$, and $\gamma_s^{\ominus}$ of the solid S:

$$(1+\cos\theta)\gamma_L = 2\left(\sqrt{\gamma_S^{LW}\gamma_L^{LW}} + \sqrt{\gamma_S^{\oplus}\gamma_L^{\ominus}} + \sqrt{\gamma_S^{\ominus}\gamma_L^{\oplus}}\right) \tag{11-9}$$

using this equation three times, with different values for θ obtained with three different liquids L.

The apolar (LW) as well as the polar (AB) components of the interfagal (IF) interaction energy $\Delta G_{132}^{IF} = \Delta G_{132}^{LW} + \Delta G_{132}^{AB}$ can then be obtained using Equations 11-2, 11-3, and 11-7.

D. DECAY WITH DISTANCE OF THE FREE ENERGIES OF INTERACTION
1. Lifshitz-van der Waals Free Energies[13]

From Equation 11-1 it can be seen that LW energies decay as the square of the distance for the flat parallel slab conformation. For spheres with a radius R the decay is

$$\Delta G^{LW} = \frac{-AR}{12\ell} \tag{11-10}$$

Equations 11-1 and 11-10 are only valid for $\ell \leq 10$ nm because at greater distances retardation effects[8] set in which strongly decrease the London-van der Waals or dispersion component of the LW forces (which is generally the most important component). At $\ell > 10$ nm Equation 11-1 decays as $1/\ell^3$ and Equation 11-10 as $1/\ell^2$.

2. Lewis Acid-Base Free Energies[13]

Polar forces do not decay with distance in proportion to the square of that distance (as is the case with LW forces; see Equation 11-1), but rather decrease at an exponential rate. The rate of decay with distance for flat parallel slabs of ΔG^{AB} is expressed as

$$\Delta G_\ell^{AB} = \Delta G_{\ell_o} \exp\left[(\ell_o - \ell)/\lambda\right] \tag{11-11}$$

where ℓ_0 is the minimum equilibrium distance (which may be assumed to be of the same order of magnitude as for LW interactions; i.e., $\ell_0 \approx 1.6$ Å) and λ is the correlation length (or decay length) typical for the solvent molecules; for liquid water λ has an empirical value of the order of about 1 nm for (hydration) repulsion or even higher, i.e., of the order of 10 nm for (hydrophobic) attraction.[17] For spheres of radius R the AB decay is expressed as

$$\Delta G_\ell^{AB} = \pi R \lambda \Delta G_{\ell_o}^{AB''} \exp\left[(\ell_o - \ell)/\lambda\right] \tag{11-12}$$

where $\Delta G_{\ell_o}^{AB''}$ is the AB free energy in the plane parallel configuration, measured at "contact" (i.e., at $\ell = \ell_0$).

3. Electrical Double-Layer Free Energies[13]

As a consequence of the shielding effect of the diffuse ionic double layers surrounding the charged sites, the effect of which varies strongly with the ambient ionic strength and with the distance between charged sites, the calculation of electrostatic interaction energies is a complicated operation. It is strongly a function of the (Debye) thickness of the diffuse ionic double layer $(1/\kappa)$, which is dependent on the ionic composition of the liquid medium.[18,20] At physiological salt concentration (i.e., 0.15 M NaCl) $1/\kappa \approx 8$ Å. In general, the rate of decay of EL interactions with distance may be expressed as

$$\Delta G_\ell^{EL} = \Delta G_{\ell_o}^{EL} \exp(-\kappa\ell) \tag{11-13}$$

It should be noted that the Debye thickness *increases* in more dilute salt solutions, e.g., the values for $1/\kappa$ are 1 nm for 0.1 M NaCl, 10 nm for 0.01 M NaCl; and 100 nm for 10^{-5} M NaCl. Thus, in more dilute salt solutions, EL interactions are measurable at much greater distances than at high ionic strengths (Equation 11-10). For the flat parallel plate configuration

$$\Delta G_\ell^{EL} = 1/\kappa\, 64 nkT\gamma_o^2 \exp(-\kappa\ell) \tag{11-14}$$

where n is the number of counterions per unit volume, k is Boltzmann's constant (= 1.38 $\times$ 10^{23} J/degree), T is the absolute temperature in degrees Kelvin, and

$$\gamma_o = \left[\exp(ve\psi_o / 2kT) - 1\right] / \left[\exp(ve\psi_o / 2kT) + 1\right] \tag{11-15}$$

where v is the valency of the counterions, e is the charge of the electron (= 4.8 $\times$ 10^{-10} ESU), and ψ_o is the potential at the surface of the particle or cell (ψ_0 can be obtained from the electrokinetic potential, ζ).[13,18-20]

For spheres of radius R

$$\Delta G_\ell^{EL} = 0.5\varepsilon R\psi_o^2 \ln\left[1 + \exp(-\kappa\ell)\right] \tag{11-16}$$

where ε is the dielectric constant of the medium (for water, $\varepsilon = 80$).

III.   LINKAGE BETWEEN ELECTRICAL DOUBLE LAYER AND LEWIS ACID-BASE INTERACTIONS

A. EFFECT OF ELECTROSTATIC NEUTRALIZATION ON LEWIS ACID-BASE INTERACTIONS

1. Electron-Donor Monopolarity of All Hydrophilic Materials

From experimental observations on a large number of biopolymers, cells, and other hydrophilic solutes and particles, it became clear that, *on a macroscopic level* (as measured by contact angle determinations on portions of surfaces with an area on the order of 0.2 cm^2), hydrophilic surfaces appear to be invariably electron-donor monopolar — i.e., they have a $\gamma_s^\ominus$ parameter which is at least a few millijoules per square meter larger than that of water (i.e., $\gamma_s^\ominus \geq 28$ mJ/m^2) and a $\gamma_s^\oplus$ which is zero, or practically zero.[12] This property allows them to repel each other, when immersed in water, even in the absence of any electrical surface potential.[12,13] The AB repulsion (which, even in the presence of EL repulsions, is usually at least an order of magnitude stronger than the latter) is the sole (or the major) driving force for polymer solubility in water, aqueous phase separation of polymer solutions, hydration pressure, stability of cell suspensions *in vivo* and *in vitro,* aqueous stability of neutral particle suspensions and of nonionic micelles, microemulsion formation, exclusion of cells and solutes by advancing ice fronts upon freezing, etc.[12-15,21]

It is important to note that, again, on a macroscopic scale, pronounced $\gamma^\ominus$ monopolarity prevails in the case of negatively charged amphoteric biopolymers (e.g., serum albumin, immunoglobulins, fibrinogen, etc.[22,23]) as well as in the case of positively charged biopolymers (e.g., lysozyme[22]).

2. "Hydrophobicity" of Charged Surfaces at Their Isoelectric Point

Upon bringing the ambient pH to the value of the isoelectric point of an amphoteric surface, its negative and positive electric charges become equal (i.e., a net zero charge exists). $\gamma^\ominus$ then tends to a very low value (and $\gamma^\oplus$ remains close to zero), making the neutral amphoteric surface strongly hydrophobic.[24,25] Thus, whilst in water both net negatively charged and net positively charged amphoteric surfaces manifest strong electron donicity and zero acceptivity, at electroneutrality (i.e., plus charges equaling minus charges) such amphoteric surfaces lose most of their electron donicity (without acquiring electron-acceptivity) and become virtually apolar, as measured macroscopically.

B. TRIPLE ROLE OF PLURIVALENT COUNTERIONS

Through the admixture of plurivalent counterions, the same "hydrophobizing" effect that arises at amphoteric surfaces at their isoelectric pH, can be obtained with electrically charged surfaces immersed in water. This effect was first reported in 1982 by Ohki,[26] who observed that negatively charged phospholipid layers became hydrophobic upon the addition of Ca^{2+} ions. Not long thereafter, Gallez and her colleagues[27] showed that the action of Ca^{2+} on negatively charged phospholipid films not only contributed to the neutralization of their electrical surface potential, but in addition exerted an inhibiting effect on the hydration pressure which originates at the surfaces of these films, immersed in water. In 1987 van Oss et al.[12] proposed that in such instances the action of Ca^{2+} is twofold: it consists of a (usually partial) neutralization of the negative surface charge, in its role as a divalent cation, together with a suppression of the electron donicity of the initially hydrophilic surfaces caused by the electron acceptivity of Ca^{2+}.* The dual role of Ca^{2+} in inducing membrane fusion was described in more detail in 1988.[28] More recently it was found that the destabilization of clay particles through the admixture of La^{3+} ions is principally (i.e., for 87%) due to a hydrophobic (AB) attraction and only slightly (for 13%) to an LW attraction.[29] Finally, even monovalent counterions (e.g., Na^+) can cause destabilization of suspensions of (in this case negatively charged) particles, partly by decreasing the ζ-potential of the particles and partly and concomitantly by rendering them more hydrophobic.[30,31] However, in accordance with Schulze and Hardy's rule,[32] much larger concentrations of monovalent than of plurivalent counterions are required to achieve flocculation.

The very low concentration of plurivalent counterions which suffices to elicit flocculation makes it possible for flocculation to occur at very low ionic strengths. For instance,

---

* As an alternative to the conjecture that the presence of Ca^{2+} gives rise to the neutralization of the electron donicity of negatively charged hydrophilic surfaces by means of its electron acceptivity, it also can be reasoned that all methods that cause the neutralization of the electrical surface charge of a charged surface (e.g., changing the ambient pH) will at the same time lead to the obliteration of all or most of its electron donicity.

flocculation of red blood cells (e.g., glutaraldehyde-fixed rabbit RBC; see below) at an ionic strength of the order of $\Gamma/2 = 0.004$, by means of ≈ 0.5 mM LaCl$_3$ (pH 7.6), takes place while the RBC still have a ψ_0-potential of -24.4 mV, i.e., a decrease of only 27.6% from the original -33.7 mV. While the presence of LaCl$_3$ causes the RBC to become quite hydrophobic, which results in a close-range attraction of about 19,000 kT, the decay of the electrostatic repulsion at the low ambient ionic strength is so gradual that at an inter-RBC distance of 3 nm a secondary maximum of repulsion of more than 200 kT still prevails. In the normal course of events this repulsion barrier should enable the cell suspension to remain stable.[33] However, in the presence of adsorbed La^{3+} ions local, microscopic attractions occur between exposed positive charges of bound La^{3+} on one cell and residual negatively charged sites on another cell, which then allow the cells to approach each other until they are less than 2 to 3 nm apart, at which time the very strong close-range "hydrophobic" attraction can prevail.

Thus, a third role played by plurivalent cations (e.g., La^{3+}), is based on their capacity to cause discrete sites of two negatively charged surfaces to attract each other to a sufficiently small distance for the relatively short-range hydrophobic attraction to take over. It would be inaccurate to state that La^{3+} ions cross-bind negatively charged surfaces directly, because the close-range electrostatic binding energy between the La^{3+} ions and negatively charged RBC is significantly smaller than the close-range hydrophobic attraction energy.

C. ESTIMATION OF ELECTRON ACCEPTICITY AND ELECTRON DONICITY BY ELECTROPHORESIS IN POLAR ORGANIC MEDIA

There exists a correlation between electrophoretic mobility in polar organic media and electron donicity or accepticity. The degree of electron donicity of a macromolecule or particle can be correlated with its electrophoretic mobility in Lewis acid organic solvents, and the degree of electron accepticity can be obtained via electrophoresis in Lewis base organic solvents[34-36] ($\oplus$ particles suspended in $\ominus$ liquids are negatively charged and $\ominus$ particles suspended in $\oplus$ liquids are positively charged). Thus, it also becomes plausible that amphoteric particles or molecules that are immersed (or dissolved) in water at a pH close to their isoelectric point have zero electron donicity *and* zero electron accepticity, which renders them apolar, or "hydrophobic".

Labib and Williams[37] tried to establish a correlation between electron donicity and the ψ_0-potential of various inorganic particles *via* electrophoresis in a number of organic $\oplus$ as well as $\ominus$ liquids. The same authors[38] studied the connection between the aqueous pH scale and the electron donicity scale. Using nonaqueous solvents, they studied the effect of moisture on the ζ-potential of inorganic particles in nonaqueous liquids; they observed that moisture could reverse the sign of charge of the solid surface and shift its electron donicity, making it more basic.[35]

Labib and Williams[35,37,38] could only establish a qualitative correlation between ζ-potential and "donicity" with respect to the Gutmann scale.[39] However, it still may be feasible to arrive at a quantitative expression for the donicity or accepticity of polar particles (1) from their ζ-potential in nonaqueous solvents (3) in terms of their $\gamma_1^{\ominus}$ (respectively, $\gamma_1^{\oplus}$) value, once one has obtained the value of ΔG_{131}^{EL}. ΔG_{131}^{EL} can then be equated with $\Delta G_{131}^{EL} = -2\gamma_{13}^{AB}$ (see Equation 11-2A), and using Equation 11-6 one can then obtain, e.g., $\gamma_1^{\ominus}$ provided $\gamma_3^{\oplus}$ of organic liquid 3 is known and provided particle 1 and solvent 3 are monopolar in the opposite sense with respect to each other.[40] For instance, if $\Delta G_{131}^{EL} = +30$ mJ/m^2, obtained from electrophoretic mobility measurements of monopolar particle 1 (for which $\gamma_1^{\oplus} = 0$), in chloroform 3 (for which $\gamma_3^{\oplus} = 3.8$ mJ/m^2 and $\gamma_3^{\ominus} = 0$), $\gamma_1^{\ominus}$ then is 14.8 mJ/m^2.

IV. CELL INTERACTIONS

A. CELL STABILITY
1. Energy vs. Distance Balance

To determine the outcome of cell-cell interactions as a function of cellular surface properties it is indispensable to elaborate energy vs. distance balances, taking into account the LW and EL as well as the AB free energies of interaction, as a function of the decay of each of these energy classes with distance. Classically this approach is known as the DLVO method (or DLVO plots, or DLVO theory), based on the work of Derjaguin and Landau[41] and Verwey and Overbeek.[42] However, the DLVO approach only takes the LW and EL energies into account, which in polar media — especially in water — usually leads to erroneous results.[33] In practically all cases one has either markedly attractive or strongly repulsive AB forces in aqueous media. It is only in the rare contingency, when a cell or particle surface has a $\gamma^{\oplus} \approx 0$ and a $\gamma^{\ominus} \approx 28$ mJ/m^2, that the net forces are neither attractive nor repulsive, leaving the LW and EL forces as the only decisive ones.

The property of the three different interaction forces to decay as a function of distance ℓ in a totally different manner, gives rise to ΔG^{TOT} ($= \Delta G^{LW} + \Delta G^{EL} + \Delta G^{AB}$) vs. ℓ plots ("extended DLVO", or XDLVO plots[33,43]), which may contain one or two maxima of repulsion and/or one or two minima of attraction at different values of ℓ. Thus, the quantitative determination of all three free energies and their mode of decay (which is especially variable in the case of ΔG^{EL} as a function of the ionic strength) is essential for assessing all possible cell-cell, cell-inert surface, and cell-biopolymer interactions.

2. Influence of the Size and Shape of Cells

From the equations given in Section II.D it is apparent that for all three types of interaction (LW, EL, and AB) the initial interaction energies as well as their mode of decay as a function of distance differ considerably between two parallel (semi-infinite) flat surfaces on the one hand and two spheres of radius R on the other. In Figure 11-1 XDLVO plots are shown for rabbit RBC, for the edge-edge interaction (R = 2 μm) as well as for the interaction between two parallel (flat-flat) biconcave discs. It can be seen that in the flat-flat configuration the energy at the secondary minimum of attraction ($\Delta G = -141$ kT) is almost 5 × that of cells in the edge-edge configuration ($\Delta G = -30$ kT). Thus, at the secondary minimum of attraction the flat/flat configuration (where cells can approach each other to a distance of about 5.5 nm) is strongly favored over the edge/edge configuration (where cells can approach each other to a distance of about 4.3 nm). This is the reason why, *in the presence of stabilizing cross-binding biopolymers* (e.g., fibrinogen, excess IgM, polymerized albumin, high molecular weight dextran) rouleau formation occurs rather than amorphous clumping.[43-45] Because the depths of energy of attraction in the flat/flat configuration at the secondary minimum of attraction differ between RBC of different species, in mixtures of RBC of different species different rouleaux occur, each separate rouleau consists of either RBC of species A or RBC of species B.[44,46]

In cell-cell interactions the radius of curvature R is also of crucial importance. From the equations in Section II.D it is clear that all three types of energies of interaction are proportional to R. Thus, other factors remaining equal, large smooth cells will interact more strongly with each other than small smooth cells. Now, on a macroscopic scale, cells which are suspended in isotonic aqueous media tend to repel each other *in vivo* as well as *in vitro*. This occurs mainly via net AB repulsions (see below). When such cells interact with each other attractively (e.g., in phagocytic engulfment, platelet clumping, hemagglutination, etc.) they can only do this via protruding processes or spicules, which end in

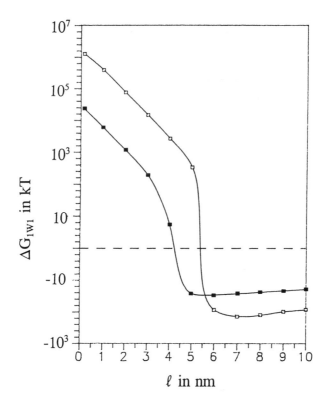

Figure 11-1 Energy vs. distance (XDLVO) diagram of glutaraldehyde-fixed rabbit erythrocytes. The distance (ℓ) between the distal surfaces of the cells' glycocalices is given on the abscissa, in nanometers. The free energy (ΔG_{1W1}) is given on the ordinate, in kT. (□) Parallel disc configuration (a contactable surface area of 25×10^{-8} cm² is assumed); (■) edge/edge configuration (for the edges a radius R = 2 μm is assumed). Conditions: phosphate-buffered saline, pH 7.2; ionic strength ($\Gamma/2$) = 0.15; $1/\kappa$ = 0.8 nm, λ = 0.6 nm. The surface tension data for the cells are: γ_1^{LW} = 40.3, $\gamma_1^{\oplus}$ = 0, and $\gamma_1^{\ominus}$ = 39.1 mJ/m². ψ_0 = −33.7mV. At the secondary minimum of attraction, ΔG_{1W1}^{TOT} = −141 kT in the parallel disc and −30kT in the edge/edge configuration. Thus, the parallel disc configuration is energetically favored, which leads to rouleau formation when the cells (which are then about 5.5 nm apart) can be stabilized through cross-binding by large asymmetrical polymer molecules.

a curved surface with a small R. By this mechanism the overall macroscopic repulsion field can be pierced and a (microscopic) local attraction (often of a specific ligand-receptor nature) can prevail, thus giving rise to cell-cell attachment or even to antigen-antibody or lectin-carbohydrate binding, etc.[47] There are numerous examples of this — e.g., facilitation of contact between phagocytic cells and bacteria by means of protruding leukocytic pseudopodia, platelet adhesion following spiculation induced by adenosine diphosphate, virus attachment to cells via spikes protruding from the virion,[48] and enhancement of hemagglutination through acanthocyte formation induced by the action of, e.g., anti-A and anti-B blood group antibodies.[49]

3. Influence of Each of the Three Primary Noncovalent Forces

Whilst net van der Waals (LW) interactions between different particles immersed in a liquid can under certain conditions be repulsive,[50] biological entities immersed in *water* always undergo an LW *attraction* which, however, is relatively feeble: it usually is on the order of $\Delta G^{LW} \approx -5$ mJ/m² at closest approach. For rabbit RBC (used in this study) ΔG^{LW} = −5.6 mJ/m².

Most biological entities, and all cells, are negatively charged. For instance, for rabbit RBC (used in this study), with a surface potential $\psi_0 = -34$mV, there is a mutual electrostatic repulsion of the order of $\Delta G^{EL} \approx +2$ mJ/m² at closest approach.

Many proteins and most polysaccharides (including those of cellular glycocalices) are strong (monopolar) electron donors (i.e., they have a $\gamma^\ominus \gg 28.3$ mJ/m² and a $\gamma^\oplus \approx 0$), so they strongly repel each other with a net hydrogen-bonding repulsion on the order of $\Delta G^{AB} \approx +10$ to $+45$ mJ/m² at closest approach. With rabbit RBC (used in this study) ΔG^{AB} = +24.3 mJ/m². Thus, typically, $|\Delta G^{AB}| > |\Delta G^{LW}| > |\Delta G^{EL}|$ at close range.

LW free energies decay as a function of distance ℓ as ℓ^{-2} in the flat/flat configuration and as ℓ^{-1} in the sphere/sphere configuration. Thus, whilst LW free energies are fairly feeble in water, they decay rather gradually (especially in the sphere/sphere configuration), so that the LW attraction between cells tends to predominate at $\ell \geq 4$ nm or thereabouts (Section II.D). This gives rise to a secondary minimum of attraction between cells, which plays a crucial role in, e.g., many varieties of hemagglutination used in connection with blood transfusion technology[43-45] (see also Figure 11-1).

Both AB and EL free energies decay exponentially as a function of ℓ, but that decay is strongly governed by the ionic strength of the suspending medium only in the case of EL forces. With EL forces even the free energy of interaction at $\ell = \ell_0$ is dependent on the ionic strength. At a very low ionic strength $\Delta G^{EL}_{\ell=\ell_0}$ is at a maximum and the decay as a function of distance is extremely gradual, while at high ionic strengths $\Delta G^{EL}_{\ell=\ell_0}$ is low and its decay with distance is very steep (see Section II.D). Figure 11-2 illustrates the difference in ΔG^{EL} and in its decay, as a function of the ionic strength, in plots of ΔG^{EL} vs. ℓ at ionic strengths ($\Gamma/2$) 0.15 and 0.015.

4. Action of Plurivalent Counterions

Whilst variations in ionic strength caused by changes in the concentration of 1-1 electrolytes such as NaCl (at concentrations up to 1 M) influence only ΔG^{EL} and not ΔG^{LW} or ΔG^{AB}, the presence of just millimolar concentrations of trivalent counterions (e.g., La^{3+}) in the aqueous suspending medium can cause a drastic decrease in ΔG^{EL} and concomitantly change the polar (AB) free energy of interaction from a repulsion (positive ΔG^{AB}) into an attraction (negative ΔG^{AB}). Only ΔG^{LW} remains virtually unaffected by the presence of plurivalent counterions. Thus, as is well known, low concentrations of $LaCl_3$ can cause RBC[51] (as well as most other suspensions of negatively charged particles) to flocculate.[29-31] However, while it is widely accepted that such plurivalent counterion-induced flocculation is caused by a strong decrease in the ζ-potential of the particles (so that $\Delta G^{EL} \to 0$), which then allows the LW attraction to become predominant,[42] this is only a minor part of the actual mechanism[29,30,33,40] (see also Section III.B above). As shown in Figure 11-3, the 0.5 mM $LaCl_3$ (at the low total ionic strength of 0.004), which easily suffices to flocculate RBC, only halves the close-range value of ΔG^{EL}. On the other hand, the (repulsive) value of ΔG^{AB} (which in the absence of La^{3+} represents about nine times the repulsive energy of ΔG^{EL} in the sphere/sphere configuration) turns into a ΔG^{AB} attraction in the presence of 0.5 mM La^{3+} which is considerably stronger than the residual ΔG^{EL} repulsion and the ΔG^{LW} attraction. Thus, the change in sign of ΔG^{AB} brought about

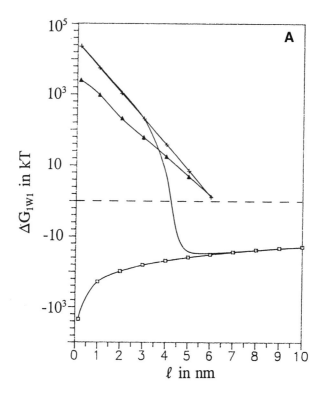

Figure 11-2 Energy vs. distance (XDLVO) diagrams of glutaraldehyde-fixed rabbit erythrocytes; for the explanation of plotted entities, see the legend to Figure 11-1. In both A and B only the edge/edge configuration is depicted. (A) Conditions as described for Figure 11-1; ionic strength ($\Gamma/2$) = 0.15. (B) Conditions as described in Figure 11-1, but $\Gamma/2$ = 0.015. In both A and B, ΔG_{1W1}^{LW} ($\square$), ΔG_{1W1}^{AB} (X), and ΔG_{1W1}^{EL} ($\blacktriangle$) and their sum, ΔG_{1W1}^{TOT} (plain line), are depicted. A differs from B in that decay of ΔG_{1W1}^{EL} with distance ℓ is much more gradual in B on account of the lower ionic strength ($\Gamma/2$) of the medium in B. This also influences the ΔG_{1W1}^{TOT} curves. In A ($\Gamma/2$ = 0.15) there is a pronounced secondary minimum of attraction of −32kT (see Figure 11-1; ΔG_{1W1}^{TOT} in Figure 11-2A is the same as the curve with the solid squares in Figure 11-1). At lower ionic strength, in B, there is no attraction at any distance; this illustrates that suspensions of charged cells or particles (of the same sign of charge) are most stable at low ionic strength (other factors remaining equal).

by the admixture of small amounts of $LaCl_3$ (changing the predominant hydrophilic repulsion into an equally predominant hydrophobic attraction) is quantitatively the primary driving force for the flocculation of these RBC. This was mainly effected by the decrease in $\gamma^\ominus$ ($\approx$ 39.1 mJ/m²) of the original RBC to $\gamma^\ominus \approx$ 15.6 mJ/m² caused by the added La^{3+} ions. These La^{3+} ions partly neutralize the negative surface potential of the RBC, still leaving one or two positive charges (i.e., electron acceptors) to neutralize, or at least depress, many of the electron-donor sites of the glycocalices of the RBC, thus changing their surface from hydrophilic ($\gamma^\ominus$ > 28.3 mJ/m²) to hydrophobic ($\gamma^\ominus$ < 28.3 mJ/m²).[52] Ca^{2+} acts on negatively charged cells in much the same manner, but up to 20 to 50 times more Ca^{2+} than La^{3+} is required to obtain the same effect.[29,30]

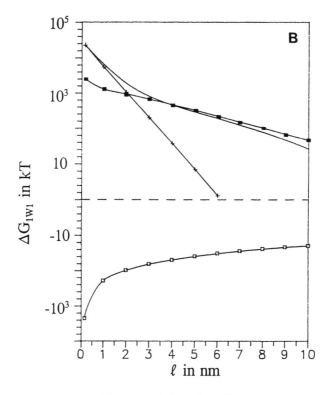

Figure 11-2 (continued)

Thus, the indirect but omnipresent linkage between EL and AB energies, as demonstrated in Figures 11-2 and 11-3, should serve as a warning that observed differences in electrophoretic mobility and, hence, in ζ-potential usually do not solely signify differences in ΔG^{EL}, but generally also indicate even greater differences in ΔG^{AB}.* The latter (usually) cannot be measured by electrophoresis, but need to be determined separately, e.g., by contact angle measurements.[12,13]

B. CELL ADHESION
1. Influence of Prior Adsorption of Biopolymers and Other Solutes
When dissolved proteins are present (which very often is the case, either through dissolution of extracellular proteins in the ambient medium or through outward diffusion of

---

* Such differences in ΔG^{AB} accompany all possible types of changes in ζ-potential, regardless of their origin. They may, for example, be caused by changes in ambient pH or by the concentration of (plurivalent) counterions, as well as by the adsorption of surfactants or polymers or by treatment with proteolytic or glycolytic enzymes (e.g., papain or neuraminidase[53]). Quantitative correlations between ζ-potential and ΔG^{AB} cannot be established with any degree of precision (see Section III.B above), but qualitatively it can be stated that the higher the ζ-potential (regardless of its sign of charge) the more hydrophilic the surface, i.e., the higher the value of $\gamma^\ominus$, its electron-donor parameter. However, it should also be remembered that there are biopolymers (especially polysaccharides — e.g., dextran) with a low or even zero ζ-potential which nevertheless have a high $\gamma^\ominus$ and, thus, are very hydrophilic in their own right. Changes in pH, etc. have very little effect on the ΔG^{AB} of such biopolymers, which usually represent an appreciable part of the glycocalices of many cell types.

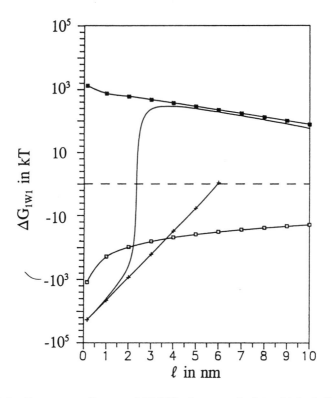

Figure 11-3 Energy vs. distance (XDLVO) diagram of glutaraldehyde-fixed rabbit erythrocytes; for the explanation of the plotted entities, see the legend to Figure 11-1. In this figure the situation is depicted where the erythrocytes are admixed with 0.5 mM LaCl$_3$ at a total ionic strength ($\Gamma/2$) of 0.004, making $1/\kappa = 3.54$ nm; $\psi_0 = -24.4$ mV. The cells interact in the edge/edge configuration (see Figures 11-1 and 11-2). The symbols used here are: ΔG_{1W1}^{LW} ($\square$); ΔG_{1W1}^{AB} (X); and ΔG_{1W1}^{EL} ($\blacksquare$). In this case, as a consequence of the hydrophobizing action of the La^{3+} ions there is a strong primary minimum of attraction at "contact" (ΔG_{1W1}^{TOT} [$\ell = \ell_0$] $= -18{,}200$ kT). There also is a secondary maximum of repulsion at $\Delta G_{1W1}^{TOT} \approx +300$ kT ($\ell \approx 4$ nm) which ideally should cause the cell suspension to be stable. In practice, however, the cells agglutinate under these conditions, owing to multiple local microscopic attractions between a bound La^{3+} ion on one cell (with still one or two available positive charges) and a negatively charged site on another cell.

intracellular proteins into the medium), the presence of protein adsorbed onto the solid surface must be taken into account; this can significantly change the nature of the surface onto which the cells will adhere. The phenomenon observed by Baier et al. with various polymer implants *in vivo*, according to which a minimum of cell adhesion occurs to solid surfaces with $\gamma_S \approx 30$ mJ/m^2,[54] while seemingly paradoxical, is therefore easily explained. Like proteins dissolved in aqueous media, cells adhere most strongly through a hydrophobic AB plus an LW attraction to medium-energy polymer surfaces such as polyethylene ($\gamma_S = 33$ mJ/m^2). With lower-energy surfaces the adhesion is somewhat weaker because of a smaller LW attraction, and with higher energy surfaces the adhesion is weaker because polymer surfaces with $\gamma_S > 33$ mJ/m^2 tend to have increasingly dominant polar sites which counteract hydrophobic AB attractions.[55] However, in the presence of protein molecules which on account of their smaller size and higher diffusion coefficient, adsorb more quickly than the much bigger cells can adhere, in practice cells usually only adhere to polymer surfaces already precoated with protein. The conditions under which *maximum protein adsorption* occurs (i.e., at $\gamma_S \approx 30$ mJ/m^2) then also become the conditions under which *minimum cell adhesion* takes place because cell adhesion is greatly decreased on surfaces precoated with hydrophilic protein.[47]

2. Specific Cell Adhesion

Cell adhesion is strongly favored on solid surfaces to which proteins or other biopolymers (adhesins) are attached for which cells have *specific receptors*. There are a great many such cell-surface-specific adhesins (see, e.g., References 56 and 57).

The same types of forces (LW, EL, and AB forces) which are sufficiently *repulsive* to keep cells and biopolymers in suspension or solution are involved in the *attraction* between the same cells and/or biopolymers when they engage in specific interactions between receptors and their corresponding ligands, such as antibody-antigen, lectin-carbohydrate, enzyme-substrate, and other ligand-receptor interactions. There naturally is a strong need for blood cells to repel each other, in order to remain in stable suspension, and for plasma proteins to repel each other enough to remain in solution. However, under certain conditions it can become necessary for individual cells and/or biopolymers to adhere to each other, e.g., to trigger a specific chain of events, such as those leading to the removal of pathogenic microorganisms. Because in the general case the interactions are repulsive, while under special circumstances they can be attractive, safeguards are needed so that macromolecules and/or cells which can attract each other do so only when it is important for the attraction to be able to override the prevailing repulsion. This is achieved *in vivo,* thanks to the fact that: (1) the nonspecific repulsive forces operate between considerably larger surface areas than the specific attractive forces; and (2) the specifically interacting moieties (e.g., antibodies) only react with their specific ligands (e.g., antigens) when the latter are present, which normally happens relatively infrequently.[47] The main advantage of the small size of the specifically reacting moieties is that they can locally overcome the generally prevailing repulsion fields as long as the nonspecific repulsion (per unit surface area) is smaller than the specific attraction (per unit surface area) at the specific attraction site.[47]

3. Positive and Negative Cell Adhesion to Advancing Ice Fronts

As a consequence of the surface thermodynamic properties of ice,[58] all blood cells are repelled by an advancing ice front at 0°C, mainly by AB and to a very minor degree by EL repulsions.[21] Thus, upon freezing of cell suspensions in aqueous media, cells (as well as microsolutes) are *excluded* by the advancing ice front. This puts the cells under considerable osmotic stress; it is the major source of cell destruction upon freezing. However, in aqueous suspensions admixed with appropriate concentrations of a cryoprotectant (e.g., glycerol), cells are attracted, and therefore *engulfed* by advancing freezing fronts: under such conditions cells do not undergo any osmotic stress and remain undamaged when frozen. This explains the destructive outcome of cell freezing under physiological conditions, as well as the protective mechanism of additives such as glycerol.[21]

C. CELL-POLYMER INTERACTIONS

Cells (e.g., peripheral blood cells in aqueous suspension) as well as biopolymers (in aqueous solution) repel one another. On a macroscopic level, cells and biopolymers also mutually repel each other, regardless of whether or not the biopolymers (microscopically) can adsorb onto the cell surface (or glycocalyx). In the presence of high concentrations of dissolved biopolymers the AB repulsion (and sometimes the AB + EL repulsion) which the polymers exert on the cells can be greater than the repulsion between the cells: this favors cell-cell interaction. On the other hand, when the repulsion between dissolved polymers and cells is smaller than the intercellular repulsion, the cells will not interact with each other and will remain monodispersed.[59] For instance, at extracellular dextran (MW $\approx$ 500,000) concentrations of 2 to 10%, erythrocytes flocculate in small spherical agglutinates: as a consequence of the higher extracellular (AB) repulsion by the polymer molecules the red cells are brought closely together, which allows them to become cross-linked by the asymmetrical dextran molecules, which bind to a red cell glycocalyx lectin (with a specificity for glucose; see Reference 60). At higher dextran concentrations (>10%) the strengths of the repulsive forces are reversed (owing to a stronger dextran adsorption at higher concentrations), so that the cells then repel each other more strongly than the repulsion between the extracellular dextran molecules and the cell surface, which allows the cells to become monodispersed. At all dextran concentrations the erythrocytes are coated with considerable amounts of dextran. It can be demonstrated that at dextran concentrations from 2 to 10%, as well as at dextran concentrations >10%, there is a thin, dextran-depleted layer between the dextran layer adsorbed onto the erythrocytes and the bulk dextran solution. It also can be shown that there is a repulsive interaction between the two layers of dextran, one adsorbed and one free. When the adsorbed dextran layer is the most concentrated, stability ensues; when the dextran in free solution is the most concentrated, flocculation occurs.[59] The depletion layer has a thickness which is approximately equal to the radius of gyration of the dissolved hydrophilic polymer molecules.[61,62] Its presence is an *effect* of the long-range AB (hydration-orientation propagated) repulsion[5-7,63] between the bulk polymer molecules and the cells' glycocalices (including adsorbed polymer molecules) and, contrary to a frequently held belief,[3] the *cause* of neither "depletion flocculation" nor "depletion stabilization".[4,59] However, the discovery of the existence of a depletion layer around erythrocytes surrounded by dissolved polymer molecules provided the explanation for the paradoxical "Brooks effect". The Brooks effect was the observation that under the influence of dissolved dextran molecules (which themselves have a zero ζ-potential[64]) the ζ-potential of erythrocytes appears to *increase* as a function of the extracellular dextran concentration.[64] On the other hand, a closer examination of the Brooks effect by Bäumler and Donath[61] and Arnold et al.[62] showed that the ζ-potential of erythrocytes in reality is *unaffected* by high dextran concentrations, but that close to the erythrocyte surface there is a layer which is depleted of dextran and, thus, has the ordinary low viscosity of the aqueous medium rather than the high viscosity of a concentrated dextran solution. In Brooks' calculations of the ζ-potential from the observed electrophoretic mobilities,[65] the presumed high viscosity of the liquid medium gave rise to the spurious conclusion of a high ζ-potential for the erythrocytes.

D. CELL FUSION

For two cells to be able to fuse they must be able to approach each other closely and their surfaces must, at least locally, attract each other more strongly than the attraction each one of them has for water (see also Reference 66):

$$W_{iwi} > W_{iw} \qquad (11\text{-}17)$$

W_{iwi} is the work of adhesion between cells in water and W_{iw} is the work of hydration of the cells, or

$$3\gamma_{iw} > \left(\gamma_i + \gamma_w\right) \qquad (11\text{-}18)$$

according to the equation

$$W_{iwi} = -\Delta G_{iwi} = 2\gamma_{iw} \qquad (11\text{-}19)$$

and

$$W_{iw} = -\Delta G_{iw} = \gamma_i + \gamma_w - \gamma_{iw} \qquad (11\text{-}20)$$

However, it should be realized that when two cells approach each other closely (driven by the W_{iwi} attraction), interstitial water molecules can be expelled even when $W_{iw} > W_{iwi}$ (expressed in energy per unit surface area). This is because for the expulsion of single water molecules of hydration the work of hydration and the work of adhesion (W_{iw} and W_{iwi}, respectively) are more properly expressed in energy units per molecule or particle, e.g., in kT units (see p. 224). Thus, when $W_{iwi} \approx 50$ mJ/m^2 and $W_{iw} \approx 120$ mJ/m^2, for the interaction between two cells the incipient contactable surface area $S_c \approx 10$ nm^2, while for individual water molecules S_c is only about 0.1 nm^2. It follows that while the intercellular energy of adhesion $W_{iwi} \approx 123$ kT per cell pair, the free energy of hydration W_{iw} is only about 3 kT per water molecule, which easily allows the water molecules of hydration to be expelled individually.[14,40]

A crucial factor for cell fusion is the presence of Ca^{2+}. As described in Section III.B above, calcium ions play several roles. For instance, (1) Ca^{2+} decreases the (negative) ζ-potential of the cells; and (2) Ca^{2+} functions as an electron acceptor, thus causing a decrease in the cells' (monopolar) $\gamma_i^\ominus$.[40] This causes the surfaces of the cells to become more hydrophobic so that instead of repelling each other the cells will attract each other[28] (see Section IV.A.4 above). The presence of extracellular polyethylene oxide further helps Ca^{2+}-induced cell fusion[62] by exerting an outward pressure on the cells which forces them more closely together (see Section IV.C above).

There is one further condition for cell fusion, i.e.,

$$W_{ii} > W_{iwi} \qquad (11\text{-}21)$$

(or the work of adhesion between the cells after expulsion of water must be greater than the work of adhesion between the cells before expulsion of water), or[40]

$$\gamma_i > \gamma_{iw} \qquad (11\text{-}22)$$

so that the surface tension of the cells must be greater than the interfacial tension between the cells and the aqueous medium. The reason for this is that for fusion to occur there must be a direct linking between the phospholipids of two cell membranes (governed by W_{ii}), rather than just an attraction between the phospholipids of two cell membranes, while still having a layer of water between them (W_{iwi}), however thin that layer of water may be.[67] In practice, the condition expressed in Equation 11-22 is satisfied for all phospholipids containing alkyl groups with $\gamma_i \geq 21.8$ mJ/m^2 ($\approx \gamma_w^{LW}$), i.e., phospholipids comprised of nonyl groups or larger. This is always the case because with phospholipids with octyl groups or smaller it also would be impossible to maintain coherent cell membranes in water.

E. HEMAGGLUTINATION

Hemagglutination, i.e., the cross-binding of (usually human) red blood cells by means of specific (usually blood group-specific) antibodies (Abs), is a phenomenon that is strongly governed by polar (AB) and, to a lesser extent, by EL and LW interactions. Under physiological conditions (*in vivo* as well as in phosphate-buffered saline) the distal ends of human RBC glycocalices cannot approach one another more closely than to about 5.0 nm (see Figure 11-1), which leaves a minimum distance of about 15.0 nm between the outer surfaces of the actual cell membranes of neighboring red cells.[45] Herein lies the cause of a great deal of experimental and theoretical work on the modification of red cell surfaces and/or on the modulation of the aqueous medium, with the aim of bringing the cells closer together. This is because IgG-class Abs, with a "reach" of only about 12 to 14 nm, cannot cross-bind two RBC, e.g., via their Rh_0 (D) epitopes, which are situated on the RBC membrane. (There is no such problem in achieving hemagglutination with IgM-class Abs, with a "reach" of 27 to 28 nm; however, in immunohematology and in blood transfusion in general, IgG Abs often are more important than IgM Abs). Thus, to make it possible to detect IgG-class blood group Abs one must either increase the "reach" of IgG or decrease the distance ℓ between RBC.[45] Only the latter approach will be considered here.

Among the various methods for decreasing the distance ℓ between RBC are centrifugation;[68,69] reduction of intercellular repulsion, e.g., by treatment with neuraminidase or papain[53] (these treatments reduce the ζ-potential of RBC, but, more importantly, they also reduce the hydrophilic (AB) repulsion of the cells by making them more hydrophobic); exertion of extracellular pressure by the admixture of, e.g., high concentrations of serum albumin or other water-soluble polymers;[45] cross-linking with large asymmetrical polymers;[44,45,60] and formation of protruding processes of a small radius of curvature, e.g., through cell spiculation caused by the admixture of anti-A or anti-B Abs or lectins[45,49] or of dextran (MW $\approx$ 40,000).[60]

ACKNOWLEDGMENTS

The author is greatly indebted to Mrs. Jeanette McGuire for her expert typographical help and to Mr. W. Wu for his assistance with the computer plotting of the XDLVO diagrams.

REFERENCES

1. **Joffe, E.W. and Mudd, S.,** A paradoxical relation between zeta potential and suspension stability in *S* and *R* variants of intestinal bacteria, *J. Gen. Physiol.,* 18, 599, 1935.
2. **Ottewill, R.H.,** Effect of nonionic surfactants on the stability of dispersions, in *Nonionic Surfactants,* Schick, M.J., Ed., Marcel Dekker, New York, 1967, 627.
3. **Napper, D.H.,** *Polymeric Stabilization of Colloidal Dispersions,* John Wiley & Sons, London, 1983.
4. **van Oss, C.J.,** Interaction forces between biological and other polar entities in water: how many different primary forces are there?, *J. Dispersion Sci. Technol.,* 12, 201, 1991.
5. **Parsegian, V.A., Fuller, N., and Rand, R.P.,** Measured work of deformation and repulsion of lecithin bilayers, *Proc. Natl. Acad. Sci. U.S.A.,* 76, 2750, 1979.
6. **Parsegian, V.A., Rand, R.P., and Rau, D.C.,** Hydration forces: what next?, *Chem. Scripta,* 25, 28, 1985.
7. **Parsegian, V.A., Rand, R.P., and Rau, D.C.,** Lessons from the direct measurement of forces between biomolecules, in *Physics of Complex and Supermolecular Fluids,* Sufran, S.A. and Clark, N.A., Eds., John Wiley & Sons, New York, 1985, 115.
8. **Israelachvili, J.N.,** *Intermolecular and Surface Forces,* Academic Press, New York, 1991.

9. **Claesson, P.M.,** Forces between Surfaces Immersed in Aqueous Solutions, Ph.D. thesis, Royal Institute of Technology, Stockholm, Sweden, 1986.

10. **Fowkes, F.,** Acid-base interactions in polymer adhesions, in *Physicochemical Aspects of Polymer Surfaces,* Vol. 2, Mittal, K.L., Ed., Plenum Press, New York, 1983, 583.

11. **Chaudhury, M.K.,** Short-Range and Long-Range Forces in Colloidal and Macroscopic Systems, Ph.D. thesis, State University of New York, Buffalo, 1984.

12. **van Oss, C.J., Chaudhury, M.K., and Good, R.J.,** Monopolar surfaces, *Adv. Colloid Interface Sci.,* 28, 35, 1987.

13. **van Oss, C.J., Chaudhury, M.K., and Good, R.J.,** Interfacial Lifshitz-van der Waals and polar interactions in macroscopic systems, *Chem. Rev.,* 88, 927, 1988.

14. **van Oss, C.J.,** Acid/base interfacial interactions in aqueous media, *Colloids Surf.,* 78, 1, 1993.

15. **van Oss, C.J. and Good, R.J.,** Prediction of the solubility of polar polymers by means of interfacial tension combining rules, *Langmuir,* 8, 2877, 1992.

16. **Visser, J.,** On Hamaker constants: a comparison between Hamaker constants and Lifshitz-van der Waals constants, *Adv. Colloid Interface Sci.,* 3, 331, 1972.

17. **Christenson, H.K.,** The long-range attraction between macroscopic hydrophobic surfaces, in *Modern Approaches to Wettability,* Schrader, M.E. and Loeb, G.I., Eds., Plenum Press, New York, 1992, chap. 2.

18. **Overbeek, J.Th.G.,** Electrochemistry of the double layer, in *Colloid Science,* Vol. 1, Kruyt, H.R., Ed., Elsevier, Amsterdam, 1952, chap. 4.

19. **Overbeek, J.Th.G. and Bijsterbosch, B.H.,** The electrical double layer and the theory of electrophoresis, in *Electrophoretic Separation Methods,* Righetti, P.G., van Oss, C.J., and Vanderhoff, J.W., Eds., Elsevier, Amsterdam, 1979, chap. 1.

20. **Hunter, R.J.,** *Zeta Potential in Colloid Science,* Academic Press, London, 1981.

21. **van Oss, C.J., Giese, R.F., and Norris, J.,** Interaction between advancing ice fronts and erythrocytes, *Cell Biophys.,* 19, 253, 1992.

22. **van Oss, C.J., Good, R.J., and Chaudhury, M.K.,** Solubility of proteins, *J. Protein Chem.,* 5, 385, 1986.

23. **van Oss, C.J.,** Surface properties of fibrinogen and fibrin, *J. Protein Chem.,* 9, 487, 1990.

24. **Holmes-Farley, S.R., Reamey, R.H., McCarthy, T.J., Dent, J., and Whitesides, G.M.,** Acid-base behavior of carboxylic groups covalently attached at the surface of polyethylene: the usefulness of contact angle in following the ionization of surface functionality, *Langmuir,* 1, 725, 1985.

25. **van Oss, C.J. and Good, R.J.,** Orientation of the water molecules of hydration of serum albumin, *J. Protein Chem.,* 7, 179, 1988.

26. **Ohki, S.,** A mechanism of divalent ion induced phosphatidylserine membrane fusion, *Biochim. Biophys. Acta,* 689, 1, 1982.

27. **Gallez, D., Prévost, M., and Sanfeld, A.,** Repulsive hydration forces between charged lipidic bilayers, *Colloids Surf.,* 10, 123, 1984.

28. **van Oss, C.J., Chaudhury, M.K., and Good, R.J.,** Polar interfacial interactions, hydration pressure and membrane fusion, in *Molecular Mechanisms of Membrane Fusion,* Ohki, S., Doyle, D., Flanagan, T.D., Hui, S.W., and Mayhew, E., Eds., Plenum Press, New York, 1988, 113.

29. **van Oss, C.J., Wu, W., and Giese, R.F.,** Linkage between ζ-potential and electron donicity of charged polar surfaces, in Abstr. 67th Colloid Surface Sci. Symp., Toronto, 1993, no. 217.

30. **Wu, W., Giese, R.F., and van Oss, C.J.,** Linkage between ζ-potential and electron donicity of charged polar surfaces. I. Implications for the mechanism of flocculation of particle suspensions with plurivalent counterions, *Colloids Surf.,* in press, 1994.

31. **Wu, W., Giese, R.F., and van Oss, C.J.,** Linkage between ζ-potential and electron donicity of charged polar surfaces. II. Repeptization of flocculation caused by plurivalent counterions by means of complexing agents, *Colloids Surf.,* in press, 1994.

32. **Overbeek, J.Th.G.,** Phenomenology of lyophobic systems, in *Colloid Science,* Vol. 1, Kruyt, H.R., Ed., Elsevier, Amsterdam, 1952, chap. 2.

33. **van Oss, C.J., Giese, R.F., and Costanzo, P.M.,** DLVO and non-DLVO interactions in hectorite, *Clays Clay Miner.,* 30, 151, 1990.

34. **Fowkes, F.M., Jinnai, H., Mostafa, M.A., Anderson, F.W., and Moore, R.J.,** Mechanism of electric charging of particles in nonaqueous liquids, *ACS Symp. Ser.,* 200, 307, 1982.

35. **Labib, M.E. and Williams, R.M.,** The effect of moisture on the charge at the interface between solids and organic liquids, *J. Colloid Interface Sci.,* 115, 330, 1987.

36. **Labib, M.E.,** The origin of the surface charge on particles suspended in organic liquids, *Colloids Surf.,* 29, 293, 1988.

37. **Labib, M.E. and Williams, R.M.,** The use of zeta-potential measurements in organic solvents to determine the donor-acceptor properties of solid surfaces, *J. Colloid Interface Sci.,* 97, 356, 1984.

38. **Labib, M.E. and Williams, R.M.,** An experimental comparison between the aqueous pH scale and the electron donicity scale, *Colloid Polym. Sci.,* 264, 533, 1986.

39. **Gutmann, V.,** *The Donor-Acceptor Approach to Molecular Interactions,* Plenum Press, New York, 1978.

40. **van Oss, C.J.,** *Interfacial Forces in Aqueous Media,* Marcel Dekker, New York, 1994.

41. **Derjaguin, B.V. and Landau, L.,** *Acta Physicochim. URSS,* 14, 633, 1941.

42. **Verwey, E.J. and Overbeek, J.Th.G.,** *Theory of the Stability of Lyophobic Colloids,* Elsevier, Amsterdam, 1948.

43. **van Oss, C.J.,** Surface free energy contribution to cell interactions, in *Biophysics of the Cell Surface,* Glaser, R. and Gingell, D., Eds., Springer-Verlag, Berlin, 1990, 131.

44. **van Oss, C.J. and Absolom, D.R.,** Influence of cell configuration and potential energy equilibria in rouleau phenomena, *J. Dispersion Sci. Technol.,* 6, 131, 1985.

45. **van Oss, C.J.,** Immunological and physicochemical nature of antigen-antibody interactions, in *Transfusion Immunobiology and Medicine,* Garratty, G., Ed., Marcel Dekker, New York, chap. 12.

46. **Sewchand, L.S. and Canham, P.B.,** Induced rouleau formation in interspecies populations of red cells, *Can. J. Pharmacol.,* 54, 437, 1976.

47. **van Oss, C.J.,** Aspecific and specific intermolecular interactions in aqueous media, *J. Mol. Recogn.,* 3, 141, 1990.

48. **van Oss, C.J., Gillman, C.F., and Good, R.J.,** The influence of the shape of phagocytes on their adhesiveness, *Immunol. Commun.,* 1, 627, 1972.

49. **van Oss, C.J. and Mohn, J.F.,** Scanning electron microscopy of red cell agglutination, *Vox Sang.,* 19, 432, 1970.

50. **van Oss, C.J.,** Mechanisms of and conditions for repulsive van der Waals and repulsive hydrogen bonding interactions, *J. Dispersion Sci. Technol.,* 11, 491, 1990.

51. **Lerche, D., Hessel, E., and Donath, E.,** Investigation of the La^{3+}-induced aggregation of red blood cells, *Stud. Biophys.,* 78, 95, 1979.

52. **van Oss, C.J.,** Hydrophilic and "hydrophobic" interactions of proteins, in *Protein Interactions,* Visser, H., Ed., VCH Verlag, Weinheim, Germany, 1992, chap. 2.

53. **van Oss, C.J. and Absolom, D.R.,** Hemagglutination and the closest distance of approach of normal, neuraminidase- and papain-treated erythrocytes, *Vox Sang.,* 47, 250, 1984.

54. **Baier, R.E., Meyer, A.E., Natiella, R.R., and Carter, J.M.,** Surface properties determine bioadhesive outcomes, *J. Biomed. Mater. Res.,* 18, 337, 1984.

55. **van Oss, C.J.,** The forces involved in bioadhesion to flat surfaces and particles — their determination and relative roles, *Biofouling,* 4, 25, 1991.

56. **Berkeley, R.C.W., Lynch, J.M., Melling, J., Rutter, P.R., and Vincent, B., Eds.,** *Microbial Adhesion to Surfaces,* Ellis Horwood, Chichester, U.K., 1980.

57. **Doyle, R.J. and Rosenberg, M., Eds.,** *Microbial Cell Surface Hydrophobicity,* American Society for Microbiology, Washington, D.C., 1990.

58. **van Oss, C.J., Giese, R.F., Wentzek, R., Norris, J., and Chuvilin, E.M.,** Surface tension parameters of ice obtained from contact angle data and from positive and negative particle adhesion to advancing freezing fronts, *J. Adhesion Sci. Technol.,* 6, 503, 1992.

59. **van Oss, C.J., Arnold, K., and Coakley, W.T.,** Depletion flocculation and depletion stabilization of erythrocytes, *Cell Biophys.,* 17, 1, 1990.

60. **van Oss, C.J., Mohn, J.F., and Cunningham, R.K.,** Influence of various physico-chemical factors on hemagglutination, *Vox Sang.,* 34, 351, 1978.

61. **Bäumler, H. and Donath, E.,** Does dextran significantly increase the surface potential of human red blood cells?, *Stud. Biophys.,* 120, 113, 1987.

62. **Arnold, K., Zchoernig, O., Barthel, D., and Herold, W.,** Exclusion of poly(ethylene glycol) from liposome surfaces, *Biochim. Biophys. Acta,* 1022, 303, 1990.

63. **van Oss, C.J.,** Hydration forces, in *Water and Biological Macromolecules,* Westhof, E., Ed., Macmillan, Basingstoke, U.K., 1993, chap. 13.

64. **van Oss, C.J., Fike, R.M., Good, R.J., and Reinig, J.M.,** Cell microelectrophoresis simplified by the reduction and uniformization of the electroosmotic back-flow, *Anal. Biochem.,* 60, 212, 1974.

65. **Brooks, D.E. and Seaman, G.V.F.,** The effect of neutral polymers on the electrokinetic potential of cells and other charged particles. I. Models for the zeta potential increase; II. A model for the effect of adsorbed polymer on the diffuse double layer; III. Experimental studies on the dextran/erythrocyte system; IV. Electrostatic effects in dextran-mediated cellular interactions, *J. Colloid Interface Sci.,* 43, 670–714, 1973.

66. **Ohki, S.,** Surface tension, hydration energy and membrane fusion, in *Molecular Mechanisms of Membrane Fusion,* Ohki, S., Doyle, D., Flanagan, T.D., Hui, S.W., and Mayhew, E., Eds., Plenum Press, New York, 1988, 123.

67. **van Oss, C.J. and Good, R.J.,** The equilibrium distance between two bodies immersed in a liquid, *Colloids Surf.,* 8, 373, 1984.

68. **Hirszfeld, L. and Dubiski, S.,** Untersuchungen über die Struktur der inkompletten Antikörper, *Schweiz. Z. Allg. Pathol. Bakteriol.,* 17, 73, 1954.

69. **van Oss, C.J.,** Stability of human red cell suspensions at $300,000 \times G$, *J. Dispersion Sci. Technol.,* 6, 139, 1985.

Chapter 12

Electrical Surface Phenomena of Endothelial Cells

Fernando F. Vargas

CONTENTS

I. INTRODUCTION

Vascular endothelial cells (ECs) form a syncytium, the endothelium, that covers the inner surface of blood vessels, and they also synthesize and secrete many factors that are involved in blood homeostasis, vasomotor regulation, inflammation, immunological reactions, angiogenesis, etc.[1-3] Besides these and other functions, vascular endothelium acts as a selective barrier between blood and tissues, restricting the passage of macromolecules and cells while allowing a high rate of fluid and small solute exchange.[4,5]

Most endothelium functions are directly or indirectly influenced by electrical charges on the cell surface. Among these functions are transport of large molecules through vesicles, transendothelial channels, intercellular junctions, etc. Charges located in these pathways impart selectivity to the passage of charged species.[5,6] Adhesion between ECs and blood cells, subcellular matrix, and artificial surfaces is influenced by electrical repulsion forces that arise between surfaces of similar polarity.[7] Adhesion of blood cells to the vascular luminal surface is of paramount importance in many pathological processes. For instance, the adhesion of leukocytes to the EC surface is part of the inflammatory response[8] and precedes their crossing the vascular barrier. Surface charge also affects enzymatic reactions[9] and ionic channel conductance and gating.[10] An overview of the electrical surface properties of endothelial cells and their measurement with electrophoresis will be presented in this chapter.

II. EVALUATION OF ENDOTHELIAL CELL SURFACE CHARGE

A. THEORETICAL

Cells have a surface layer — the glycocalyx — formed by charged polysaccharides attached to proteins. A preponderance of ionogenic groups of one polarity (usually negative) give rise to a net charge on the cell surface, which is divided by the surface area to obtain the charge density, σ. When the chemical group bearing the charge has been identified, σ can be determined as the product of the charge concentration and the glycocalyx thickness. However, when several chemical species contribute to the surface charge, measurement of charge concentration alone cannot be directly correlated with electrical surface properties of cells, since the effect of different ionogenic groups on these properties will depend on their pKs and their location in the glycocalyx. An alternative approach is to measure the overall effect of the charged groups by using cell electrophoresis. In this method the velocity imparted to saline-suspended single cells by application of an electrical field is measured. Cells move toward the electrode with a polarity opposite to that of its surface charge and with a velocity that depends on the cell σ, the electrical field intensity, hydrodynamic friction, and other factors to be discussed later in this chapter. Cell velocity (in micrometers per second) per unit driving force (in volts per centimeter) is electrophoretic mobility (EPM). Whereas EPM has been studied extensively in free blood cells, like erythrocytes,[11,12] lymphocytes,[13] and monocytes,[14] there is a paucity of EPM data on ECs.

EPM measurements can be used to calculate electrical surface parameters like σ through the use of an appropriate model like the one developed by Levine et al.,[15] which is based on RBC electrophoretic data. This model treats the cell as a sphere covered by a layer of charged polymers, moving under the driving force of an electrical field. Charge distribution is assumed to be homogeneous throughout the glycocalyx. Surface potential, charge density, and distance from the cell surface are related through the linearized form of the Poisson-Boltzmann equation. Glycocalyx polymers are assumed to be equivalent to chains of spheres, such that hydrodynamic resistance to fluid flowing through the glycocalyx is calculated from the Stokes expression for a sphere moving through a newtonian fluid. The force acting on the cell is the electrical field times the cell surface charge. At steady state this force must be exactly balanced by the hydrodynamic frictional resistance to cell movement, as expressed mathematically by the Navier-Stokes equation. A solution of this equation was derived by Levine et al.[15] and used to calculate EPM for RBCs. σ was obtained as the product of the concentration of sialic acid, which is the main contributor to the surface charge in this cell, and the glycocalyx thickness, β. Theoretical curves of EPM vs. ionic concentration fit the experimental EPM data of RBCs fairly well, showing the model to be a good approximation.

Vargas et al.[16] applied the model of Levine et al. to the investigation of electrical surface parameters of ECs and calculated σ and surface potential from EPM measurements. In the study described in this chapter the same model was applied in the opposite direction — that is, to calculate EPM from σ. The latter was obtained from the charge density of sulfated polysaccharides, which are the main charge carriers in ECs.[17] A simplified form of Levine et al.'s solution of the Navier-Stokes equation was used to relate EPM with σ, and with hydrodynamic frictional parameters as follows:

$$EPM = \left\{ \sigma / 2\eta\kappa^2\beta \right\} \left\{ -1 - (\kappa/\gamma) - 2(\kappa^2/\gamma^2) + (\kappa^2/[\kappa^2 - \gamma^2])([\kappa/\gamma] - 1) \right\} \quad (12\text{-}1)$$

where η is the fluid viscosity, κ is the reciprocal of the Debye length ($\kappa = 1.08 \times 10^{15} \times$ C), β is the glycocalyx thickness, and γ is a frictional term defined by $\gamma^2 = 6\pi r N$, with r the radius and N the volume density of polymer segments in the glycocalyx. Surface charge density, σ, in electrostatic units of charge (ESU) per unit area (square centimeter),

Table 12-1 **Electrophoretic mobility and surface charge density of attached and free cells**

Cell	Sialic Acid Molecules/cm² ($\times 10^{-13}$)	Mobility, (mm/s)/(V/cm)		Charge, ESU/cm² ($\times 10^{-4}$)
		Control	NANADase	
RBC	2.22[32]	1.03[16]	0.00[16]	0.91[16,a]
Lymphocyte[13]	4.34	0.89	0.49	0.81
EC[31]	240–510	—	—	—
EC[12]	130	0.74	0.79	2.62[a]
Neuron[29]	—	1.40	1.30	—

Note: NANADase = neuraminidase; EC = endothelial cell. Superscript numbers are references.

* Calculated using the model of Levine et al.[15]

is the product of the charge concentration in the glycocalyx (ρ) and β. If ρ, instead of σ, is used in Equation 12-1, β cancels out and EPM becomes independent of glycocalyx thickness.

B. EXPERIMENTAL MEASUREMENTS AND CALCULATIONS

Cell electrophoresis is done by placing a cell suspension inside a capillary. An electric field is applied between both ends of the capillary, and cell movement under the electrical driving force is observed with a microscope. To obtain single cells that can freely move, Vargas et al.[16] separated cultured bovine pulmonary artery ECs with a rubber policeman and measured EPM after 1 h of incubation in saline at pH 7.4 and 37°C. EPM measured immediately after the cells were transferred from culture medium to saline was 0.94 ± 0.07 (μm/s)/(V/cm) (N = 27). On the other hand, frequent measurements of EPM during saline incubation showed a gradual EPM decrease, which reached a plateau at 0.74 ± 0.07 (μm/s)/(V/cm) (N = 29) in about 1 h. This is probably associated with the loss of heparan sulfate that has been observed in ECs exposed to protein-free solutions.[18] For comparison purposes, 0.74 ± 0.07 (μm/s)/(V/cm) was considered as the standard EPM of bovine pulmonary artery ECs in saline. However, the initial EPM must be closer to the true EPM of intact ECs. The only other reported EPM value for ECs is that of rat aortic ECs, 1.24 ± 0.07 (μm/s)/(V/cm).[19]

EPM of ECs was lower while σ (calculated from Equation 12-1) was higher than control RBC values (Table 12-1). This is expected since the EC glycocalyx is thicker than that of RBCs (300 Å thick[20,21] vs. 50 to 75 Å thick[15]) and, consequently, exerts a stronger hydrodynamic drag than that of the latter.

It is of great interest to compare experimental EPM of EC with EPM calculated from the model of Levine et al. For this purpose σ was taken as the product of charge concentration of sulfated GAGS, 26.5 meq/L,[22] times Avogadro's number (6.02×10^{23} M^{-1}) and electronic charge (4.8×10^{-10} ESU) to obtain $\rho = 7.66 \times 10^9$ ESU/cm³. For the glycocalyx thickness of ECs in Ringer's, 280 Å,[20] σ was 2.14×10^4 ESU/cm². This chemically determined σ is very close to the value obtained from Equation 12-1 and experimental EPM, which was 2.62×10^4 ESU/cm²,[16] showing that the model correctly predicted this surface parameter. η and κ^2 were calculated for the ionic concentrations of the solutions used, while N was taken from RBC data[15] since the author is not aware of equivalent data for ECs.

Theoretical curves for EPM as a function of NaCl concentration were derived from Equation 12-1 using a computer program. For this calculation η, κ^2, and γ were functions of the ionic concentration, while β was held constant although it might change with ionic strength. Results from this computation showed that EPM of ECs decreased with NaCl

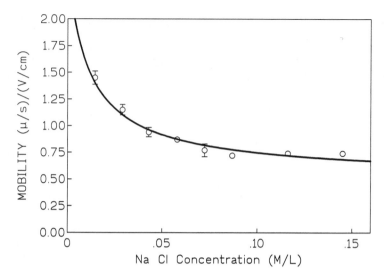

Figure 12-1 Endothelial cell mobility decrease caused by increasing NaCl concentration. The thick line represents the theoretical electrophoretic mobility value calculated from Equation 12-1 using a computer program.

concentration, as expected from ionic shielding and binding to the glycocalyx charge, and reached a plateau at approximately physiological electrolyte concentration (Figure 12-1). Experimental EPM were measured at different concentrations of NaCl while keeping osmolarity constant by substituting sorbitol for the electrolyte. The computer-derived theoretical curves shown in Figure 12-1 closely fit the experimental data, indicating that the model is a good approximation of the real physical system.

In spite of the excellent results obtained applying the model of Levine et al. to ECs, it is advisable to keep in mind that some of the assumptions the model is based on are just approximations of the real physical conditions of the system. For instance, charge distribution on the surface of ECs is not uniform, since differentiated microdomains of charge are revealed by cationic ferritin marking of the luminal vascular surface.[23,24] In addition, opposite sides of ECs in intact endothelium and confluent monolayers differ in protein and glycoprotein composition.[25,26] This latter asymmetry is assumed to diminish in the 2-h lapse between cell dispersion and experimental measurement, since membrane protein redistribution after EC separation takes less than 1 h.[25] Asymmetry of heparan sulfate distribution on the two sides of ECs also exists, but it may disappear with cell separation since it depends on EC attachment to the subendothelial matrix.[27,28]

III. GLYCOCALYX COMPONENTS CARRYING THE SURFACE CHARGE ON ENDOTHELIAL CELLS

Ionogenic groups responsible for the surface charge of blood cells have been identified by measuring EPM changes after enzymatic treatment. Neuraminidase, which lyses sialic acid (SA), induces a marked decrease in EPM in RBCs[11,12] and lymphocytes,[13] indicating that most of the surface charge resides in this acid. In contrast, EPM of ECs[16] and neurons[29] is not diminished by neuraminidase (Table 12-1), whose ability to remove SA was chemically verified. This is apparently in contradiction with data showing that ECs contain much higher amounts of SA than do RBCs.[30-32] Measurement of SA released by neuraminidase, using spectrophotometric methods, has shown that bovine pulmonary ECs used in the present calculations release at least 50 times more SA per cell than human

Table 12-2 **Electrophoretic mobility changes caused by enzymes**

| Enzyme | Mobility, $(\mu m/s)/(V/cm)$[a] | | p |
	Control	Treated	
Heparinase and heparitinase	0.84 ± 0.076 (94)	0.64 ± 0.14 (102)	<0.05
Chondroitinase on luminal side	0.85 ± 0.14 (31)	0.71 ± 0.07 (42)	<0.05
Chondroitinase on both sides	0.79 ± 0.13 (61)	0.00 (72)	<0.05
Collagenase	0.81 ± 0.04 (35)	0.63 ± 0.08 (31)	<0.05
Papain	0.65 ± 0.26 (10)	0.00 (10)	<0.05

[a] Mean $\pm$ SD; number in parenthesis denote number of experiments.

erythrocytes (Table 12-1). Since the pK of sialic acid is 2.6, its carboxylic groups must be dissociated at the physiological pH. Under these circumstances the only possible explanation for the lack of effect of SA removal on the surface charge of ECs, is that SA is preferentially located in glycoproteins near the plasmalemma, since charges that are too deep in the glycocalyx exert very little or no effect on surface electrical phenomena.[33] Moreover, the thick and heavily charged layer of sulfated GAGS formed by heparan and chondroitin sulfates, would cover sialic acid and dominate charge effects. Extensive work on the effect of enzymes on the distribution and density of cationized ferritin binding sites on the luminal surface of the microvasculature showed that neuraminidase caused only a moderate decrease of binding sites in the plasmalemma proper,[24] suggesting that most of the charge resides in ionogenic groups other than SA. These groups are sulfated GAGS, as shown by electrophoretic measurements combined with enzymatic removal of EC surface components. In these measurements, lysis with heparinase, heparitinase, and chondroitin ABC lyase produced a marked drop in EPM.[17]

The chemical difference between charged species in blood cells and ECs is probably a reflection of the complexity and thickness of the glycocalyx of matrix-attached cells, which besides SA contain additional carboxylic groups as well as sulfated groups residing on polysaccharide polymers. Another important difference between free and attached cells is that surface charge distribution is uniform on the former, while the latter show membrane[25,26] and glycocalyx[27,28] asymmetry. GAGS asymmetry was investigated by Vargas et al.,[17] who measured EPM changes caused by enzymatic lysis of EC GAGS. Heparinase and heparitinase caused a small EPM reduction when applied over the whole cell surface (Table 12-2). This small effect was expected since heparan sulfate was partially lost during incubation with saline. In some experiments, chondroitin sulfate was enzymatically removed from the cell side facing the culture medium (luminal side), and ensuing EPM changes were compared with those from cells whose total surface was exposed to enzymes. Removal of chondroitin sulfate from only the luminal side was accomplished by adding chondroitin ABC lyase on top of intact EC monolayers. Enzymatic action on the abluminal side was assumed to be negligible, since passage of proteins across endothelium is a slow process. Therefore, the abluminal side should have been only briefly exposed to a low enzyme concentration. After samples of cells treated on one side had been taken and EPM measured, the remaining cells were mechanically dispersed and suspended in enzyme-containing saline. EPM was measured after 1 h of incubation in saline. Results from the above-mentioned differential treatment (Table 12-2) showed that chondroitin sulfate lyase applied to the whole cell surface completely abolished EPM, while only a moderate change was observed when it was applied to the luminal side, suggesting that chondroitin sulfate is preferentially located on the abluminal side.

IV. EFFECT OF MONO- AND DIVALENT CATIONS ON ELECTROPHORETIC MOBILITY

Ionic binding and shielding of surface charges cause a decrease in endothelial cell EPM, as seen when the NaCl concentration is increased (Figure 12-1). Addition of $CaCl_2$ to the 0.145 M NaCl control solution further reduces EPM from 0.74 (μm/s)/(V/cm) down to zero at 60 mM Ca^{2+}. This complete suppression of EPM is probably caused by Ca^{2+} binding to GAGS.

Since Ca^{2+} binding neutralizes the surface charge, the observed EPM decrease can be used to estimate binding parameters. Seaman et al.[34] applied Stern's treatment for molecular adsorption to a surface in order to estimate Ca^{2+} binding site density and change in adsorption free energy from EPM changes with Ca^{2+}. At equilibrium

$$\frac{nCa^{++}}{N} = \frac{K[Ca^{2+}]}{1+K[Ca^{2+}]} \tag{12-2}$$

where n_{Ca2+} is the number of occupied binding sites, n is the total number of sites, K is the equilibrium constant for ion adsorption, and $[Ca^{2+}]$ is calcium concentration in moles per liter.

The change in net surface charge caused by cation binding, $\Delta\sigma$, is equal to twice the electron charge (2e) times the number of sites occupied by Ca^{2+}. n_{Ca2+} can be replaced by its value in Equation 12-2 to obtain $\Delta\sigma$ as a function of $[Ca^{2+}]$.

$$\Delta\sigma = 2e\,nCa^{++} = \frac{2e\,N\,K[Ca^{2+}]}{1+K[Ca^{2+}]} \tag{12-3}$$

A linear form is obtained by inversion of Equation 12-3:

$$\frac{1}{\Delta\sigma} = \frac{1}{2e\,N} + \frac{1}{2e\,N\,K[Ca^{2+-}]} \tag{12-4}$$

Plotting $1/\Delta\sigma$ vs. $1/[Ca^{2+}]$ yields a straight line with intercept $1/2en$ and slope $1/2enK$ (Figure 12-2). Seaman et al. estimated n as 4.97×10^{12} sites per square centimeter for RBCs and 3.83×10^{12} sites per square centimeter for lymphocytes.[34] The free energy change, ΔF, for Ca^{2+} adsorption on RBCs was calculated from the equilibrium constant K as -3.82 kcal/mol for control RBCs and -4.47 kcal/mol for neuraminidase-treated RBCs. The higher ΔF of the latter as compared to the untreated cell was explained as being indicative of a second binding site with a stronger affinity for Ca^{2+} than sialic acid.[34] Vargas et al.[35] plotted $1/\Delta\sigma$ from ECs exposed to different concentrations of Ca^{2+} in the bathing medium. n for ECs was 4.6×10^{13} sites per square centimeter, which is about ten times higher than that of RBC's. This is consistent with σ for the former being higher than that of the latter (Table 12-1). ΔF for Ca^{2+} adsorption on ECs was -4.25 kcal/mol, which is higher than that for control RBCs but close to that of neuraminidase-treated RBCs in the experiments of Seaman et al.[34] The postulated second Ca^{2+} binding site of the RBC resembles that of the EC not only in the magnitude of ΔF, but also in the absence of charge reversal concentration with Ca^{2+}. The latter was observed in untreated RBCs and lymphocytes whose EPM changed sign at high Ca^{2+} concentration,[34] but not in ECs.[16] This suggests that each Ca^{2+} ion binds to two charges on ECs, since binding to only one charge would leave one positive charge free for each bound Ca^{2+}. Hence, Ca^{2+} concentration over

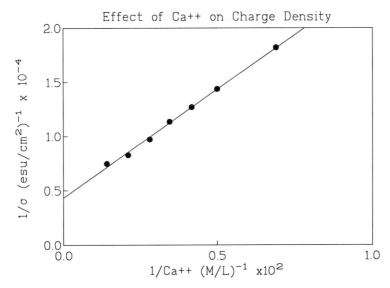

Figure 12-2 $1/\Delta\sigma$ vs. the reciprocal of Ca^{2+} concentration. σ at each Ca^{2+} concentration was calculated from Equation 12-1 and its difference with σ at zero Ca^{2+} ($\Delta\sigma$) was inverted and plotted. Mobilities measured in 30 to 50 cells were used for the calculation of σ.

that needed to neutralize the surface charges would make the cell surface positive and invert mobility direction. Absence of polarity inversion in ECs is consistent with its high charge density allowing Ca^{2+} binding to two neighbor charges. This is in agreement with equilibrium dialysis measurements of Ca^{2+} binding on GAGS, which show a binding stoichiometry consistent with a binding site for Ca^{2+} consisting of two anionic groups.[36]

V. EFFECT OF ASCORBIC ACID ON ELECTROPHORETIC MOBILITY

Ascorbic acid (AA) has many effects on ECs, among which collagen synthesis,[37] vasodilatation,[38] and scavenging of oxygen radicals[39-42] are the best known. It is therefore of interest to investigate the effect of AA on EC surface properties. EPM of ECs incubated in AA solutions increased from the reference level, 0.8 (μm/s)/(V/cm), to 0.97 (μm/s)/(V/cm) in saline.[43] This may be due to AA binding onto the endothelial surface and increasing its negativity. Binding of dextran sulfate to arterial ECs increases EPM from 1.24 to 2.32 (μm/s)/(V/cm).[19] However, the need for about 1 h of incubation time with AA suggests that the surface charge increase is caused by AA promoting secretion of negatively charged collagen. Collagen contributes to the surface negativity as shown by the decrease in EPM caused by collagen lysis (Table 12-2). Moreover, AA stimulates collagen secretion in cultured cells with a half-time of 1 to 1.5 h.[37,44] Although AA binds calcium,[45] the relatively low concentration used would exclude any effect of this nature.

EPM reached a maximum at about 60 min of incubation time and 20 μM AA concentration (Figure 12-3). Since the latter concentration is in the range normally found in human plasma (20.3 to 99.6 μM),[46] surface charge of ECs would be directly influenced by the plasmatic level of vitamin C. Besides increasing EC EPM, ascorbic acid counteracts the effect of low Ca^{2+} concentration on EPM, but it does not prevent EPM from reaching zero at about 70 mM Ca^{2+} (Figure 12-4).

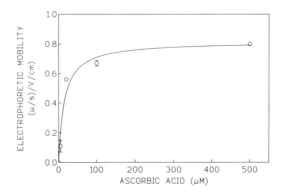

Figure 12-3 Electrophoretic mobility increase caused by 8-h preincubation of endo-thelial cells (ECs) with different concentrations of ascorbic acid. Measurements were carried out in saline containing 0.36 mM CaCl$_2$. Mean ± SEM.

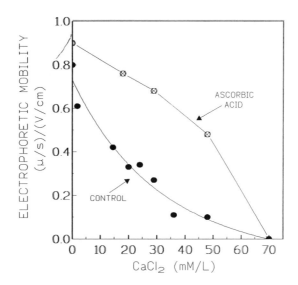

Figure 12-4 Electrophoretic mobility vs. Ca^{2+} concentration in control cells (lower curve) and cells preincubated with 500 μM ascorbic acid. Notice that values at zero Ca^{2+} are already higher than control, but both drop to zero at about the same Ca^{2+} concentration.

VI. EFFECT OF SURFACE CHARGE ON ENDOTHELIAL CELL FUNCTIONS

Quantification of EC electrical parameters opens the way to an evaluation of transendothelial transport selectivity to charged molecules, EC adhesion to cells and surfaces, and other phenomena of physiological interest.

A. TRANSPORT SELECTIVITY FOR CHARGED SOLUTES

Transport of charged solutes across the vessel wall is subject to the influence of charges located at different points in the pathway. Restriction to the passage of anions and facilitation for cations is expected from negatively charged pathway structures. Organ or tissue studies suffer from the problem that there are several charged structures in the path of the solutes being measured. Adamson et al.[47] avoided this problem by using single capillaries. They showed that polycharged cationic lactalbumin crossed the capillary wall twice as fast as anionic ribonuclease, in spite of their similar size (radius = 20 Å). O'Donnell and Vargas[48] used isolated blood vessels to measure the passage of small ions *in vitro.* They found that transport numbers for the transendothelial movement relative to those in free solution were the same for NaCl, but smaller when Na_2SO_4 was the test solute. This suggests selectivity for a cation as small as sulfate. The effect of charge on transvascular transport in different organs and territories has been reviewed recently by Curry.[6]

The different levels of the transendothelial pathway possibly involved in selectivity are: (1) the glycocalyx surface, where the charged solute concentration may be quite different from that in the solution; (2) the inside of the glycocalyx, which acts as an ionic exchange resin, with partition and diffusion coefficients different from those in the external solution; and (3) the intercellular junction channels, where charges on the channel walls interact with the moving solutes.

1. Surface Concentration of Charged Solutes

The surface potential modifies ionic concentrations at the glycocalyx surface by attracting ions of opposite polarity (counterions) and repelling those of the same sign (coions). Higher cation concentration and lower pH on the cell surface are therefore associated with negative surface potentials. The surface potential is defined as the electrical potential difference between the EC glycocalyx surface and the bulk surrounding fluid — i.e., at the edge of the glycocalyx — $\psi(0)$. Surface potential can be calculated from the following equation derived from the model of Levine et al.:

$$\Psi(0) = \left\{4\pi\sigma / 2 E\kappa(\kappa\beta)\right\}\left\{1 - e^{-2\kappa\beta}\right\} \tag{12-5}$$

where E is the dielectric constant.

The surface concentration of charged solutes depends on the magnitude of $\psi(0)$ through the Boltzmann equation, as follows:

$$c / c^0 = \exp(-zF\Psi / RT) \tag{12-6}$$

where c and c^o are the ionic concentrations at the surface and in the bulk solution, respectively, z is the valence, F the Faraday number, Ψ the electrical potential difference between the luminal surface and the bulk solution, R the gas constant, and T the absolute temperature.

The surface potential $\Psi(0)$ calculated from Equation 12-5 is –1.5 mV for ECs[16] and –3 mV for RBCs.[15] Hence, the concentration of monovalent cations on the EC surface, as calculated from Equation 12-6, is 5% higher than that of the bulk solution. This is a very minor contribution to selectivity.

2. Role of Glycocalyx in Selectivity

Since the net charge in the glycocalyx is negative, a partition coefficient higher than 1 for cations should be expected, while the inverse should hold for anions. Diffusion coefficient for cations, on the other hand, would be lower than in the bulk solution. These parameters depend on the fixed charged density and on the mobile solute charge.

Zamparo and Comper[22] investigated selectivity of membranes made of sulfated polysaccharides, which should be a good simile of the EC glycocalyx. As described earlier in this chapter, the charge density of EC glycocalyx ($\rho = 7.66 \times 10^9$ ESU/cm^3), estimated from data obtained by Zamparo and Comper for sulfated polysaccharides,[22] was very close to the value obtained from EPM data. It is therefore reasonable to expect that their analysis of charge selectivity in polysaccharide membranes would also apply to the EC glycocalyx. These membranes contained heparan and chondroitin sulfates at different concentrations, with a maximum at 40 meq/l. The latter is about twice the polysaccharide concentration that they measured in the kidney glomerulus. However, even at the highest concentration used, selectivity for polycharged solutes, like dextrans, was negligible. Given these results it may be concluded that in spite of its net charge the EC glycocalyx is not a selectivity barrier for charged molecules moving across the vascular wall.

3. Selectivity at the Intercellular Channel

Since neither the surface potential nor the glycocalyx ionic exchange properties can explain the selectivity for charged solutes passing across the endothelium, charge screening must reside in the intercellular junction channels. These junctions form narrow passages or channels about 100 Å wide; moving solutes must come close to the fixed charges in order to pass through these channels. This situation has been analyzed by Curry[6] for solutes with a radius of 10 to 70 Å and with approximately 20 electron charges that enter a cylindrical pore with 100 Å radius and with a wall charge similar in sign and magnitude to that of the solute. For a solute with a 20 Å radius and a Debye length of 10 Å, the calculated rise in solute exclusion would reduce the partition coefficient between the pore and the bulk solution from 0.64 to 0.44. These calculations are in good agreement with experimental results obtained by Garby and Areekul reported in the same paper.[6] Data from Michel and Turner on charged vs. uncharged myoglobin transport (cited by Curry[6]) also showed selectivity, but their magnitude was lower than that calculated.

B. ENDOTHELIAL CELL ADHESION TO CELLS AND SURFACES

Adhesion of leukocytes to the luminal surface of microvascular vessels is frequently followed by cell passage across the vessel wall. Several intercellular forces play a role in cell adhesion.[7,49] These include nonspecific electrostatic forces, represented by van der Waals attraction forces between hydrophobic surfaces, and repulsion forces between negatively charged cell surfaces. A nonspecific mechanical force, the steric repulsive force, arises when the distance between two interacting cells becomes less than the sum of their glycocalyx thickness. This causes bending and compression of the surface polymers and, consequently, a steric force similar to an elastic recoil.[7]

Since the magnitude of electrostatic and steric forces depends on the intercellular distance of approximation, the glycocalyx thickness plays an important role in adhesion. For blood cells adhering to ECs glycocalyx thickness of 100 and 300 Å, respectively, may be assumed, giving a combined thickness of 400 Å. However, cells can come closer under an external force like a centrifugal force or an attraction force generated by specific ligand-to-ligand binding. Bongrand and Bell[7] calculated the interaction between two similar cells having ρ twice that of ECs and a glycocalyx thickness of 100 Å each and found that electrostatic forces were negligible at distances greater than 300 Å. At cell-to-cell distances under 300 Å the sum of electrostatic and steric repulsion forces is stronger than the van der Waals attraction forces.[7] Hence, if these were the only acting forces no adhesion would probably occur. The extra attraction force required to overcome repulsion forces is provided by specific ligand/receptor binding. The magnitude of the free energy of binding has been estimated as 6 to 12×10^{-13} ergs/bond.[7] Assuming a bond density of 10^{13} bonds per square centimeter (one bond per 1000 Å^2), the interaction energy for binding is 10 ergs/cm^2, which is higher than the total repulsive energy calculated by Bongrand and Bell for distances over 200 Å.[7] Calculated energy for specific binding is time dependent, since after initial adhesion unbound sites diffuse toward the adhesion area, thus increasing site density and adhesion force. This makes the contact duration an important variable.

Microvilli are cellular projections present in some cells. While the cells are still apart they can establish connections with other cells at distances longer than the combined sum of their glycocalyx thickness.[7,49] This increases the probability of adhesion, and it is particularly relevant for ECs, which have surface projections 5×10^3 Å in diameter and 5 to 30×10^3 Å in length.[50] These extremely long microvilli should markedly increase EC adhesion to any surface, since at these distances the electrostatic and steric forces are essentially zero.

Cell-cell adhesion can be modified by acting on the repulsion forces. One example of this is the reduced adhesion caused by increasing steric repulsion, since it occurs when either charged or uncharged polymers are bound to the cell surface.[7] An increase in calcium concentration also reduces adhesion.[7] It is of physiological interest that adhesion and diapedesis of leukocytes takes place at the postcapillary venules, where repulsion forces are low because anionic sites density is the lowest among microvascular segments.[23]

The EPM increase caused by AA at physiological Ca^{2+} levels suggests that adhesion of ECs to other cells or surfaces may be decreased by the stronger repulsion forces associated with the higher surface charge caused by AA. Consistent with this hypothesis is the finding that AA significantly reduces adhesion of neutrophils to ECs.[40]

Surface charge also influences adhesion of ECs to artificial materials. This is important for the preparation of vascular grafts and for EC culture on different types of plastic materials.[50,51]

REFERENCES

1. **Jaffe, E. A.,** Physiologic functions of normal endothelial cells, in *Vascular Medicine. A Textbook of Vascular Biology and Diseases,* Loscalzo, J., Creager, M. A., and Dzav, V. J., Eds., Little, Brown, Boston, 1992, 19.

2. **Cotran, R. S.,** New roles for the endothelium in inflammation and immunity, *Am. J. Pathol.,* 129, 407, 1987.

3. **Fajardo, L. F.,** The complexity of endothelial cells. A review, *Am. J. Clin. Pathol.,* 92, 241, 1989.

4. **Parker, J. C., Perry, M. A., and Taylor, A. E.,** Permeability of the microvascular barrier, in *Edema,* Staub, N. C. and Taylor, A. E., Eds., Raven Press, New York, 1984, chap. 7.

5. **Simionescu, M. and Simionescu, N.,** Endothelial transport of macromolecules: transcytosis and endocytosis. A look from cell biology, *Cell Biol. Rev.,* 25, 1, 1991.

6. **Curry, F. E.,** Mechanics and thermodynamics of transcapillary exchange, in *Handbook of Physiology,* Vol. 4, Cardiovascular System, Renkin, E. M. and Michel, C. C., Eds., American Physiological Society, Bethesda, MD, 1984, chap. 8.

7. **Bongrand, P. and Bell, G. I.,** Cell-cell adhesion: parameters and possible mechanisms, in *Cell Surface Dynamics: Concepts and Models,* Perelson, A. S., DeLisi, C., and Wiegel, F., Eds., Marcel Dekker, New York, 1984, chap. 14.

8. **Gimbrone, M. A., Brock, A. F., and Shafer, A. I.,** Leukotriene B_4 stimulates polymorphonuclear leukocyte adhesion to cultured vascular endothelial cells, *J. Clin. Invest.,* 74, 1552, 1984.

9. **Wojtczak, L. and Nalecz, M. J.,** Surface charge of biological membranes as a possible regulator of membrane-bound enzymes, *Eur. J. Biochem.,* 94, 99, 1979.

10. **Latorre, R., Labarca, P., and Naranjo, D.,** Surface charge effects on ion conduction in ion channels, *Methods Enzymol.,* 207, 471, 1992.

11. **Cook, G. M. W., Heard, D. H., and Seaman, G. V. F.,** Sialic acid and electrokinetic charge of the human erythrocyte, *Nature,* 191, 44, 1961.

12. **Seaman, G. V. F., Knox, R. J., Nordt, J., and Regan, D. H.,** Red cell aging. I. Surface charge density and sialic acid content of density fractionated human erythrocytes, *Blood,* 50, 1001, 1977.

13. **Shimizu, M. and Iwaguchi, T.,** Effect of sialic acid on the electrophoretic mobility of mouse splenic lymphocytes, *Electrophoresis,* 8, 556, 1987.

14. **Bauer, J. and Hannig, K.,** Relationship between electrophoretic mobility and activation of human monocytes, *Electrophoresis,* 6, 301, 1985.

15. **Levine, S., Levine, M., Sharp, A. K., and Brooks, D. E.,** Theory of the electrokinetic behavior of human erythrocytes, *Biophys. J.,* 42, 127, 1983.

16. **Vargas, F. F., Osorio, H. M., Ryan, U. S., and De Jesus, M.,** Surface charge of endothelial cells estimated from electrophoretic mobility, *Membr. Biochem.,* 8, 221, 1989.

17. **Vargas, F. F., Osorio, H. M., Basilio, C., De Jesus, M., and Ryan, U. S.,** Enzymatic lysis of sulfated glycosaminoglycans reduces the electrophoretic mobility of vascular endothelial cells, *Membr. Biochem.,* 9, 83, 1990.

18. **Shimada, K. and Ozawa, T.,** Release of heparan sulphate proteoglycans from cultured aortic endothelial cells by thrombin, *Thromb. Res.,* 39, 387, 1985.

19. **Suemasu, K., Watanabe, K., and Ichikawa, S.,** Contribution of polyanionic character of dextran sulfate to inhibition of cancer metastasis, *Gann,* 62, 331, 1971.

20. **Adamson, R. H. and Clough, G.,** Plasma proteins modify the endothelial cell glycocalyx of frog mesenteric microvessels, *J. Physiol. (London),* 445, 473, 1992.

21. **Simionescu, N.,** Cellular aspects of transcapillary exchange, *Physiol. Rev.,* 63, 1536, 1983.

22. **Zamparo, O. and Comper, W. D.,** Model anionic polysaccharide matrices exhibit lower charge selectivity than is normally associated with kidney ultrafiltration, *Biophys. Chem.,* 38, 167, 1990.

23. **Simionescu, M., Simionescu, N., Santoro, F., and Palade, G. E.,** Differentiated microdomains of the luminal plasmalemma of murine muscle capillaries: segmental variations in young and old animals, *J. Cell Biol.,* 100, 1396, 1985.

24. **Simionescu, M. Simionescu, N., Silbert, J. E., and Palade, G. E.,** Differentiated domains on the luminal surface of the capillary endothelium. II. Partial characterization of their anionic sites, *J. Cell Biol.,* 90, 614, 1981.

25. **Beer, D. M., Baldwin, L. A., Piszczkiewicz, S., and Jacobson, B. S.,** Asymmetry in endothelial cells: post synthesis targeting of proteins to apical and basolateral membrane domains, *J. Cell. Biol.,* 107, 506A, 1989.

26. **Muller, W. A. and Gimbrone, M. A.,** Plasmalemmal proteins of cultured vascular endothelial cells exhibit apical-basal polarity: analysis by surface-selective iodination, *J. Cell Biol.,* 103, 2389, 1986.

27. **Keller, R., Pratt, B. M., Furthmayr, H., and Madri, J. A.,** Aortic endothelial cell protoheparan sulfate. II. Modulation by extracellular matrix, *Am. J. Pathol.,* 128, 299, 1987.

28. **Silbert, J. E., Gill, P. J., Humphries, D. E., and Silbert, C. K.,** Cell-surface protoheparan sulphate structure-function relationship, *Ann. N.Y. Acad. Sci.,* 556, 51, 1989.

29. **Mironov, S. L. and Dolgaya, E. V.,** Surface charge of mammalian neurons as revealed by electrophoresis, *J. Membr. Biol.,* 86, 197, 1985.

30. **Vargas, F. F., Osorio, H. M., Ryan, U. S., and Toro, A.,** Surface charge in cultured endothelial cells from bovine pulmonary artery, *FASEB J.,* 2, A934, 1988.

31. **Born, G. V. R. and Palinski, W.,** Unusually high concentrations of sialic acids on the surface of vascular endothelia, *Br. J. Exp. Pathol.,* 66, 543, 1985.

32. **Streichman, S., Segal, E., Tatarsky, I., and Marmur, A.,** Moving boundary electrophoresis and sialic acid content of normal and polycythaemic red blood cells, *Br. J. Haematol.,* 48, 273, 1981.

33. **Pasquale, L., Winisky, A., Oliva, C., Vaio, G., and McLaughlin, S.,** An experimental test of new theoretical models for the electrokinetic properties of biological membranes, *J. Gen. Physiol.,* 88, 697, 1986.

34. **Seaman, G. V. F., Vassar, P. S., and Kendall, M. J.,** Calcium ion binding to blood cell surfaces, *Experientia,* 25, 259, 1969.

35. **Vargas, F. F., Osorio, H. M., Ryan, U. S., and Toro, A.,** Calcium adsorption on endothelial cells from bovine pulmonary artery, *FASEB J.,* 2, 1882, 1988.

36. **Hunter, G. K., Wong, K. S., and Kim, J. J.,** Binding of calcium to glycosaminoglycans: an equilibrium dialysis study, *Arch. Biochem. Biophys.,* 260, 161, 1988.

37. **Peterkofsky, B.,** Ascorbate requirement for hydroxylation and secretion of procollagen: relationship to inhibition of collagen synthesis in scurvy, *Am. J. Clin. Nutr.,* 54, 1135, 1991.

38. **Chang, K. C., Chong, W. S., Sohn, D. R., Kwon, B. H., Lee, I. J., Kim, C. Y., Yang, J. S., and Joo, J. I.,** Endothelial potentiation of relaxation response to ascorbic acid in rat and guinea pig thoracic aorta, *Life Sci.,* 52, 37, 1993.

39. **Niki, E.,** Action of ascorbic acid as a scavenger of active and stable oxygen radicals, *J. Am. Clin. Nutr.,* 54, 1119, 1991.

40. **Jonas, E., Dwenger, A., and Hager, A.,** In vitro effect of ascorbic acid on neutrophil-endothelial cell interaction, *J. Biolumin. Chemilumin.,* 8, 15, 1993.

41. **Washko, P. W., Wang, Y., and Levine, M.,** Ascorbic acid recycling in human neutrophils, *J. Biol. Chem.,* 268, 15531, 1993.

42. **Buettner, G. R.,** The pecking order of free radicals and antioxidants: lipid peroxidation, alpha tocopherol and ascorbate, *Arch. Biochem. Biophys.,* 300, 535, 1993.

43. **Osorio, H. M.,** Efecto del Acido Ascorbico sobre la Carga Electrica Neta de la Superficie de Celulas Endoteliales, M.Sc. thesis, School of Public Health, University of Puerto Rico, 1987.

44. **Pollard, B. H. and Heldman, E.,** personal communication.

45. **Tajmir-Riahi, H. A.,** Coordination chemistry of vitamin C. I. Interaction of L-ascorbic acid with alkaline earth metal ions in the crystalline solid and aqueous solution, *J. Inorg. Biochem.,* 40, 181, 1990.

46. **Dhariwal, K. R., Hartzell, W. O., and Levine, M.,** Ascorbic acid and dehydroascorbic acid measurements in human plasma and serum, *Am. J. Clin. Nutr.,* 54, 712, 1991.

47. **Adamson, R. H., Huxley, V. H., and Curry, F. E.,** Single capillary permeability to proteins having similar size but different charge, *Am. J. Physiol.,* 254, 304, 1988.

48. **O'Donnell, M. P. and Vargas, F. F.,** Electrical resistance of isolated rabbit aorta, *Fed. Proc.,* 41, 1598, 1982.

49. **Bell, G. I., Dembo, M., and Bongrand, P.,** Cell adhesion: competition between nonspecific repulsion and specific bonding, *Biophys. J.,* 45, 1051, 1984.

50. **Smith, U., Ryan, J. W., Michie, D. D., and Smith, D. S.,** Endothelial projections as revealed by scanning electron microscopy, *Science,* 173, 925, 1971.

51. **Gourevitch, D., Jones, C. E., Crocker, J., and Goldman, M.,** Endothelial cell adhesion to vascular prosthetic surfaces, *Biomaterials,* 9, 97, 1988.

52. **Van Wachem, P. B., Hogt, A. H., Beugeling, T., Feijen, J., Bantjes, A., Detmers, J. P., and Van Aken, W. G.,** Adhesion of cultured human endothelial cells onto methacrylate polymers with varying surface wettability and charge, *Biomaterials,* 8, 323, 1987.

Chapter 13

Application of Cell Electrophoresis for Clinical Diagnosis

Wolfgang Schütt, Nabuya Hashimoto, and Motomu Shimizu

CONTENTS

I. GENERAL REMARKS

The electrophoretic mobility of a cell is directly correlated with the surface charge, which in turn is based on ionized chemical groups forming parts of the membrane macromolecules and on ions adsorbed at the surface. Because of this relationship between surface macromolecules and the charge, the electrophoretic mobility can be regarded as an image of the cell and thus be used to discriminate different cell types.[1,2]

The electrophoretic mobility may be changed if cells are incubated *in vitro* with substances which bind to the surface or which influence the cell membrane in another way. A second important application of cell electrophoresis is, therefore, the detection and analysis of interactions of various substances with the cell membrane. Furthermore, since exposure of cells to ionizing radiation leads to changes in the membrane, an analysis of the effects of potential biochemical protectors is possible. The electrophoretic mobility also can be used to characterize nonbiological particles and the interaction of substances with these particles. In living cells these processes are very specific and often connected with receptors, whereas artificial particles involve no specific adsorption processes.

Interest in particle electrophoresis has remained at a moderate level for over 50 years, with a steady stream of applications in biomedicine, chemistry, pharmacy, brewing, soil technology, and engineering.[1] The more widespread consideration of cell electrophoresis as an aid in biomedical research and clinical diagnostics depends on the availability of equipment capable of performing the necessary measurements on many cells quickly and accurately. During the 1970s many interesting applications of cell electrophoresis in the hematology, cardiology, and clinical immunology fields aroused increased interest in the technique.[2,3]

In particular, Field's development of the so-called macrophage electrophoretic mobility (MEM) test,[4,5] used for the immunological detection of cancer, stimulated several groups to modify and simplify the test method and to develop an automated measuring system for cell electrophoresis. The use of tanned sheep red blood cells (tSRBC) became a standard test for the detection of the charge-changing lymphokines produced *in vitro* by sensitized lymphocytes of tumor patients after antigen contact.[6,7] Concerning this test, several research teams in the East and the West euphorically reported highly significant

0-8493-8918-6/94/$0.00+$.50
© 1994 by CRC Press, Inc.

results in connection with tumor tests (80 to 95% correct diagnosis),[8-10] which made it worthwhile and necessary to automate the procedure for measuring the electrophoretic mobility of the indicator cells. These results and the expectations attached to this tumor test accelerated the development of electrophoresis equipment that could perform the measurements objectively and quickly.[11]

The development of a corresponding automatic measuring device[12,13] made jointly with Carl Zeiss Jena in 1978 was intended from the very start to provide a tool for teams concerned with the application of the tSRBC test (accurate estimation of mean value) and to develop a system for biophysical and biomedical applications in general (production of high-resolution histograms). At that time the available devices, based on laser doppler electrophoresis, analytical free-flow electrophoresis, and image transduction electrophoresis, were more suitable for measuring particles and cells suspended in solutions with low ionic strength. (For a review see References 2 and 14.) Frequency distributions may contain artifacts, since the intensity of light scattered at a given angle is influenced by the size, shape, orientation, and refractive index of a cell. Studies of cells with different sizes in one suspension cause such problems. Finally, no absolute mobility values from individual cells can be obtained. Attempts have, therefore, been made to compensate for these disadvantages by projecting the cells moving in a classical electrophoretic chamber onto a television (TV) screen and using computer analysis. The automated analytical electrophoretic microscope, developed under NASA contract, and the system developed by Hashimoto permit rapid multiple velocity measurements per cell. In the Parmoquant system, produced by Carl Zeiss Jena since 1979 and licensed to Kureha Chemical Ltd., Tokyo, an electronic tracking system enables the automatic objective measurement of many individual cells in a short period of time.[15]

The principle of the particle speed measurement of single cells by the Parmoquant system is shown in Figure 13-1. By choosing a suitable contrast technique the particles are brightly displayed on a dark background. The optical image is converted by the TV camera into an electronic image. After filtering, the electronic signal can be digitized using an adjustable discriminator. The discriminator responds whenever the optical signal exceeds a certain brightness level. After scanning the TV picture, the microprocessor calculates from this information in the center of mass and area of any cells.

Cells within a depth range of ± 5 to 8 μm around the stationary layer are selected optically. In order to estimate the migrating speed, the position and area of all sharply imaged cells with a constant area are recorded simultaneously every $\frac{1}{3}$ ($\frac{1}{15}$ in PQ-L) of a second after the cells have stopped accelerating (1 s). The method of the least mean square is used for calculation of every cell speed.

The persuance algorithm contains many test criteria that exclude those particles which do not migrate in one plane, which undergo changes during migration, or whose paths have been disturbed by others. The mobility of each particle followed during the entire forward and backward run is calculated by associating the velocity with the voltage applied during the migration time and actually measured by special electrodes inside the chambers. The particle diameters for automatic measurement can vary from 2 to 20 μm. Using a cell concentration of 2×10^6 cells per milliliter, about 20 of the migrating cells in view are tracked by the image processing unit at the same time. One such measuring cycle takes nearly 10 s. The results of measurements of the electrophoretic mobility of each measured single cell, including the mean cell mobility, the standard deviation, and the histogram, as well as the conditions of electrophoresis are displayed on a monitor and a printout. An external computer can be connected for processing the data.

With this high-precision equipment it is now possible to reproduce the original Field MEM test with macrophages as indicator particles, but not the simpler and frequently published tSRBC version of the tumor test.[16]

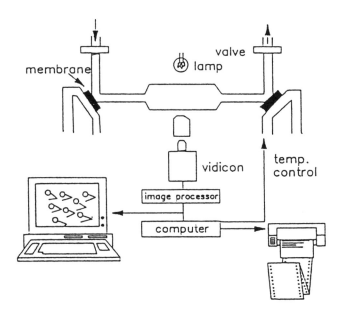

Figure 13-1 The automated cell electrophoresis system, Parmoquant.

In 1990 it was concluded by Preece and Brown[17] that a major difficulty in promoting cell electrophoresis as an advantageous technique for cellular biology was that it had to overcome the unfavorable press it had received when it was employed in the tumor test. For many scientists in biomedicine the mere term "cell electrophoresis" conjures up a picture of a piece of apparatus that caused considerable controversy when used as a test for cancer. Results generated at that time were often inconsistent, irreproducible, and of marginal significance because of a combination of factors, such as the physical limitations of the apparatus, poor experimental design, and occasionally an unintentionally biased subjectivity by the operator of the cytopherometer. These problems, combined with the forceful claims and counterclaims of certain groups, only served to discredit the technique and its applicability. Fortunately, that particular period failed to discourage the enthusiastic and determined biophysicists from developing and constructing analytical and preparative cytopherometers which are currently often automated and rapidly provide reproducible results. At present the development of these machines has advanced at a faster pace than their biological application.[14-16]

During the international meetings on cell electrophoresis in Rostock, Germany in 1982, 1984, 1988, and 1990 attempts have been made to overcome these difficulties through discussions between groups from the East and the West. New measuring systems and biomedical applications have been presented and compared during electrophoresis meetings in Bristol in 1979 and 1981, in London in 1986, and in Tokyo in 1983, 1990, and 1992.

II. OBJECTS OF INVESTIGATION

The basis for the acceptance of cell electrophoresis as a powerful tool in the biomedical sciences is the relation between the expression of a particular surface charge by leukocytes, red blood cells, or platelets and their state of function, differentiation, or their phenotypic identification and the interaction with exogenic factors. Many fields of applications have become very competitive during the last 40 years. The cells and

Table 13-1 **Cells favored as objects in practical application of cell electrophoresis**

Cells (untreated)	
Erythrocytes	Anemia, stress, incorporation of parasites or viruses, storage
Thrombocytes	Bernard-Soulier syndrome, von Willebrand disease, hyperlipidemia, acute myeloid leukemia
Lymphoid cells	Percentage of B- and T-cells and other subpopulations in cancer, transplantation, immunopathies, dialysis, irradiation
Leukemic cells	Differentiation
Keratinocytes	Psoriasis
Bacteria, algae, parasites	Differentiation, pathogenicity, virulence
Liposomes	Prenatal lung maturity, lung lavage
Cells influenced by	
Irradiation	Sensitors, protectors, cancer treatment
Pharmaceuticals	Neuroleptics, narcotics, endotoxin, antibiotics, chemotherapeutics, immunomodulators, aggregating substances
Antibodies	Blood group antibodies, antilymphocyte sera, monoclonal antibodies
Lectins	Receptor identification
Mitogens	Stimulation
Toxic substances	Toxicology
Biomaterials	Selective adherence, lymphokine production
Nonbiological particles	
Latex	Detection of pathological changes in body fluids: cerebrospinal fluid (meningitis), sera (cystic fibrosis, LDL/HDL relation), secretion (RDS), lavage (lung disease)

Note: LDL = low-density lipoproteins; HDL = high-density lipoproteins; RDS = respiratory distress syndrome.

particles favored as objects in practical applications are described in several reviews[1-3,16-26] and summarized in Table 13-1.

III. LYMPHOCYTE ELECTROPHORESIS

Electrophoretic measurements of density gradient-separated human mononuclear cells reveal a histogram of heterogeneously distributed cells. Two populations are clearly distinguishable, and it was first shown by rosette formation that the high and low mobility populations correlate with T- and B-lymphocytes, respectively.[27] In the Rostock laboratory a dissection of such an electrophoretic histogram into immunologically defined lymphocyte subpopulations was obtained using monoclonal antibodies to lymphocyte surface markers in a complement-dependent lysis assay combined with electrophoretic measurements.[28] In this way a correlation between electrophoretic mobility and different lymphocyte subpopulations as well as monocytes was obtained. Interestingly, CD4+ and CD8+ cells also differed in their electrophoretic mobility.

Therefore, the analysis of electrophoretic histograms determined from human mononuclear cells can be used to evaluate the populations of lymphocytes in disease and other

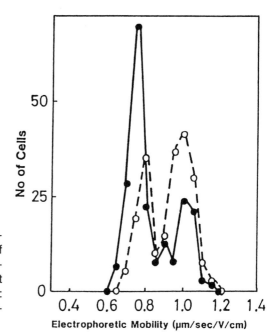

Figure 13-2 Typical electrophoretic mobility histograms of whole peripheral blood mononuclear cells in cancer patient and normal subject. ●——●: cancer patient and ○——○: normal control.

clinical disorders. For the determination of changes in the percentage of subpopulations a borderline was set between high mobility cells (T-lymphocytes) and low mobility cells (B-lymphocytes and monocytes) at the electrophoretic mobility of 0.96×10^{-8} m^2 s^{-1} V^{-1} (mean value of freshly drawn human erythrocytes: 1.08×10^{-8} m^2 s^{-1} V^{-1}). In this way changes were determined in the percentage of T-lymphocytes following kidney transplantation and open heart surgery, and during pregnancy, dialysis, irradiation, and in the case of cancer.[3,16-20]

It has been reported that the electrophoretic mobility of peripheral blood mononuclear cells (PBMC) in cancer patients is lower than that of healthy controls.[29-31] However, the characteristics of low mobility cells still have not been clarified. Since the electrophoretic mobility can be estimated easily by means of the Parmoquant, extended studies have been performed in Japan to find a correlation between immunological characteristics, clinical significance, and electrophoretic mobility. Typical histograms of whole PBMC in a cancer patient and a normal subject are shown in Figure 13-2. The heights of the two peaks in the histogram are reversed in cancer patients with solid tumors, such as rectal, gastric, breast, colon, and uterus tumors. The ratio of low to high mobility cells at the borderline at 0.96 was defined to show these changes quantitatively. As shown in Figure 13-3, this ratio was 81 ± 15 (mean and standard deviation) in healthy controls. The ratio attained very high values and incidences with the development of the stages of a patient and was especially high in recurrent cancer.

To analyze the low mobility cells, adherent and nonadherent PBMC were separated on plastic dishes (Figure 13-4). In the pattern of the electrophoretic histogram of adherent cells in cancer patients, the high mobility peak disappeared and the low mobility one significantly increased. The ratio of low to high mobility PBMC in the patients increased with the increase in monocytes. Adherent cells were shown to be monocytes by morphological examination (Wright-Giemsa staining) and by flow cytometry using monoclonal antibodies (CD14+). On the other hand, since B-cells are also low mobility cells, changes in B-cells (CD19+) did not increase and showed a constant percentage. These results indicate that low mobility cells in cancer patients are monocytes.

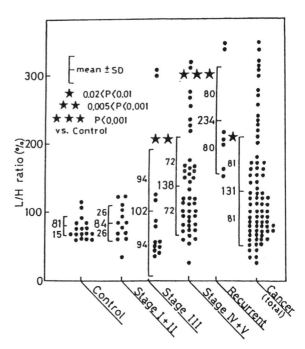

Figure 13-3 The ratio of low (L) and high (H) mobility cells of whole peripheral blood mononuclear cells in cancer patients and normal control. The statistical significance was determined by student *t*-test between patient and normal control.

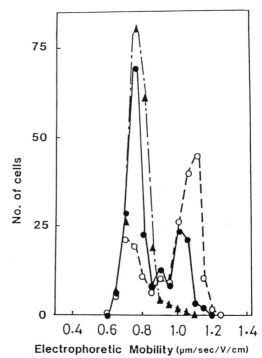

Figure 13-4 Typical electrophoretic mobility histograms of whole, nonadherent and adherent peripheral blood mononuclear cells (PBMC) in cancer patient. Adherent and nonadherent cells were separated from whole PBMC using a fetal bovine serum-coated dish. ●——●: whole cells; ○——○: nonadherent cells; ▲——▲: adherent cells.

The relationship between suppressor T-cells (CD8+/CD11b+) and monocytes (CD14+) was studied in whole PBMC of cancer patients. The percentage of suppressor T-cells correlated significantly with that of monocytes. This result suggests that the appearance of suppressor T-cells and of monocytes are closely related. The production of interleukin 1 and prostaglandin E2 was determined to clarify the function of monocytes. The production of both decreased in stage IV and V and recurrent cancer patients.

The ratio of low and high mobility PBMC closely correlated with total serum protein and the number of peripheral lymphocytes,[31] which are a measure of general and immunological status of cancer patients. The ratio reflects the performance and immunological status of cancer patients much better than other previously reported parameters.[31] The ratio can be applied to all cancer patients with a solid tumor. Using the automatic electrophoresis system Parmoquant it is very easy to determine the ratio and obtain highly reproducible data.[32] Thus, the ratio may be a new and useful parameter for monitoring cancer.

Another parameter may be the electrophoretic mobility of monocytes of cancer patients. Human monocytes were isolated from the peripheral blood of five control donors and eight cancer patients and measured.[30] Histograms of the monocytes from the five control persons showed one sharp peak at 0.85×10^{-8} m^2 s^{-1} V^{-1}. The monocytes from two patients with intermediate-stage cancer formed broad and plural divided peaks. The monocytes from four patients with terminal-stage cancer again formed a single peak, but it obviously was located on the lower mobility side.

Also, after treatment of patients with a biological response modifier, an activator of monocytes, the peak of low mobility cells changed to a lower electrophoretic mobility zone, although the absolute number of monocytes was constant. Thus, measurement of electrophoretic mobility offers many advantages over routine monocyte counts in peripheral blood. Monocytes possess suppressor, helper, and cytotoxic activities in the tumor-bearing host. On the other hand, a significant relationship between suppressor T-cells and monocytes suggests that monocytes regulate the number and function of suppressor T-cells, and vice versa.[33] Interleukin 1 enhances the immune response, but prostaglandin E2 suppresses it.[34] Although monocyte content increased in PBMC of patients, interleukin and prostaglandin production per cell decreased in stage IV and V and in recurrent cancer patients (decrease in monocyte function).[34] It was reported that monocytes in cancer patients release a decreased level of interleukin 1[35] and an increased level of prostaglandin E2.[36] This discrepancy in prostaglandin may be due to the differences in preparation of monocytes and the assay method. Taken together with these results, the ratio of low to high mobility cells increases whereas interleukin and prostaglandin production from monocytes decreases in cancer patients; these findings indicate that these parameters are a good index for the follow-up of patients and show the changes in the immune system in tumor-bearing hosts.[37]

The PBMC in the autoimmune disease lupus nephritis exhibited a high ratio of low to high mobility cells, and the changes in the ratio correlated with clinical status. These results suggest that the percentage of low mobility cells is dependent on specific immunological changes such as tumors and autoimmune disease.[38,39]

IV. LEUKEMIA CELLS

During research on human blood cells interest was focused on the characterization of leukemic cells by particle electrophoresis and use of this method in the diagnosis of leukemias. The first experiments in this field were performed by Sabolovic et al.,[40] but since then much progress has been achieved in the diagnosis of leukemias by immunological phenotyping. This method has revealed that leukemias are more heterogeneous

Table 13-2 **Methods used in clinically relevant applications**

Changes in subpopulations of lymphoid cells after biomaterial-blood contact
In vitro: spleen cells (low mobility cells, adherent)
In vivo: peripheral blood mononuclear cells of uremic patients before, during, and
after dialysis
Cytokine release after *in vitro* biomaterial-mononuclear cell contact as an assay for
cellular immunity[42]
Protein adsorption behavior on latex particles of different surface structure

with respect to cell type than previously expected. Despite this progress it has been concluded that a multiparameter analysis of cells using different methods is necessary for an exact diagnosis. Groups in Rostock, Bratislava, and Stettin have investigated the diagnostic suitability of electrophoretic mobility measured by the Parmoquant system as a cell marker for differentiation of leukemias.[3,18,20]

In acute lymphatic leukemias, subtyping by means of cell electrophoresis is possible. This subtyping correlates largely with the help of monoclonal antibodies. Acute lymphatic leukemia (T-ALL) can be differentiated clearly from the non-T-ALL forms. This differentiation was of particular importance for treatment strategies and prognosis of leukemia in children. In a number of cases Knippel and Rychly[18] have ascertained the method to be well suited also for monitoring the rate of leukemic blast cell formation during the course of therapy. For example, differentiation of normal T-lymphocytes and leukemic T-cells in T-ALL is not always possible with monoclonal antibodies, but the electrophoretic mobilities of the two forms of cells can differ. In acute myelocytic leukemias, differentiation between the Fab subtypes M2, M4, and M5 can be made electrophoretically. Knippel and Rychly consider this a valuable capability since the Fab classification, based on morphology and cytochemistry. is relatively subjective and, thus, can be supported by electrophoretic mobility as an objective parameter. Distinction between myelocytic and lymphatic leukemias remains important for the prognosis and therapy of acute leukemia. Electrophoretic mobility can, above all, be useful in the highly immature forms that are often difficult to differentiate because in such instances myeloid cells exhibit high electrophoretic mobility in comparison with lymphatic cells.

Cell electrophoresis can be recommended for diagnosis of acute leukemia as a part of the multiparameter analysis of leukemic cells. In many cases it allows more distinct diagnosis.[18,20]

V. BIOCOMPATIBILITY

A further field for the application of cell electrophoresis including such histogram analysis is that concerned with investigations of blood compatibility at the level of blood-biomaterial interactions. Concerning the assessment of selected parameters of blood compatibility, there are several means by which this relatively simple method can be applied. The methods used by Thomaneck et al.[41] in clinically relevant applications are summarized in Table 13-2.

Thomaneck et al. found a good correlation to the biomaterial surface condition, which is very important to characterize before clinical use. Protein adsorption on latex can be used as a model system for the first step of the interaction between biomaterials and blood.

Table 13-3 **Plasma protein adsorption phenomena providing the basis for body fluid investigation by means of particle electrophoresis**

- Important plasma proteins have different electrophoretic plateaus
- Thus, electrophoretic measurements can detect adsorption of competing proteins on solid particles
- Adsorption studies at different ionic strengths
- Development of extended electrokinetic theory
- Conclusion regarding orientation and tertiary structure
- Antigenicity of proteins after adsorption

VI. BODY FLUID INVESTIGATION

The understanding and control of the interaction of proteins with solid surfaces is important in a number of fields, including pharmacy, protein processing, biotechnology, diagnostic products, and biochips. Particle electrophoresis is a very simple method providing experimental data with regard to this field which have been poorly applied to date. In contrast to other methods, such as radioisotope labeling, particle electrophoresis allows rapid, direct measurements without modification of the molecules involved. The adsorption of proteins on synthetic particles is made manifest by changes in the electrical surface charge density of the particles compared to bare particles and, thus, by altered electrophoretic mobilities.

Knippel et al.[18,20,43] have opened a new application field for particle electrophoresis based on the results which have already been described extensively and which are summarized in Table 13-3. Based on the adsorption phenomena mentioned above, a simple method for detecting pathological changes in body fluids has been developed. The procedure is simple and rapid, and it requires only small amounts of body fluids. The following diseases were considered in a study involving more than a 1000 subjects: cystic fibrosis (serum), respiratory distress syndrome in the newborn (pharyngeal secretions), lung diseases (bronchoalveolar lavage fluid), meningitis (cerebrospinal fluid), and fetal lung maturity.[3,18,19,44]

Using electrophoretic fingerprinting it was determined that only certain types of indicator particles respond in the test. Electrophoretic fingerprinting is a new *in situ* analytical technique for characterizing the surface of polymer microspheres.[45] The technique requires the measurement of the electrophoretic mobility of synthetic particles over all pertinent electrochemical variables. This allows one to produce surface representations of electrophoretic mobility as a function of both potential determining ion (pdi) concentration and overall ionic transport properties measured to a first approximation by the specific conductivity. Analysis of the data produces a mobility-pdi concentration profile from which it is possible to differentiate surface functional groups and ion adsorption correctly. Using this technique of electrophoretic fingerprinting, differences in surface chemistry have been shown between two batches of polysterene latex particles. One batch was responsive while the other was unresponsive. The analysis suggests that the surface must contain amphoteric groups in order for the latex particle to be responsive to serum components from cystic fibrosis gene carriers. Polymer microspheres that have surfaces containing carboxylic acid groups (modified or unmodified) will probably not respond. Investigation of other surfaces using electrophoretic fingerprinting will determine the suitability of this technique for predicting whether a certain type of indicator particle will respond in body fluid tests. The technique also may help clarify the mechanisms involved in the binding of proteins and other biomolecules contained in body fluids to synthetic materials.[18] The electrophoretic test for detecting cystic fibrosis gene carriers may become valuable in clinical practice in addition to DNA diagnosis owing to its easy inexpensive handling and low cost. However, many details of the methodology remain to be clarified.

ACKNOWLEDGMENTS

We would like to thank Drs. U. Thomaneck, E. Knippel, and J. Rychly (Rostock) for longlasting cooperation and Prof. D. Sabolovic (Paris) and Prof. P. Todd (Boulder, CO) for continual scientific exchange.

REFERENCES

1. **Sherbet, G. V.,** *The Biophysical Characterization of the Cell Surface,* Academic Press, London, 1978.
2. **Preece, A. W. and Sabolovic, D.,** *Cell Electrophoresis: Clinical Application and Methodology,* Elsevier/North-Holland, Amsterdam, 1979.
3. **Schütt, W. and Klinkmann, H., Eds.,** *Cell Electrophoresis,* Walter de Gruyter, Berlin, 1985.
4. **Field, E. J. and Caspary, E. A.,** Lymphocyte sensitization: an in vitro test for cancer?, *Lancet,* 2, 1337, 1970.
5. **Field, E.J.,** The macrophage electrophoretic mobility test: clinical applications — a guide, in *Cell Electrophoresis,* Schütt, W. and Klinkmann, H., Eds., Walter de Gruyter, Berlin, 1985, 747.
6. **Porzsolt, F., Tautz, C., and Ax, W.,** Electrophoretic mobility test. I. Modifications to simplify the detection of malignant diseases in man, *Behring Inst. Mitt.,* 57, 128, 1975.
7. **Jenssen, H. L. and Shenton, B. K.,** Electrophoretic mobility test for lymphocyte sensitization using tanned sheep erythrocytes, *Acta Biol. Med. Ger.,* 34, K29, 1975.
8. **Tautz, C., Laier, E., and Schneider, W.,** Der EM—Test, ein hochsensibler Malignom-Test, *Monatsschr. Kinderheilkd.,* 125, 456, 1977.
9. **Jenssen, H. L., Köhler, H., and Werner, H.,** Die Anwendung der Zellelektrophorese zur Untersuchung verschiedener Immunphänomene, Habilitationsschrift, Wilhelm-Pieck-Universität, Rostock, Germany, 1977.
10. **Müller, M., Irmscher, I., Fischer, R., Friemel, H., Jenssen, H. L., Köhler, H., Seyfarth, M., von Broen, B., and Pasternak, C.,** Zellelektrophorese-Mobilitätstest in der Geschwulstdiagnostik: simultan an mehreren Meßgeräten durchgeführt — methodisch orientierte Blindversuche, *Dtsch. Gesundheitswes.,* 32, 1057, 1977.
11. **Jenssen, H. L., Köhler, H., Werner, H., Meyer-Rienecker, H., Seyfarth, M., Büttner, H., Schütt, W., Günther, M., Jennsen, R., and Friemel, H.,** Ergebnisse und Stand der Entwicklung eines immunologischen Tumortestes für die Praxis auf der Grundlage der Zell-Elektrophorese-Mobilitäts-Methode, *DDR Med. Rep.,* 5, 876, 1976.
12. **Pohl, A. and Schütt, W.,** Increased reliability and productivity of cell electrophoresis by objectivated and automated measurements and data processing, *Jenaer Rundsch.,* 6, 270, 1978.
13. **Schütt, W.,** PARMOQUANT — ein automatisches Meßmikroskop für die Partikelelektrophorese, *Laborpraxis Med. (München),* 3, 16, 1981.

14. **Schütt, W., Grümmer, G., Preece, A., and Goetz, P.,** Principles of cell electrophoresis, in *Physical Characterization of Biological Cells,* Schütt, W., Klinkmann, H., Lamprecht, I., and Wilson, T., Eds., Verlag Gesundheit GmbH, Berlin, 1991, 215.

15. **Schütt, W., Thomaneck, U., Knippel, E., Rychly, J., Klinkmann, H., Hayashi, H., Toyoama, N., Fujii, M., Hirose, F., Yoshikumi, C., Onodera, C., Kotagawa, K., Endo, K., Nishimimura, Y., and Kawai, Y.,** Automated single cell electrophoresis realized in PARMOQUANT 2 and PARMOQUANT L, in *Cell Electrophoresis,* Schütt, W. and Klinkmann, H., Eds., Walter de Gruyter, Berlin, 1985, 55.

16. **Schütt, W.,** Beiträge zur Entwicklung eines automatischen Zellelektrophoresemeßsystems für biomedizinische Anwendungen, Habilitationsschrift, Universität Rostock, Rostock, Germany, 1986.

17. **Preece, A. W. and Brown, K. A.,** *Recent Trends in Particle Electrophoresis,* Elsevier/North-Holland, Amsterdam, 1990.

18. **Knippel, E. and Rychly, J.,** Anwendungen der Partikel-elektrophorese in Biologie und Medizin, Habilitationsschrift, Universität Rostock, Rostock, Germany, 1988.

19. **Schütt, W., Thomaneck, U., Knippel, E., Rychly, J., and Klinkman, H.,** Biomedical and clinical applications of automated single cell electrophoresis, *Electrophoresis,* 11, 970, 1990.

20. **Knippel, E., Rychly, J., Thomaneck, U., and Schütt, W.,** Particle electrophoresis in medicine and biology, *Med. Focus Int.,* 3, 8, 1992.

21. **Hashimoto, N.,** Clinical application of cell electrophoresis, *Physico-Chem. Biol. (Jpn.),* 36, 323, 1992.

22. **Sabolovic, D. and Hashimoto, N.,** Alterations of uninfected red blood cell membranes during in vitro and in vivo parasite growth, *Phys. Chem. Biol. (Jpn.),* 36, 347, 1992.

23. **Knopf, B. and Wollina, U.,** Electrophoretic mobilities of keratinocytes from normal skin and psoriatic lesions, *Arch. Dermatol. Res.,* 284, 117, 1992.

24. **Shimizu, M., Sabolovic, D., Ozawa, H., and Iwaguchi, T.,** Cyclophosphamide-induced suppressor cells in nude mice, *Anti-Cancer Drugs,* 3, 427, 1992.

25. **Schütt, W., Klinkmann, H., Lamprecht, I., and Wilson, T.,** *Physical Characterization of Biological Cells,* Verlag Gesundheit GmbH, Berlin, 1991.

26. **Schütt, W.,** New trends in cell electrophoresis, *Phys. Chem. Biol. (Jpn.),* 36, 329, 1991.

27. **Sabolovic, D., Sabolovic, N., and Dumond, F.,** Identification of T and B cells in mouse and man, *Lancet,* 2, 927, 1972.

28. **Sabolovic, D., Knippel, E., Thomaneck, U., Rychly, J., and Schütt, W.,** Electrophoretic mobility and monoclonal antibody combined in the study of human peripheral blood lymphocytes, in *Cell Electrophoresis,* Schütt, W. and Klinkmann H., Eds., Walter de Gruyter, Berlin, 1985, 333.

29. **Shimizu, M., Iwaguchi, T., Sekine, K., Mori, T., and Hayashi, H.,** Lymphocyte electrophoresis of tumor-bearing host in basic and clinical diagnosis, in *Physical Characterization of Biological Cells,* Schütt, W., Klinkmann, H., Lamprecht, I., and Wilson, T., Eds., Verlag Gesundheit GmbH, Berlin, 1991, 245.

30. **Mori, T. and Kosaki, G.,** The changes of lymphocyte electrophoretic mobility in cancer patients, in *Cell Electrophoresis,* Schütt, W. and Klinkmann, H., Eds., Walter de Gruyter, Berlin, 1985, 356.

31. **Mori, T., Shimizu, M., and Iwaguchi, T.,** Immunological characterization and clinical significance of low mobility cells appearing in the peripheral blood mononuclear cells of cancer patients, *Eur. J. Cancer Clin. Oncol.,* 2, 1463, 1988.

32. **Hayashi, H., Toyama, N., Fujii, M., Yoshikumi, C., Kawai, Y., and Iwaguchi, T.,** Determination of cell mixtures by an automated cell electrophoretic instrument and monoclonal antibody, *Electrophoresis,* 8, 224, 1987.

33. **Kunkel, S. L., Chensue, S. W., and Phan, S. H.,** Prostaglandins as endogenous mediators of interleukin 1 production, *J. Immunol.,* 136, 186, 1986.

34. **Shimizu, M., Sekine, K., Iwaguchi, T., Kataoka, T., and Mori, T.,** Diagnosis of cancer using lymphocyte electrophoresis: electrophoretic mobility, interleukin 1 and prostaglandin E2 production of monocytes in peripheral blood of cancer patients, *Jpn. J. Cancer Chemother.,* 17, 639, 1990 (in Japanese).

35. **Yokota, M., Sakomoto, S., Koga, S., and Ibayashi, H.,** Decreased interleukin 1 activity in culture supernatant of lipopolysaccharide stimulated monocytes from patients with liver cirrhosis and hepatocellular carcinoma, *Clin. Exp. Immunol.,* 67, 335, 1987.

36. **Nara, K., Odagiri, H., Fujii, M., Yamanaka, Y., Yokoyama, M., Morita, T., Sasaki, M., Kon, M., and Abo, T.,** Increased production of tumor necrosis factor and prostaglandin E2 by monocytes in cancer patients and its unique modulation by their plasma, *Cancer Immunol. Immunother.,* 25, 126, 1987.

37. **Shimizu, M. and Iwaguchi, T.,** Analysis of lymphocyte electrophoresis, *Phys. Chem. Biol. (Jpn.),* 36, 329, 1992.

38. **Koide, K., Thoyama, J., and Terada, M.,** Analysis of electrophoretic mobility histogram of peripheral blood lymphocytes in patients with lupus nephritis, in Report on Progressive Nephritis Research, Japanese Ministry of Health and Welfare, Tokyo, 1989, 299 (in Japanese).

39. **Shimizu, M., Sekine, K., Matsuzawa, A., and Iwaguchi, T.,** Cell electrophoretic characterization of abnormally expanded lymphocytes in autoimmune lpr, gld and Yaa mice, and of thymocyte subsets, *Electrophoresis,* 13, 136, 1992.

40. **Sabolovic, D., Dumont, F., Chollet, P., and Amiel, J. L.,** Electrophoretic mobility of blood lymphocytes in patients with chronic lymphocytic leukemia, *Biomedicine,* 19, 222, 1973.

41. **Thomaneck, U., Schütt, W., Behrend, D., Falkenhagen, D., and Klinkmann, H.,** in *Immune and Metabolic Aspects of Therapeutic Blood Purification Systems,* Smeby, L., Jorstad, S., and Wideroe, T., Eds., S. Karger, Basel, 1986, 163.

42. **Thomaneck, U. and Hashimoto, N.,** Current application of the electrophoresis mobility test, in *Physical Characterization of Biological Cells,* Schütt, W., Klinkmann, H., Lamprecht, I., and Wilson, T., Eds., Verlag Gesundheit GmbH, Berlin, 1991, 265.

43. **Knippel, E., Schütt, W., Thomaneck, U., Rychly, J., Klinkmann, H., Goetz, P. J., Marlow, B., and Fairhurst, D.,** Cell electrophoresis applied to biological research and clinical diagnosis, in *Proc. 6th Congr. Int. Electrophoresis Society,* Schäfer-Nielsen, E., Ed., VCH-Verlag, Weinheim, 1988, 208.

44. **Schütt, W., Thomaneck, U., Knippel, E., Rychly, J., and Klinkmann, H.,** Biomedical and clinical applications of automated single cell electrophoresis, *Phys. Chem. Biol. (Jpn.),* 34, 293, 1990.

45. **Marlow, B., Fairhurst, D., and Schütt, W.,** Electrophoretic fingerprinting and the biological activity of colloidal indicators, *Langmuir,* 4, 776, 1988.

Chapter 14

The Negative Surface Charge Density of Cells and Their Actual State of Differentiation or Activation

Johann Bauer

CONTENTS

I. INTRODUCTION

Several decades ago it was observed that cells suspended in an isotonic salt solution start to move toward a positively charged anode upon application of an electrical current.[1,2] The speed of the movement is proportional to the density of the negative charges located on the cell surfaces. It is called the electrophoretic mobility (EPM) and has been defined as the distance (in micrometers) covered by a cell in 1 s at a voltage of 1 V/cm (10^{-4} cm^2/Vs). Since this time the method of cell electrophoresis has been developed further and extensively applied. The efforts in this field have been aimed in two directions: (1) to analyze the negative surface charge density of distinct cell populations, and (2) to separate cell populations with different EPM values on a preparative scale.

During analysis of the negative surface charge densities the EPM values of many kinds of animal and human cells have been determined (Table 14-1).[3-42] The predominant aim of these studies was to detect correlations between an EPM value of a cell and its additional biological or biochemical parameters. Such a correlation was claimed when cancer cell lines were investigated.[3-7] The results of these studies led to the speculation that the metastasizing capacity of cancer cells might depend on their negative surface charge density. However, it was soon realized that metastasis is a very complex process which cannot be explained by just one cellular parameter, such as the negative surface charge density of cancer cells.[43]

Around 1970 it was detected that the body fluids of cancer patients contain soluble factors which influence the EPM of macrophages.[44] On the basis of this observation many efforts were made to use the EPM of macrophages for detection of tumors in a very early

0-8493-8918-6/94/$0.00+$.50
© 1994 by CRC Press, Inc.

Table 14-1 **EPM values of several kinds of mammalian cells**

First Author	Kind of Cells	EPM, μm · cm/Vs	EPM of Standard Cells, μm · cm/Vs[a]		Ref.
Ambrose	Hamster erythrocytes	1.35			3
	Hamster kidney cells	0.7			
	Hamster kidney tumor cells	1.15			
Purdom	Mouse sarcoma cells from				4
	Ascites fluid	2.0			
	Solid tumors	1.1			
Forrester	Hamster fibroblasts	1.17			5
Arnold	Rat erythrocytes	1.24			6
	Rat lymphocytes	1.06			
	Rat granulocytes	0.87			
	Rat ascites (parenchyma cells of liver)	1.21			
Mayhew	Ehrlich ascites cells	0.7, 1.2	1.02	he	7
Lichtman	Human granulocytes from				8,9
	Peripheral blood	1.38	2.13	he	
	Bone marrow (mature)	1.51			
	Bone marrow (immature)	1.84			
Zeiller	Cells from rat bone marrow and lymphatic organs	3.41–2.16	3.41	re	10
Suemasu	Rat endothelial cells	1.24			11
Zeiller	Rat antibody-forming cells	2.7, 2.4	3.41	re	12
Hannig	Human eosinophils	1.0	1.10	hly	13
Mehrishi	Human epidermoid cancer cells (HeP 2) cultured	1.07	1.08	he	14
Bosman	Murine melanoma cells	2.07, 2.77	2.79	he	15
Blume	Murine T-cells	1.0, 1.3	1.08	he	16
Bamberger	Rat erythrocytes	1.04, 1.3	1.10	he	17
	Rat thrombocytes	0.87	1.10	he	
Cocker	Chick erythrocytes	1.3, 1.8			18
Hannig	Sheep erythrocytes	1.25, 0.95	1.10	he	19
	Guinea pig erythrocytes	1.20			
	Rabbit erythrocytes	0.55			
Bauer	Human monocytes	0.95			20
	Human macrophages (*in vitro* maturation)	1.10, 0.95			
Mironov	Rat ganglion cells	1.40	1.80	he	21
Cohly	Mouse epithelial cells	1.28	1.57	hly	22
Bauer	Resting human B-cells	0.90			23
Rychly	Human T-cells, CD4+	1.18			24
	Human T-cells, CD8+	1.09			
	Human B-cells	0.78			
Hyrc	Bomirski hamster melanoma cells	0.89, 1.25			25
Easty	Mouse epidermal keratinocytes	0.72, 1.08			26

Table 14-1   **EPM values of several kinds of mammalian cells (continued)**

First Author	Kind of Cells	EPM, $\mu m \cdot cm/Vs$	EPM of Standard Cells, $\mu m \cdot cm/Vs$[a]		Ref.
Price	Mouse kidney cells	1.26	1.21	he	27
Silva-Filho	Mouse peritoneal macrophages:				28
	Resident	0.96			
	Elicited	1.01			
	Activated	1.10			
Gommans	Human keratinocytes	1.26			29
	Psoriatic human keratinocytes	1.48			
Bauer	Mature human B-cells	0.92, 1.06	1.10	he	30
	Activated human B-cells	0.92			
Nair	Mouse leukemic cells		1.40	he	31
	P388/SS	1.07			
	P388/R	1.35			
	Human lymphoblastoid cells CCRF-CEPM	1.32			
Yamamoto	Human fibroblast cell line TIG-1	1.65, 1.17	1.10	he	32
Carter	Rat prostatic adenocarcinoma cells	1.28, 1.96	1.25	he	33
Vargas	Bovine endothelial cells	0.74	1.03	he	34
Bauer	Murine myeloma cells	0.94, 1.03	1.10	he	35
Crawford	Human neutrophils	0.8, 1.2			36
Shimizu	Mouse thymocytes	1.03, 0.95, 0.75			37
Knopf	Human keratinocytes	0.59			38
	Psoriatic human keratinocytes	0.76			
Dolgaya	Mouse T-lymphocytes	1.02, 1.30			39
Hyrc	Hamster melanoma cells	0.95, 1.24	1.08	he	40
Crook	Human platelets	0.93, 0.88, 0.83			41
Slivinsky	Parenchymatous rat liver cells	1.02, 1.09	1.04	he	42

[a] Abbreviations: he = human erythrocytes, hly = human T-cells, re = rat erythrocytes.

phase of malignancy.[45,46] Later on it was claimed that, during *in vitro* activation of T-cells, factors appear in the supernatant which modify the EPM of these lymphocytes.[16] All these attempts to establish a link between the physical cell parameter of EPM and other biological parameters are still being controversially discussed. Nevertheless, as a whole they suggest that the EPM value of a cell may provide information about its biological state. Thus, it is still of interest to compare EPM values of cells with their other cellular parameters.

In this chapter studies on human monocytes and B-cells are described which suggest that cytokines prevent these cells from changing their EPM value during *in vitro* culturing.

II. ELECTROPHORETIC MOBILITIES OF MAMMALIAN CELLS

In order to see whether the electrophoretic mobilities of different kinds of mammalian cell populations have something in common, the EPM values of the human and animal cell populations shown in Table 14-1 were compared. The data were found in the literature written over 35 years of research. To the author's knowledge they represent the greater part of cell populations whose EPM values have been determined so far. The collection reveals two interesting aspects.

A. RANGE OF THE ELECTROPHORETIC MOBILITY VALUES

Table 14-1 shows only EPM values which are in the range between 0.5 and 3.5×10^{-4} cm²/ Vs. Since the studies were performed using several types of electrophoresis devices and different analysis conditions, it appears necessary to compare the EPM values of the standard cells used in many studies. The standard cells used most are human and rat erythrocytes. Each of these standard cells showed various EPM values when measured in different laboratories. If one assumes that each of these standard cells has only one distinct EPM in nature, one may suppose values of 1.10 and 1.24×10^{-4} cm²/Vs for human and rat erythrocytes, respectively, since the majority of the experimentaters have found these or similar values. Based on this assumption the EPM values of the cell populations can be recalculated. Such a recalculation reveals that most of the cells have electrophoretic mobilities between 0.7 and 1.40×10^{-4} cm²/Vs. This is an astonishingly narrow range.

The very narrow range of naturally occurring EPM values is a serious drawback for the application of cell electrophoresis devices for cell purification. However, it suggests that the negative surface charge density of a cell is a very important biological parameter. Since cells with an EPM outside of this range have rarely been found so far, it appears likely that such cells do not, or even cannot, exist in animals and humans. Of course, determinations of further kinds of cells are required in order to prove this speculation. Especially the EPM of tissue cells has been poorly investigated so far. Nevertheless, the data available so far suggest that the EPM of cells is a strictly controlled cellular parameter and that the possibilities for variations of the overall densities of the negative cell surface charges are restricted.

B. THE BIMODALITY OF ELECTROPHORETIC MOBILITY DISTRIBUTION CURVES OF MORPHOLOGICALLY DEFINED CELL POPULATIONS

The results of electrophoretic analyses collected in Table 14-1 show another very interesting aspect: a considerable number of cell populations belonging to one cell lineage contain cells with two distinct electrophoretic mobilities. The differences between the two EPM values are small, and both of these values are within the range of 0.7 to 1.4×10^{-4} cm²/Vs. However, they can be determined and distinguished in a reproducible manner.

The studies on the bimodality of EPM distribution curves of morphologically defined cell populations have a long history. Almost 40 years ago, when cells of cultured cancer cell lines were analyzed, it was observed that morphologically defined cell populations sometimes contained cells with different electrophoretic mobilities.[3-7] The frequency of cells showing either one or the other EPM value could vary within a population. It was thus suggested that cells might be able to switch from one EPM value to the other. The EPM changes usually occurred during the aging of a cell line or during the transition of a solid tumor cancer cell to a fluid-phase cancer cell.

About two decades later it was shown that not only malignant cell populations but also normal cell populations such as B-cells may contain cells with variable EPM values. Zeiller et al. noted an increase in EPM of B-cells from rat lymph nodes when the lymphocytes started secreting antibodies.[12] Lichtman and Weed observed a decrease in granulocyte EPM shortly before the cells left the bone marrow.[9] Meanwhile, a considerable number of human and animal cell populations with cells showing various EPM values have been found (Table 14-1). The studies clearly demonstrate that cells of healthy mammals also can alter their EPM. This means that normal physiological processes occur when cells switch from one EPM value to another.

Recently, biological and biochemical parameters which change concomitantly with the EPM were studied. For this purpose, populations of human neutrophils and platelets were fractionated electrophoretically; thereby fractions were obtained containing either low or high mobility neutrophils or platelets, respectively. Then the cells in the various fractions were characterized. Functional assays revealed that low mobility neutrophils had higher chemotactic, phagocytic, and oxygen burst activities than the high mobility neutrophils if both types of cells had been isolated from one person.[36] Furthermore, high mobility platelets expressed more alpha-2-adrenoreceptors than low mobility platelets.[41] The studies suggested a correlation between the actual state of activation of a cell and its EPM value.

An attempt was also made to obtain information about the biological relevance of EPM values. For this purpose, isolated human monocytes and B-cells were cultured in such a way that EPM alterations were possible.[20,30,35,47] The EPM alterations could be influenced by modifying the incubation techniques or by adding drugs to the culture supernatant. The early findings of these studies suggest that EPM changes occur during cell differentiation.[20,30] Further work has shown that differentiation and EPM alteration are not strictly linked to each other and that differentiation also may occur without EPM alteration.[47,48] In the following section the author's efforts to investigate the process of EPM alteration shall be summarized. The results suggest that the interaction of a distinct cytokine (monokine) with its receptor prevents monocytes, and probably also B-cells, from switching from an EPM below 1×10^{-4} cm^2/Vs to an EPM above 1×10^{-4} cm^2/Vs.

III. REGULATION OF ELECTROPHORETIC MOBILITY

The aim of these studies was to look for mechanisms which regulate alterations of the negative surface charge density of cells and to find out why the alterations occur frequently during cell activation and differentiation. For this purpose, cell electrophoresis studies were performed on human peripheral blood monocytes and B-cells. Both kinds of cells were electrophoresed either immediately after isolation or after they had been cultured for a few days after blood drawing. It was assumed that results obtained from freshly isolated cells or cells cultured for a short time might give more information about a normal, healthy *in vivo* process than results obtained from immortalized cell lines. However, one drawback was that monocytes, as well as B-cells, are minor cell populations of the human peripheral blood. Hence, it was necessary to make sure that an EPM value measured by an apparatus was correctly assigned to a distinct kind of cell. This meant that the desired cell populations had to be purified.

A. CELL ISOLATION

When the desired cells were enriched an attempt was made to perform the purification procedure without changing the current *in vivo* status of the cells. Therefore, multistep physical isolation procedures were applied for the enrichment of the isolated cells.[49] In this way purification steps which include cell adherence to surfaces or the attachment of antibodies to the cells were avoided; such steps are known to deliver activation or differentiation signals to the cells.[50-53]

Ficoll-Hypaque gradient centrifugation was the first step of cell enrichment using peripheral blood as starting material.[54] Then the mononuclear cells of the interface were washed and separated in a countercurrent centrifugal elutriator[55] (Beckman Instruments, Palo Alto, CA). With this machine several cell fractions were obtained containing cells differing in size.[49,56,57] The various cell populations were either fractionated by free flow electrophoresis or incubated for a few days.

In some experiments cells were electrophoresed which had been cultured for a few days after isolation. Cultured cells were preseparated only by countercurrent centrifugal elutriation. Since the number of cells harvested from culture flasks was usually small, it was not possible to apply the well-known standard elutriator of Beckman Instruments, which was used for fractionation of the mononuclear leukocytes obtained from the Ficoll-Hypaque gradient. Consequently, a smaller device was developed, i.e., a table-top elutriator.[58] Its main features are small outer dimensions (20×40 cm), a small separation chamber (0.5 ml), and small pieces of tubing leading the cells into and out of the chamber.

If the cells were preseparated, as described above, considerable enrichment was achieved. Nevertheless, it was not possible to perform analytical electrophoretic measurements of cells because the concentrations of the desired cells, especially of the B-cells, were still too low to reliably assign an EPM value recorded by the electrophoresis device to the cell which caused the value. Therefore, the cell suspension was fractionated with the help of the semipreparative free-flow electrophoresis apparatus ACE-710 (Hirschmann Gerätebau, Unterhaching, Germany). This machine was developed at the Max-Planck-Institut für Biochemie, Martinsried, Germany, in the department of Professor Hannig. Its technical details have already been described extensively.[19] Its main features are a small separation chamber and very precisely working accessories, such as the cooling system and the injection or elution pumps. Furthermore, an optical detection system allows the observation of the migration of the cells within the electrical field and the control of the stability of the cell flow during a preparative separation experiment.

Cells which had passed through the electrophoresis chamber were fractionated with the help of 96 pieces of tubing placed side by side at the bottom of the chamber. One piece of tubing collected cells with an EPM range of 0.05×10^{-4} cm^2/Vs. Monocytes, resting B-cells, and activated B-cells were enriched up to 98%, 70%, and 10%, respectively, after cell electrophoresis. At this concentration they could be identified by flow cytometric and histological methods.[57] The presence of preactivated or mature B-cells in an electrophoresis fraction was proved by antibody detection assays.[23,30] For this purpose the electrophoresed cells were incubated for 1 to 7 days; then cell supernatants were analyzed for human antibodies.

B. STUDIES ON THE ELECTROPHORETIC MOBILITY ALTERATIONS OF HUMAN B-CELLS

Earlier analytical studies on B-lymphocytes of human peripheral blood had shown that even the peripheral blood of healthy human beings contains resting B-lymphocytes and preactivated B-cells[59] as well as antibody-producing plasma cells.[60] These three types of human B-cells were fractionated and enriched on a preparative scale (Figure 14-1). They could be identified and distinguished by the following characteristics:

1. Resting B-cells had a small volume and left an elutriator at low counterstream rates. They did not produce antibodies unless they were stimulated by mitogens or heterologous leukocytes. If they were stimulated, they required at least 5 days of culturing before they started antibody secretion. They had a low EPM of 0.90×10^{-4} cm^2/Vs.

2. Activated B-cells had an increased volume and left an elutriator at intermediate counterstream rates. They produced antibodies after 3 days of culturing, even if no mitogens or heterologous leukocytes were added to the cultures. They produced higher quantities of antibodies in the presence of interleukins[30] or drugs.[56] They had a low EPM of 0.90×10^{-4} cm^2/Vs.

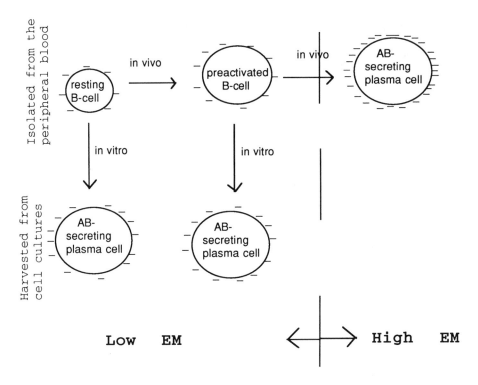

Figure 14-1 Electrophoretic mobility (EPM) alterations of human B-cells. AB = antibody.

3. Mature plasma cells left the elutriator at high counterstream rates together with the monocytes. They started antibody production on the first day of culturing. Their antibody secretion capability could not be enhanced either by mitogens or by interleukins[30] or drugs.[56] They had a high EPM of 1.05×10^{-4} cm^2/Vs.

During the characterization studies on freshly isolated B-cells neither resting or activated B-cells, with EPM values around 1.05×10^{-4} cm^2/Vs nor mature plasma cells, with an EPM around 0.90×10^{-4} cm^2/Vs, were detected. Consequently, it was concluded that human B-cells increase their EPM when they mature to antibody-secreting plasma cells. In order to prove this conclusion, B-cell differentiation behavior was investigated *in vitro*. For this purpose, resting and activated B-cells were cultured under suitable conditions until they secreted antibodies into the culture supernatant. When resting and activated B-cells had been cultured for 5 and 7 days, respectively, they showed maximal antibody secretion activities. At this time the cells were harvested, preseparated by the tabletop elutriator, and fractionated in the ACE-710. Each electrophoresis fraction was incubated for another 24 h; then their culture supernatants were screened for human antibodies. The tests revealed that the B-cells had retained their low EPM of 0.90×10^{-4} cm^2/Vs if they had matured to antibody-secreting plasma cells *in vitro* (Figure 14-1). The results accordingly suggest that no EPM alterations occur during B-cell maturation *in vitro*.

The studies showed that human antibody-secreting B-cells can have either low (0.90×10^{-4} cm^2/Vs) or high (1.05×10^{-4} cm^2/Vs) EPM, depending on whether they have matured *in vivo* or *in vitro*. Since there was interest in learning more about the EPM alteration of human B-cells during their maturation *in vivo*, possible ways to trigger a

B-cell EPM alteration also during maturation *in vitro* were investigated. So far no cytokine, drug, or incubation condition has been found which triggers B-cells isolated from the human peripheral blood to increase their EPM during maturation *in vitro.*

In another approach to this problem, the electrophoretic behavior of the mouse myeloma cell line Ag8 has been studied. This cell line contains very mature B-cells which have been immortalized and arrested at a step of differentiation where antibody secretion begins.[35] During cell electrophoresis the cell line showed a very broad EPM distribution curve with a distinct shoulder at the right edge. The cells could be separated into four fractions on a preparative scale. If cells from the four different fractions were immediately re-electrophoresed, four different narrow, unimodal EPM distribution curves were found. Their peaks were at these four positions, which could be calculated for each fraction, respectively. If the fractionated cells were incubated for 2 days and then analyzed, four identical very broad distribution curves were found, all of which looked like the starting distribution curve. The results, therefore, show that a fast-growing Ag8 myeloma cell population consists of a defined mixture of fast and slowly moving very mature B-cells.

In further experiments an attempt was made to change the ratio of fast- and slowly moving cells within a population. It was found that the quantity of fast-moving cells in a culture — i.e., the shoulder at the right edge of the distribution curve — was smaller if cells from a monolayer or from a colony were measured. On the other hand, the number of fast-moving cells increased if cells from a culture flask with low cell density were analyzed. The result that myeloma cells growing at high cell density have a predominantly low EPM suggested that either cell-cell contact or a cytokine released into the supernatant might be involved in the EPM regulation of mouse myeloma B-cells.

C. STUDIES ON THE ELECTROPHORETIC MOBILITY ALTERATIONS OF HUMAN MONOCYTES/MACROPHAGES

Monocytes are a very small population among the cells of human peripheral blood. They could be enriched up to 70% by the aforementioned preseparation procedure. This was sufficient to study the EPM of the monocytes using the semipreparative cell electrophoresis apparatus ACE 710.

Several buffers were used for electrophoresis of monocytes. It proved useful to have a considerable quantity of sodium and chloride ions within the separation buffer. When a buffer mixture of 50% low ionic strength medium (as described by Zeiller and Hannig[10]) and 50% Hanks' balanced salt medium was used, monocytes showed the narrowest EPM distribution curve, with a peak at 0.95×10^{-4} (cm^2/Vs). With this buffer system it was possible to purify the monocytes to more than 98%.

The degree of activation of the isolated monocytes was very low. This was proved by testing the interleukin 1 (IL-1) production activity, the oxygen reduction activity, the accessory cell function in concanavalin A (Con A) stimulation assays, and the helper activity in antibody production experiments.[20,61] Nevertheless, monocytes differentiated to macrophages if they were incubated in plastic dishes. The differentiation occurred irrespective of whether the monocytes were incubated after the Ficoll-Hypaque gradient (i.e., at a concentration of 20% among the other populations of the mononuclear leukocytes), after the centrifugal elutriation (i.e., 70 to 80% purity), or after the final electrophoresis step (i.e., >95% homogeneity). During the differentiation process the negative surface charge density of the monocytes/macrophages either changed or remained constant, depending on the method of cell culturing (Figure 14-2). Under suitable conditions an EPM enhancement from 0.95×10^{-4} to 1.10×10^{-4} cm^2/Vs could occur within 10 days of incubation. After this time the monocytes reached a normal macrophage size of about 2000 µm^3 irrespective of whether or not they had increased EPM.

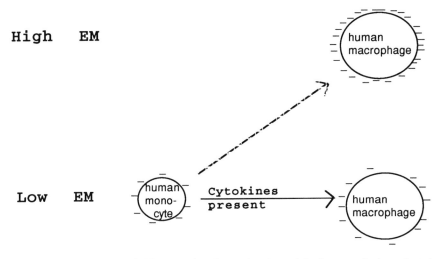

Figure 14-2 Electrophoretic mobility (EPM) alterations of human monocytes/macrophages.

The culture conditions which directed the monocyte differentiation to macrophages with either low (0.95×10^{-4} cm²/Vs) or high (1.10×10^{-4} cm²/Vs) EPM were determined. Macrophages with low EPM were found after 10 days of incubation if

1. Monocytes had been incubated after gradient centrifugation (i.e., together with lymphocytes) in a medium supplemented with human serum.[20]
2. Monocytes had been incubated after gradient centrifugation in a medium supplemented with fetal calf serum (FKS) and Con A or heterologous leukocytes.[61]
3. Monocytes had been incubated after any step of purification (i.e., with or without lymphocytes) in a medium supplemented with 12.5 µg/ml BAY R 1005 and either FKS or human serum. In the presence of the synthetic muramyl dipeptide analogon the volume increase stopped at 800 to 1000 µm³.[47]

Macrophages with high EPM were found after 10 days of incubation if

1. Monocytes had been incubated after each step of purification in a medium supplemented with FKS.[20]
2. Monocytes had been incubated after centrifugal elutriation or after final cell electrophoresis (i.e., after removal of the greater part of the lymphocytes) in a medium supplemented with human serum or with FKS.[20]

In an attempt to understand the biological relevance of the enhancement of the surface charge density, several activities of macrophages with low and with high EPM were compared.[47] The tests revealed that macrophages with low EPM could differ from macrophages with high EPM with regard to some activities.[20,47,61] However, macrophages with low EPM also showed varying activities (Table 2). The differences in the activities of the macrophages obtained by various culturing methods, therefore, seem to be caused by the culture conditions but are not related to a decreased electrophoretic mobility.[47] Only one culture component has so far been found to be required in order to obtain macrophages with low EPM: cytokines had to be present in the culture flasks at elevated

Table 14-2 **Comparison of human macrophages with low electrophoretic mobility (EPM)**

Features	Macrophages with Low EPM Obtained by Culturing Monocytes for 10 Days with		
	Human Serum and Lymphocytes	Fetal Calf Serum after Preactivation	Fetal Calf Serum and BAY R 1005
Volume	2000 μm³	2000 μm³	800 μm³
Accessory function in concanavalin A assays	Low	High	Low
Spreading	Yes	Yes	No
Helper activity in antibody production	ND	Low	High
Plaque-forming activity	Yes	ND	ND
Cytokines	Added	Produced	Produced

Note: nd = not done.

levels. They had, either to be added together with human serum, or to be produced by the monocytes after activation.[47] This observation strongly suggested that distinct monokines may be involved in the regulation of the negative surface charge density of human monocytes/macrophages.

IV. SUMMARY AND CONCLUSIONS

Quantitative determinations of electrophoretic mobilities of many kinds of cells suggest that most mammalian cells have very similar negative surface charges densities. Some of the cells are able to vary their electrophoretic mobilities in a small but reproducibly detectable manner. Our studies on the EPM alterations of human B-cells and monocytes showed that EPM changes are not strictly coupled to cell activation or differentiation. They seem to be regulated by cytokines which are predominantly produced during cell activation or differentiation.

The conclusion is strongly supported for monocytes by the data shown in Table 2. Monocytes maintain their EPM during differentiation only if cytokines are available.[47] Whether or not B-cells maintain their EPM in a similar way is not certain at the moment, but some data suggest it: During B-cell maturation many cytokines are produced.[62] B-cells which had matured in vitro could not be completely separated from the cytokines in the supernatant and from the cells (monocytes, T-cells), which produced the cytokines. *In vivo,* mature B-cells found in the peripheral blood have already left their environment of maturation (i.e., lymph nodes). Furthermore, the fast and continuous growth of myeloma cells is due to an autocrine interleukin 6 (IL-6)/IL-6 receptor loop.[63] Such a loop may be weakened if the cells grow at low concentration, which may trigger the cells to change their EPM.

Overall, it seems to be worthwhile to consider cytokines when cellular EPM alterations are further investigated.

ACKNOWLEDGMENT

The author wishes to thank Mrs. S. Bauer for help in preparing this manuscript.

REFERENCES

1. **Abramson, H. A. and Mayer, L. S.,** *The Electrophoresis of Proteins,* Reinhold, New York, 1942, 44.
2. **Fuhrmann, G. F. and Ruhenstroth-Bauer, G.,** Cell electrophoresis employing a rectangular measuring cuvette, in *Cell Electrophoresis,* Ambrose, E. J., Ed., J. & A. Churchill, London, 1965, 22.
3. **Ambrose, E. J., James, A. M., and Lowick, J. H. B.,** Differences between the electrical charge carried by normal and homologous tumor cells, *Nature,* 177, 576, 1956.
4. **Purdom, L., Ambrose, E. J., and Klein, G.,** A correlation between electrical surface charge and some biological characteristics during stepwise progression of a mouse sarcoma, *Nature,* 181, 1586, 1958.
5. **Forrester, J. A., Ambrose, E. J., and Macpherson, J. A.,** Electrophoretic investigations of a clone of hamster fibroblasts and polyoma-transformed cells from the same population, *Nature,* 196, 1068, 1962.
6. **Arnold, R.,** Pathological haemocytopherograms of rats and mice, in *Cell Electrophoresis,* Ambrose, E. J., Ed., J. & A. Churchill, London, 1965, 37.
7. **Mayhew, E.,** Electrophoresis of Ehrlich ascites carcinoma cells grown in vitro and in vivo, *Cancer Res.,* 28, 1590, 1968.
8. **Lichtman, M. A. and Weed, R. I.,** Alteration of the cell periphery during granulocyte maturation: relationship to cell function, *Blood,* 39, 301, 1972.
9. **Lichtman, M. A. and Weed, R. I.,** Electrophoretic mobility and N-acetyl neuramic acid content of human normal and leukemic lymphocytes and granulocytes, *Blood,* 35, 12, 1970.
10. **Zeiller, K. and Hannig, K.,** Evidence for specific organ distributions of lymphoid cells, *Hoppe-Seyler's Z. Physiol. Chem.,* 352, 1162, 1971.
11. **Suemasu, K., Watanabe, K., and Ichikawa, S.,** Contribution of polyanionic character of dextran sulfate to inhibition of cancer metastasis, *Gann,* 62, 331, 1971.
12. **Zeiller, K., Pascher, G., and Hannig, K.,** Preparative electrophoretic separation of antibody forming cells, *Prep. Biochem.,* 2, 21, 1972.
13. **Hannig, K., Heidrich, H. G., Klofat, W., Pascher, G., Schweiger, A., Stahn, R., and Zeiller, K.,** Separation of cells and particles by continuous free flow electrophoresis, in *Techniques of Biochemical and Biophysical Morphology,* Glick, D. and Rosenbaum, R. M., Eds., John Wiley & Sons, New York, 1972, 191.
14. **Mehrishi, J. N.,** Molecular aspects of mammalian cell surface, *Prog. Biophys. Mol. Biol.,* 25, 1, 1972.
15. **Bosman, H. B., Bieber, G. F., Brown, A. E., Gersten, D. M., Kimmerer, T. W., and Lione, A.,** Biochemical parameters correlated with tumor cell implantation, *Nature,* 246, 487, 1973.
16. **Blume, P., Malley, A., Knox, R. J., and Seaman, G. V. F.,** Electrophoretic mobility as a sensitive probe of lectin-lymphocyte interaction, *Nature,* 271, 378, 1978.
17. **Bamberger, S., Storch, F., Valet, G., and Ruhenstroth-Bauer, G.,** Postnatale Entwicklung von Rattenthrobozyten: Die Volumenverteilung und elektrophoretische Beweglichkeit von Rattenthrombozyten während des ersten Lebensmonats, *Blut,* 40, 421, 1980.
18. **Cocker, E. N., Perry, M. C., and Plummer, D. T.,** Changes in the zeta potential of chick erythrocytes after the addition of insulin, *Biochem. Soc. Trans.,* 9, 89, 1981.
19. **Hannig, K.,** New aspects in preparative and analytical continuous free flow cell electrophoresis, *Electrophoresis,* 3, 235, 1982.

20. **Bauer, J. and Hannig, K.,** Changes of the electrophoretic mobility of human monocytes are regulated by lymphocytes, *Electrophoresis,* 5, 269, 1984.

21. **Mironov, S. L. and Dolgaya, E. V.,** Surface charge of mammalian neurones as revealed by microelectrophoresis, *J. Membr. Biol.,* 86, 197, 1985.

22. **Cohly, H., Albini, B., Weiser, M., Green, K., and van Oss, C.,** Microelectrophoresis of epithelial cells and lymphocytes, in *Cell Electrophoresis,* Schütt, W. and Klinkmann, H., Eds., Walter de Gruyter, Berlin, 1985, 611.

23. **Bauer, J. and Hannig, K.,** Human antibody secreting cells enriched by free flow electrophoresis, *Electrophoresis,* 7, 367, 1986.

24. **Rychly, J., Knippel, E., Thomaneck, U., Schütt, W., Sabolovic, D., Babusikova, O., Eggers, G., Anders, O., and Klinkmann, H.,** Suitability of electrophoretic histograms for the characterization of human blood lymphocytes, in *Electrophoresis '86,* Dunn, M. J., Ed., VCH Verlagsgesellschaft, Weinheim, Germany, 1986, 56.

25. **Hyrc, K. L. and Cieszka, K.,** Electrophoretic properties of pigmented hamster melanoma cells derived from tumours and cultures of different age, in *Electrophoresis '86,* Dunn, M. J., Ed., VCH Verlagsgesellschaft, Weinheim, Germany, 1986, 94.

26. **Easty, D. J., Crook, M. A., Crawford, N. V., and Evans, D. J.,** The isolation of subpopulations of keratinocytes from epidermal suspensions using free flow electrophoresis, in *Electrophoresis '86,* Dunn, M. J., Ed., VCH Verlagsgesellschaft, Weinheim, Germany, 1986, 108.

27. **Price, J. A. R., Pethig, R., Lai, C. N., Becker, F. F., Gascoyne, P. R. C., and Szent-Györgyi, A.,** Changes in the cell surface charge and transmembrane potential accompanying neoplastic transformation of rat kidney cells, *Biochim. Biophys. Acta,* 898, 129, 1987.

28. **Silva-Filho, F. C., Santos, A. B. S., DeCarvalho, T. M. U., and DeSouza, W.,** Surface charge of resident, elicited, and activated mouse peritoneal macrophages, *J. Leukocyte Biol.,* 41, 143, 1987.

29. **Gommans, J. M. and DeJongh, G. J.,** Electrophoretic mobilities of keratinocytes from psoriatic lesions and normal epidermis, *Epithelia,* 1, 101, 1987.

30. **Bauer, J., Kachel, V., and Hannig, K.,** The negative surface charge density is a maturation marker of human B-lymphocytes, *Cell. Immunol.,* 111, 554, 1988.

31. **Nair, S., Horton, A., Leif, R. C., and Krishan, A.,** Electrophoretic mobility studies on doxorubicin-resistant and -sensitive murine P388 leukemic cells, *Cytometry,* 9, 232, 1988.

32. **Yamamoto, K., Yamamoto, M., and Ooka, H.,** Changes in the negative surface charge of human diploid fibroblasts, TIG-1, during in vitro aging, *Mech. Ageing Dev.,* 48, 183, 1988.

33. **Carter, H. B. and Coffey, D. S.,** Cell surface charge in predicting metastatic potential of aspirated cells from the Dunning rat prostatic adenocarcinoma model, *J. Urol.,* 140, 173, 1988.

34. **Vargas, F. F., Osorio, H. M., Basilio, C., DeJesus, M., and Ryan, U. S.,** Enzymatic lysis of sulfated glycosaminoglycans reduces the electrophoretic mobility of vascular endothelial cells, *Membr. Biochem.,* 9, 83, 1990.

35. **Bauer, J. and Kachel, V.,** The increase of electrophoretic mobility and alkaline phosphatase activity are parallel events during B-cell maturation, *Immunol. Invest.,* 19, 57, 1990.

36. **Crawford, N., Eggleton, P., and Fischer, D.,** Population heterogeneity in blood neutrophils fractionated by continuous flow electrophoresis (cfe) and by partitioning in aqueous polymer 2-phase systems (paps), in *Separation Technology,* Kampala, D. and Todd, P., Eds., American Chemical Society, Washington, D.C., 1991, 190.

37. **Shimizu, M., Sekine, K., Matsuzawa, A., and Iwaguchi, T.,** Cell electrophoretic characterization of abnormally expanded lymphocytes in autoimmune Ipr/cg, Ipr, gld and Yaa mice and of thymocyte subsets, *Electrophoresis,* 13, 136, 1992.

38. **Knopf, B. and Wollina, U.,** Electrophoretic mobilities of keratinocytes from normal skin and psoriatic lesions, *Arch. Dermatol. Res.,* 284, 117, 1992.

39. **Dolgaya, E. V., Krylova, I. V., and Pinchuk, G. V.,** The role of adenylate cyclase system in concanavalin A mediated changes in the electrophoretic mobility of murine T-lymphocytes, *Biol. Membr.,* 5, 1138, 1992.

40. **Hyrc, K., Wilczek, A., and Cieszka, K.,** Electrophoretic heterogeneity of pigmented hamster melanoma cells, *Pigment Cell Res.,* 6, 100, 1993.

41. **Crook, M. and Crawford, N.,** Alpha$_2$-adrenoreceptor status of human platelet subpopulations separated by continuous flow electrophoresis, *Thromb. Haemostasis,* 69, 60, 1993.

42. **Slivinsky, G. G.,** Simultaneous two-parameter measurements of electrophoretic features of subpopulations of cells and their different sedimentation characteristics, in *Cell Electrophoresis,* Bauer, J., Ed., CRC Press, Boca Raton, FL, 1994, chap. 10.

43. **Mehrishi, J. N.,** Cell surface charge increase by both tumor inhibitory and tumor growth promoting polyanions, *Nature,* 228, 364, 1970.

44. **Field, E. J. and Caspary, E. A.,** Lymphocyte sensitization: an in-vitro test for cancer?, *Lancet,* 2, 1337, 1970.

45. **Schütt, W. and Klinkmann, H., Eds.,** *Cell Electrophoresis,* Walter de Gruyter, Berlin, 1985.

46. **Schütt, W., Hashimoto, N., and Shimizu, M.,** Application of cell electrophoresis for clinical diagnosis, in *Cell Electrophoresis,* Bauer, J., Ed., CRC Press, Boca Raton, FL, 1994, chap. 13.

47. **Bauer, J., Stuenkel, K. G. E., and Kachel, V.,** Linkage between monokine production and regulation of the negative surface charge density of human monocytes, *Immunol. Invest.,* 21, 507, 1992.

48. **Bauer, J.,** Correlation between site of antibody release and negative surface charge density, *FASEB J.,* 2, 665, 1988.

49. **Bauer, J. and Hannig, K.,** Free flow electrophoresis: an important step among physical cell separation procedures, in *Electrophoresis '86,* Dunn, M. J., Ed., Verlag Chemie, Weinheim, Germany, 1986, 13.

50. **Van Wauwe, J. P., De May, J. R., and Goosens, J. G.,** OKT3: a monoclonal anti-human T-lymphocyte antibody with potent mitogenic properties, *J. Immunol.,* 124, 2708, 1980.

51. **Chiorazzi, N., Fu, S. M., and Kunkel, H. G.,** Stimulation of human B-lymphocytes by antibodies to IgM and IgG; functional evidence for the expression of IgG on B-lymphocyte surface membranes, *Clin. Immunol. Immunopathol.,* 15, 380, 1980.

52. **Gonwa, T. A., Frost, J. P., and Karr, R. W.,** All human monocytes have the capability of expressing HLA-DQ and HLA-DP molecules upon stimulation with interferon-gamma, *J. Immunol.,* 137, 519, 1986.

53. **Larson, E. L., Anderson, J., and Couthino, A.,** Functional consequences of sheep red blood cell rossetting for human T-cells: gain of reactivity to mitogenic factors, *Eur. J. Immunol.,* 8, 693, 1978.

54. **Böyum, A.,** Isolation of mononuclear cells and granulocytes from human blood. Isolation of mononuclear cells by one centrifugation and of granulocytes by combining centrifugation and sedimentation at 1g, *Scand. J. Clin. Lab. Invest.,* 21 (Suppl. 97), 77, 1968.

55. **Meistrich, M. L.,** Experimental factors involved in separation by centrifugal elutriation, in *Cell Separation: Methods and Selected Applications,* Vol. 1, Pretlov, T. G. and Pretlov, T. P., Eds., Academic Press, New York, 1982, 33.

56. **Bauer, J. and Stünkel, K. G.,** Isolation of human B-cell subpopulations for pharmacological studies, *Biotechnol. Prog.,* 7, 391, 1991.

57. **Bauer, J. and Kachel, V.,** Separation accuracy of free flow electrophoresis as proved by flow cytometry, *Electrophoresis,* 9, 62, 1988.

58. **Bauer, J. and Hannig, K.,** Countercurrent centrifugal elutriation in a table-top centrifuge, *J. Immunol. Methods,* 112, 213, 1988.

59. **Anderson, K. C., Roach, J. A., Daley, J. F., Schlossman, S. F., and Nadler, L. M.,** Dual fluorochrome analysis of human B-lymphocytes: phenotypic examination of resting, anti-immunoglobulin stimulated, and in vivo activated B cells, *J. Immunol.,* 136, 3612, 1986.

60. **Fauci, A. S., Whalen, G., and Burch, C.,** Activation of human B-lymphocytes. XV. Spontaneously occurring and mitogen induced anti-sheep red blood cell plaque forming cells in normal human peripheral blood, *J. Immunol.,* 124, 2410, 1980.

61. **Bauer, J. and Hannig, K.,** Relationship between electrophoretic mobility and activation of human monocytes, *Electrophoresis,* 6, 301, 1985.

62. **Kishimoto, T.,** B-cell stimulatory factors (BSFs): molecular, biologic function, and regulation of expression, *J. Clin. Immunol.,* 5, 343, 1987.

63. **Jourdan, M., Zhang, X. G., Portier, M., Boiron, J. M., Bataille, R., and Klein, B.,** IFN-alpha induces autocrine production of IL-6 in myeloma cell lines, *J. Immunol.,* 147, 4402, 1991.

V
Cell Electrophoresis in Space

Chapter 15

Cell Electrophoresis in Microgravity: Past and Future

Dennis R. Morrison

CONTENTS

I. INTRODUCTION

Researchers that developed preparative free-fluid electrophoresis systems recognized that many technical problems were due to gravity-dependent phenomena.[1,2] Theoretical modeling predicted that much larger chambers, higher field strengths, improved resolution, and greater sample throughput could be achieved in the absence of gravity.[3] The designs of apparatus for early space experiments focused on two types of systems and objectives: static column free-zone electrophoresis (FZE) systems aimed at demonstrating maximum resolution and continuous flow electrophoresis (CFE) systems aimed at preparative-scale throughput and improved resolution. Experiments on Apollo and Apollo-Soyuz Test Project (ASTP) missions demonstrated several of the advantages of operating both static column and CFE systems in microgravity. Economic studies of potential health care impacts led to commercial feasibility studies that identified multi-million-dollar markets for selected biological products obtainable from space electrophoresis.[4] FZE and CFE system experiments conducted by NASA and McDonnell-Douglas Aeronautics Corporation (MDAC) on early space shuttle missions demonstrated the feasibility of purifying commercial quantities of certain pharmaceuticals and isolating subsets of mammalian cells that had specific secretory functions. These experiments demonstrated that microgravity electrophoresis offered selective advantages for isolating live cells that could be used for transplantation or seeding commercial bioreactors to produce valuable pharmaceuticals in Earth-based plants. A large preparative-scale commercial CFE device, equivalent to 24 of the Shuttle middeck units, was designed and partially constructed for use in the Shuttle cargo bay; however, the Challenger accident in 1986 effectively curtailed the MDAC commercial space electrophoresis effort. Meanwhile, new experimental free-fluid electrophoresis space systems have been developed by companies in Germany, France, Japan, Russia, and the U.S. The German CFE system has flown on the Texus sounding rocket, two Russian electrophoresis systems have flown on the Mir space station, and the Japanese free-flow electrophoresis unit (FFEU) has flown on the Spacelab-J mission. The French RAMSES and American USCEPS systems are manifested for space flights in 1994 and 1995. Current plans for a biotechnology research facility (BTF) on the Space Station Freedom include several types of electrokinetic separation systems. This chapter will review the rationale for electrokinetic separation experiments in microgravity, the basic design features of the different flight systems, results of flight experiments, lessons learned, and future experimental systems that are planned for Spacelab and Space Station research. Cell electrophoresis is emphasized; however, some protein electrophoresis experiments are included since current plans include protein and macromolecular separation experiments that are already manifested on near-term Shuttle, Spacelab, and Spacehab missions.

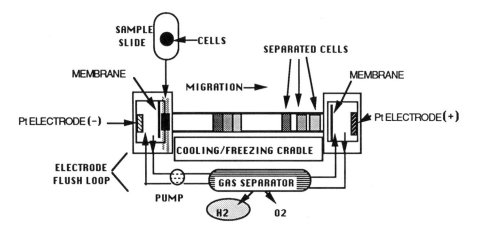

Figure 15-1 Schematic of free-zone, static column electrophoresis system (MA-011) used on Apollo-Soyuz Test Project and STS-3. The sample slide is inserted at the cathode end of the column, thawed, and then an electric field is applied. Cells migrate toward the anode (left to right).

II. PRINCIPLES OF MICROGRAVITY ELECTROPHORESIS
A. GRAVITY-INDUCED LIMITATIONS

Three gravity-dependent phenomena affect free-fluid electrophoresis: particle sedimentation, droplet sedimentation, and thermal convection.[1,2,5] In a free zone system the buffer is essentially a static fluid column between the electrode chambers, and the cells or particles migrate in the electric field toward the anode (see Figure 15-1). Sedimentation is a function of the density difference between the particles and the fluid buffer. On Earth, sedimentation is so dominant that free zone electrophoresis of cells is mostly restricted to horizontal operations.[6] In a CFE system where carrier buffer flow is upward, the cells or particles migrate orthogonal to the carrier flow (see Figure 15-2) and sedimentation will either increase the effective residence time (at low concentrations) or inhibit sample flow entirely (at high concentrations). This is illustrated in both density gradient electrophoresis and CFE separation of mammalian cells, where the more dense cells are found in fractions different from the less dense cells even when they have the same electrophoretic mobility (EPM).[6,7] In a CFE system which flows downward, particle sedimentation can cause the cells to descend faster through the chamber, thus reducing residence time in the electric field.[8] Also, a significant sedimentation potential can be created when a charged particle moves relative to another charged particle. This "counterstreaming potential" is directly proportional to the ionic strength of the surrounding medium and inversely proportional to the distance over which a charge can exert a force in the fluid. This potential can be as much as 20 mV, which is comparable to the net cell surface charge.[8]

Zone or droplet sedimentation can occur when a high concentration of large molecules or cells is surrounded by a peripheral zone in which the concentration is much lower. Cells or macromolecules only diffuse out of their zone slowly, while small molecules can diffuse rapidly into the zone, making it denser than the surrounding solution. This results in aggregation properties among the cells or particles which cause the zone to sediment as a droplet within the fluid.[5,8]

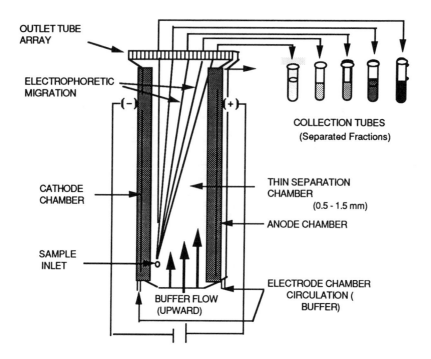

Figure 15-2 Conceptual schematic of continuous flow electrophoresis system with upward-flowing carrier buffer. Sample and carrier buffer are pumped at different rates. The electrode chambers are flushed independently of the separation chamber to remove evolved gases. Cell migration is toward the anode (right).

Although free-fluid electrophoresis systems use low conductivity buffers, thermal convection is produced from electric current flow and joule heating of the buffer. On Earth, CFE field strengths required to achieve migration rates of several milliliters per minute require flow chambers less than 1 mm thick which must be cooled vigorously at the walls to prevent thermal convection from remixing the separands.[3,5] Higher electric fields require faster carrier flow and, thus, reduce residence time in the electric field. A typical commercial instrument,* having a chamber 0.15 cm thick, 5 cm wide, and 50 cm long, is limited to field strengths of 100 V/cm and carrier flow rates of only 10 ml/min in unit gravity.[3]

Fluid dynamics within a free-zone, static column system are characterized mainly by sample movement in the center of the chamber and both return flow (opposite direction in a closed system) and electroosmotic flow produced along the walls by the electric field.[2,8] Complex fluid dynamics within a CFE chamber include two-dimensional parabolic Poiseuille flow, which is imposed on the carrier stream by viscous drag at the chamber walls, and electroosmotic flow, due to high zeta potential on the walls, causing flow along the walls perpendicular to the carrier flow. As a result the sample stream becomes crescent shaped in cross section (see Chapter 4, Figure 4-13), which can cause overlapping sample migration patterns and presents difficulties with harvesting the separands at the outlet.[3,8,9] Electroosmosis is not gravity dependent and must be negated by appropriate coating of the chamber walls to eliminate the zeta potential.[1,10] However, crescent-shaped sample streams can be avoided only if the sample stream diameter is kept small compared to the thickness of the

---

* Continuous Particle Electrophoresis, Beckman Instruments, Anaheim, CA.

chamber[3] or if low CFE carrier flow rates or special designs (e.g., deflected-lamina electrophoresis)[11] are used to obviate the effects of Poiseuille flow.

B. MICROGRAVITY DESIGN CONSIDERATIONS

Electroosmosis flow (electroendoosmosis) of the buffer solution along the stationary chamber wall produces a parabolic fluid flow profile. This spreads out the band of separands, making sample collection more difficult and effectively restricting the resolution of the electrophoresis process. Microgravity does little to modify or eliminate electroosmosis. It must be remedied either by using a wall material which bears no surface charge or else by coating the wall with a polymer having a low zeta potential, such as silane or covalently linked cellulose derivatives.[10]

Gas formation at the electrodes results from hydrolysis of water, leading to an accumulation of hydrogen at the cathode and oxygen at the anode. Coating the electrodes with platinum or palladium can reduce this problem, but not eliminate it, especially at the high current densities required for large-scale, high-resolution separations. In microgravity the evolved gas is not bouyant; therefore, it must be removed by circulating the buffer solution through the electrode compartments and then removing the gases by membrane or phase separation downstream from the electrodes.[5] This requires an apparatus design that maintains the separation between the electrodes and the electrophoresis chamber by using semipermeable membranes and gas bubble separation methods that operate efficiently in microgravity. Finally, the recovered hydrogen gas and oxygen gas must be contained or catalyzed.

CFE systems also must withstand launch vibrations and reentry loads up to 4 g on the space shuttle. Automated CFE systems flown on sounding rockets must accommodate lift-off accelerations of 13 *g* or more.[3] Sounding rocket experiments are are typically limited to 6 to 15 min duration and must be completely automated. Upon reentry the system must withstand accelerations up to 25 *g*.

Experiments flown on shuttle and Spacelab missions have utilized some crew participation, usually for sample loading, unloading, stowage, and monitoring during the runs. The free-zone (static column) MA-011 electrophoresis unit flown on the ASTP and STS-3 missions required the cell samples to be frozen preflight, thawed for the separation run, and then refrozen. Astronauts transferred the frozen sample slides, started the electrophoresis runs, took pictures of the columns during the runs, and then stowed the frozen columns in an onboard liquid N_2 freezer. CFE experiments utilized microprocessor controllers to operate the hardware, but crew members loaded the sample injection syringes and changed the fraction collection modules between experiments. After each run the cell fractions were stowed in a temperature-controlled middeck locker for the duration of the flight. Man-tended operations allowed several different types of experiments on the same flight.

As yet it has not been practical to change out carrier buffers in orbit. Therefore, the separation of proteins and living cells on the same flight has been difficult because different buffers, thermal tolerance, carrier and sample flow rates, and other specific operating conditions are required. Protein separations and limited cell separations using dilute sample concentrations can be carried out on Earth and then compared with experiments in microgravity. Computer modeling of the electrohydrodynamics within the chamber, ground-based tests, and initial flight data enabled computer projections of the expected efficiencies when gravity was reduced to less than 10^{-2} *g*. On initial flights the electrophoresis apparatus was used first to separate concentrated solutions of proteins, which demonstrated both resolution and throughput improvements in microgravity. Later, cell separations were conducted on succeeding flights once the microgravity operating characteristics of the particular device were confirmed.

III. EARLY SPACEFLIGHT EXPERIMENTS

Early flight experiments were designed to demonstrate improved resolution using increased sample concentration whenever sample sedimentation and convective mixing were eliminated from the electrophoretic chamber. Therefore, the first demonstrations used free zone, static column electrophoresis apparatus. Later, continuous flow systems were developed to characterize the interdependent relationships and to demonstrate the magnitude of increased resolution and sample throughput in microgravity.

A. FREE ZONE ELECTROPHORESIS
1. Apollo 14 and Apollo 16 Experiments

Electrophoresis in a free fluid was first demonstrated in space during the Apollo 14 mission using deoxyribonucleic acid (DNA), hemoglobin, and soluble dyes.[13] Sample bands were to be photographed periodically during electrophoresis in cylindrical chambers that were 10 cm long by 0.64 cm in diameter. The electrodes were separated from the chamber by semipermeable membranes and rinsed continually to remove any electrolytic products. The samples to be separated were confined in sample chambers with impermeable membranes which were removed to release each sample before the voltage gradient was applied. The dyes separated as expected, but the protein separation was less than expected, presumably because of sample changes during the long time between loading the apparatus and initiating the experiment in orbit. Particular problems encountered during these demonstrations included a misaligned sample insertion assembly that injected the samples near the column walls, poor photography from a hand-held camera, and bacterial degradation of the DNA and hemoglobin because of the long storage time. However, a separation was measured between two dyes, and operation of the fluid and electrical systems was normal.

The Apollo 16 electrophoresis demonstration was performed using the same basic apparatus, but with 0.23- and 0.80-μm-diameter monodispersed polystyrene latex particles as models for living cells.[14] Three chambers were used; one chamber contained the 0.23-μm particles, one contained the 0.80-μm particles, and the third contained a mixture of the two different-size particles. During electrophoresis all three samples migrated as elongated parabolic (bullet-shaped) profiles. The elongated profiles were attributed to the strong electroosmotic flow (10 μm cm V^{-1} s^{-1}) of the medium moving in the opposite direction along the wall of the chamber. The leading 0.80-μm particles moved at the same rate as the particle mixture; however, the 0.23-μm particles moved more slowly. Analysis of the migration of the latex particles was complicated by the appearance of bubbles in the electrophoresis chambers, which dramatically reduced the nominal voltage gradient of 26 V/cm. Densitometric scans of the photographs clearly showed the parabolic cone of 0.23-μm particles nested inside the parabola of the 0.80-μm particles, which agreed with the calculated particle velocities.

This Apollo 16 experiment demonstrated the feasibility of electrophoretic separations in space. Unfortunately, the separation of the different-size particles was not totally successful because of the strong electroosmotic flow counter to the direction of electrophoretic migration of the latex particles and because of bubbles which formed in each column due to electrolytic gas generation.

2. Apollo-Soyuz MA-011 Experiments

In 1975 the Electrophoresis Technology Experiment (MA-011) was conducted during the ASTP flight. These experiments were performed to demonstrate static free-zone electrophoresis and isotachophoresis of biological cells under microgravity conditions.[15]

Preflight each 0.06-ml sample slide was loaded with 5.22×10^6 rabbit red cells, 3.44×10^6 human cells, and 7.26×10^6 horse cells and then frozen. In microgravity the astronaut loaded each frozen sample slide (see Figure 15-1) into a separate electrophoresis column for each run. The cell sample was thawed and the electric field was applied to start the experiment. The cells migrated toward the anode at rates proportional to their EPM. At the end of each electrophoresis period the column was frozen and stowed in a liquid N_2 freezer. Postflight harvesting was accomplished by extruding the frozen sample cylinder from the glass column and slicing it into some 28 fractions which were then analyzed separately for viability, EPM, and function.[15,16]

Results from the ASTP indicated that the electroosmotic flow was effectively eliminated due to use of an interior wall coating that had a zeta potential of nearly zero. Fixed horse red blood cells (RBCs) were separated from a mixture of human, rabbit, and horse RBCs; however, the human and rabbit cells did not separate completely as expected. The recovery of cells was 70 to 90% of the theoretical yield. Timed inflight photographs of the RBC band migration showed no band spreading or parabolic curvature due to electroosmosis, but densitometry showed that the separation between horse and human cells was greater than anticipated. EPMs of the cells from sample slices taken from each cylinder were measured by analytical particle microelectrophoresis to determine the relative distribution of each type of RBC.

Human kidney cells were successfully separated into 28 subsets, some of which produce a valuable enzyme called urokinase that dissolves blood clots. One of the separated kidney cell fractions produced six to seven times more urokinase than ever before possible under ground-based tissue culture conditions.[15] However, the small number of cells returned from orbit limited the postflight analysis and precluded statistical analyses of the surface charge character of the target cells producing high levels of urokinase.

Another experiment attempted to separate 1.5×10^6 human lymphocytes; however, cell migration was not detected because of blockage in a fluid line leading to the electrodes, which caused accumulation of gas bubbles and a high resistance. Only 6% of the recovered lymphocytes were viable, and the buffer was unexpectedly very acidic.

3. Isotachophoresis

Free-fluid isotachophoresis is achieved by placing a solution of charged sample particles (cells) in a column containing in addition two electrolytes, one of which has coions which are all more mobile than any sample coions; the other electrolyte has coions which are all less mobile than any sample coions. All of the electrolytes share a common counterion. When the electric field is applied, the sample coions begin moving at different velocities until a steady state is reached, wherein the sample ions are separated into contiguous zones in order of their mobilities. The zones migrate at the same velocity and adjust themselves into compartments of various lengths, concentrations, and conductivities such that the product of the coion mobility and the field gradient is the same within every zone. The boundaries are self-restoring and will reform if perturbed by convection or other factors. Once the separation has been achieved, the concentration of each substance within a compartment will remain constant. The microgravity applications of isotachophoresis have been reviewed by Bier and co-workers.[17]

On ASTP the isotachophoresis columns (Pyrex® glass, 0.97 cm O.D., 0.64 cm I.D., and 15.24 cm length) had electrode chambers at each end containing electrodes of either silver (anode) or palladium (cathode). The buffer system chosen for the isotachophoresis columns consisted of a lead buffer of 10 mM phosphate, pH 7.4 and a terminator buffer of 200 mM L-serine, pH 8.2 with Tris as the common counterion. This buffer avoided cell aggregation and also contained 4.2% dextrose as an osmotic contributor and 3 M glycerol as a cryoprotective. One experiment used a mixture of frozen rabbit and human erythrocytes, and the other utilized formalin-fixed human and rabbit red cells. The final concentration in the sample slide was 4.5×10^8 cells per 0.098 ml.

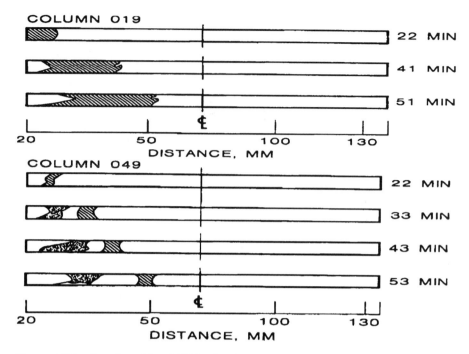

Figure 15-3 Drawing of fixed RBC migration patterns vs. time of free zone electrophoresis on STS-3. Upper panel (column 019) shows a single wide band of RBCs. The lower panel (column 049) shows two bands of RBCs with the human cells leading the rabbit erythrocytes. Note the concave shape of the leading edges of the human RBCs with respect to the migration (left to right). (Redrawn from NASA photos.)

Upon electrophoresis the frontal boundary moved more than 1 mm/min, and the fixed cell boundary was flat and sharp as predicted by isotachophoresis models using an antielectroosmotic coating. However, the unfixed red cells migrated only 60% of the predicted distance. No clear separation between the species of red cell was observed in either of the two experiments.[15]

4. STS-3 Experiments

On the third shuttle mission the electrophoresis equipment verification test used the MA-011 static column hardware to conduct two successive separations of fixed RBCs and six separations of human embryonic kidney (HEK) cells. The objective was to separate cells at very high concentrations that normally could not be resolved in unit gravity. A new buffer (D-1) was developed for STS-3, and preflight the formalin-fixed mobilities of the human and rabbit erythrocytes were found to be -2.05 ± 0.11 and -1.46 ± 0.09 μm cm $V^{-1}s^{-1}$, respectively, in the D-1 buffer. Photographs of the RBC migrations were recorded during the flight and analyzed by microdensitometry. The leading edge of the foremost cell bands in both columns migrated at more than 1 mm/min.[18] Figure 15-3 illustrates the relative positions of the RBCs for each column. During the 60-min electrophoresis run the fastest cell column (#019) migrated to only 71 mm in a field of 13.8 V/cm. It was predicted that the human RBCs should have been visibly separated from the rabbit RBCs before the human cells reached 70 mm; however, this was not apparent in the photos. Densitometry scans of the photographs indicated some cell-to-cell interactions between the human and rabbit erythrocytes in column #019, and only one large band was evident. It also was shown that the leading and trailing edges corresponded to the mean EPM of the human and rabbit RBCs, respectively.[19] The cell migration pattern observed in column #049, however, did indicate two bands. The effective electric field applied to these cells was about 14.2 V/cm at an average temperature gradient of 11 to 12°C. Inflight photographs showed one leading band of RBCs followed by a diffuse second band. The separation of human and rabbit RBCs on STS-3 was not as distinct as predicted, possibly due to cell-to-cell interactions or aggregations,[20] which also were suspected in the ASTP experiment. It was concluded, however, that cells could be separated at much higher concentrations under microgravity conditions. Follow-up experiments using a rotating tube, free-zone electrophoresis system confirmed that the patterns observed on STS-3 could occur if enough electroosmosis was present when the ratio of human to rabbit cells was 2:1.[5,6]

The STS-3 HEK cells were selected from 30 lots of primary cells that were characterized for viability, attachment, outgrowth, proliferative life span, karyology, EPM, and urokinase production of electrophoretic fractions. Inflight photos of the HEK separations lacked enough contrast to measure the effective EPM, and a postflight sample freezer mishap precluded analysis of urokinase secretion. However, the extensive characterization of the HEK cells was invaluable in the design of CFE separations of HEK cells flown on STS-8.[18]

5. Salyut-7 Experiments

In 1983 the Soviet free-fluid electrophoresis system "Tavriya" flown on Salyut-7 was used to separate bone marrow cells into five distinct fractions. It had a unique fraction collection system using lateral outlets from the separation chamber whereby separated fractions could be harvested simultaneously using buffer injected from opposite ends.[21] Tavriya was also operated as an isoelectric focusing system, using an artificial pH gradient composed of borate-polyol, to separate five variants of human serum albumin which could not be resolved in the ground-based unit.

6. Results of Free Zone Electrophoresis Experiments In Space

Although the early FZE microgravity experiments did not accomplish very high-resolution separations of cells, they did demonstrate that the normal operating ranges of electric field strength and sample concentration could be expanded several orders of magnitude. These experiments also confirmed that cell samples could be adequately separated in microgravity at concentrations 100 to 1000 times greater than possible on Earth. Figure 15-4 illustrates the effects of increased cell concentration (droplet sedimentation) on the variability of EPM as determined in horizontal microelectrophoresis (orthogonal), upward density gradient column (antiparallel), and 0 g (STS-3) systems. Preparative electrophoresis at cell concentrations above 2×10^7/ml clearly has an advantage in microgravity. Table 15-1 also shows a comparison of EPMs of fixed erythrocytes separated by FZE in space and on Earth. The EPMs of human and rabbit RBCs were the same at 10^9 cells per milliliter in space as they were at 10^6 cells per milliliter in 1 g.

Once convective mixing and particle sedimentation were eliminated the pervasive role of electroosmosis became dramatically apparent.[14] Development of the MA-011 hardware included new coatings that could adhere to the chamber walls even in the presence of flowing buffers.[10] The MA-011 experiments demonstrated that proper coating of the chamber walls to reduce the zeta potential largely eliminated the electroosmotic flows; however, cell migration profiles did not always match predictions. It was suspected that zone sedimentation affected the local mobility of cells, especially in very concentrated samples, and that conductivity mismatches between the sample and the carrier buffer could have significant consequences in FZE.[8,20] Later this principle also was illustrated in CFE experiments on STS-7. The advantages of isotachophoresis also were demonstrated. At this point the commercial interest for preparative-scale electrophoresis of

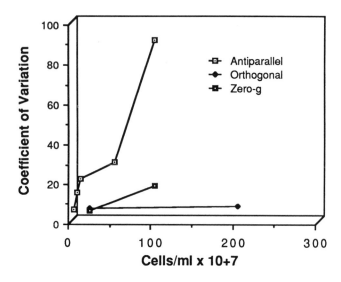

Figure 15-4 Graph of the effect of droplet (zone) sedimentation on the variation in electrophoretic mobility measurements from RBCs conducted in three different systems. Cell concentration effects are manifested as rapid increases in the coefficient of variation when the cells are electrophoresed upward in a density gradient (antiparallel) system in suspensions greater than 10^6 cells per milliliter. Orthogonal data are taken from horizontal migration in a microelectrophoresis system. The 0-*g* data are from the EEVT on STS-3. (Adapted from Reference 5.)

protein hormones and of cells with specific secretory functions changed the focus of space bioprocessing research toward CFE systems.

B. CONTINUOUS FLOW ELECTROPHORESIS

The laboratory use of commercial CFE systems (Beckman Instruments, Bender and Hobein, Hirschmann Geratebau, etc.) gave researchers insight into the unique relationship between cell surface charge and cell function. As the limitations of the ground systems were realized and microgravity FZE results began to illuminate the electrohydrodynamic principles that imposed these limitations, new interest was generated for studying preparative CFE of pharmaceuticals and live cells under microgravity conditions.

Table 15-1 **Mobility of RBCs in free zone (static column) electrophoresis**

Buffer	Gravity ($\times g$)	Concentration (cells/ml)	RBC EPM (μm s^{-1} V cm^{-1}) Human	Rabbit
A-1	1	10^6	-1.94 ± 0.10	-1.56 ± 0.10
	0	8.7×10^7	-1.96	-1.70
D-1	1	10^6	-1.42 ± 0.18	-0.94 ± 0.12
	0	10^9	-1.32	-0.83

Note: EPM = electrophoretic mobility.

1. Apollo-Soyuz MA-014 Experiments

German CFE experiments on ASTP were designed to characterize the effects of carrier buffer flow rates and temperature gradients in a wide-gap rectangular chamber. The objectives included demonstrations of increased resolution and throughput. The MA-014 automated system had a chamber 3.8 mm thick, 2.8 cm wide, and 18 cm long which could operate at 60 V/cm. It could not collect fractions, but the separands were monitored by light passing through the thickness of the chamber to be recorded by a 128-photodiode array. The four experiments included rat bone marrow, mixed rabbit and human RBCs, rat spleen cells, and rat lymph node cells. Astronauts had to remove the frozen samples and exchange the sample containers for each run. Cells were injected at 10^8 cells per milliliter and compared to ground separations at 10^7 cells per milliliter.[23] Sample flow rates were stepped from 5 ml/h to 3 ml/h; carrier flow rates were 16.5 ml/min and 11.1 ml/min with a voltage field of 60 V/cm and 40 V/cm, respectively. Each separation run continued for almost 6 min.

The continuous electrophoretic separation of human and rabbit cells failed to give a record of the separation. It is believed that the separation was satisfactory, but the ultraviolet (UV) light that illuminated the separated particle streams was so intense (halogen lamps burn brighter in microgravity due to lack of thermal convection) that it overpowered the photodetector, giving only intermittent traces of the particle streams when spacecraft power levels periodically dropped.[22] Postflight analysis did indicate that all of the space separations, with the exception of the bone marrow, were comparable in band spread and resolution to the 1 g controls.[23] However, the bone marrow cell run gave a band spread approximately twice as wide in space as in the ground controls. Based on other experiments, discussed below, it appears that the EPM of the bone marrow cells is dramatically narrowed at concentrations of 10^8 cells per milliliter in 1 g. At fast flow rates (16.5 ml/min) 7×10^6 cells per minute were separated; however, the improved resolution expected at reduced sample flow rates was not demonstrated on this flight.[22]

2. Sounding Rocket Systems

a. Advanced Applications Flight Experiment (G.E.)

In an attempt to improve resolution over ground-based CFE systems, General Electric Co. (G.E.) developed the Advanced Applications Flight Experiment (AAFE) Continuous Flow Electrophoretic Separator designed specifically for microgravity operations. The AAES instrument had a chamber 5 mm deep, 5 cm wide, and 10 cm in length.[3] It was designed to give the same throughput as the Beckman CPE at a substantially improved resolution of 0.01 μm cm V^{-1} s^{-1} or a tenfold greater throughput at the same resolution as the CPE (0.04 μm cm V^{-1} s^{-1}). It could separate up to 50 fractions and had a 512-channel UV detector array near the outlet. Ground test samples included hepatitis vaccine, sperm cells, lymphocytes, antihemophilic factor, influenza viral antigen, and erythropoietin. The AAFE was valuable in determining the regions of stable and unstable operations as a function of power and residence time (Figure 15-5). It also helped validate computer models which showed a potential increased resolution of 5 to 10 × and an increased throughput of 50 to 100 × as a function of the increased chamber thickness and sample concentration that was possible in microgravity. The AAFE was designed to fly on the Black Brandt rocket; however, the unit was never tested in space due to changes in the NASA microgravity sciences program.

b. TEM06-13 (MBB/ERNO)

A new CFE device, based on the lessons learned with the MA-014 system, was designed by MBB/ERNO for the German "Texus" sounding rocket and the "Biotex" research program for the Spacelab. It had a separation chamber 4 mm thick, 7 cm wide, and 20 cm

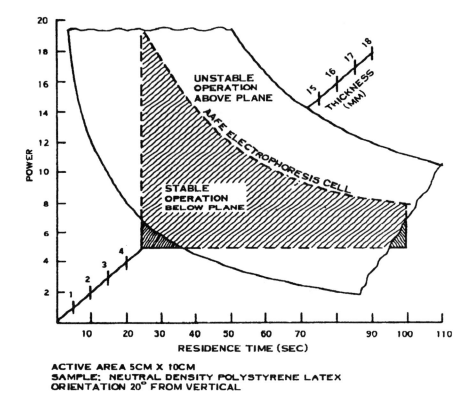

ACTIVE AREA 5CM X 10CM
SAMPLE: NEUTRAL DENSITY POLYSTYRENE LATEX
ORIENTATION 20° FROM VERTICAL

Figure 15-5 Graphic representation of regions of stable and unstable operation of the Advanced Applications Flight Experiment (AAFE) chamber plotted as a function of electric field power and residence time. Cross-hatched area shows limits of stable operations for a 5-mm-thick chamber. (Adapted from Reference 3.)

long and operated at electric fields up to 143 V/cm. It had a sample volume of 10 ml, a video camera to record bubbles and visible sample streams, and a UV detector at the outlet and could collect seven fractions.[12] It first flew on the Texus rocket in May 1988 as the TEPE-1 experiment. It was designed to begin steady-state operations on the ground, perform the 1 g reference experiment, then continue operations throughout the launch (13 g) and perform the space experiment during the 6 min of microgravity. Operations ceased just prior to reentry. The unit successfully separated rat, guinea pig, and rabbit erythrocytes.[23]

IV. CONTINUOUS FLOW ELECTROPHORESIS EXPERIMENTS ON THE SPACE SHUTTLE

A. CONTINUOUS FLOW ELECTROPHORESIS SYSTEM EXPERIMENTS (STS-4 TO STS-61B)

In 1981 MDAC negotiated the first Joint Endeavor Agreement with NASA for a series of joint experiments using a continuous flow electrophoresis system (CFES) installed in the middeck of the shuttle cabin. The first flight was on STS-4 in 1982. The MDAC flight CFES device had a chamber 3 mm thick, 16 cm wide, and 120 cm long. The ground control unit had a chamber 1.5 mm thick, 16 cm wide, and 120 cm long. Both operated at fields strengths up to 40 V/cm and had a 197-tube array to collect fractions. On STS-

4, 6, 7, and 8 the CFES used a low-conductivity triethanolamine-potassium acetate carrier buffer, pH 7.25, 296 mOsm/l for cells (or barbital buffer for proteins) flowing upward through the separation chamber at a rate of 20 ml/min. The electrode and separation chambers were pumped separately. Computerized flow balancing eliminated transverse flow and pressure gradients in the sample chamber, which enhanced cell stream stability and reproducibility of successive runs. Cells at 10^7 to 10^8/ml were injected into the bottom of the chamber using an infusion pump (4 ml/h). Cells were exposed to the electric field for 4 min (ground) or 30 min (flight) before they exited the top of the flow chamber through 197 fraction collection ports. A comparative measure of the separation capacity, E_t, expressed in volt-minutes per centimeter, is the product of the electric field strength, E, and the residence time, t. The MDAC space CFES was used for separations of test proteins on shuttle flights STS-4 and 6, of latex particles on STS-7, and of mammalian cells on shuttle flight STS-8. Flights STS-41D, 51D (1984), and 61D were devoted to purification of erythropoietin to obtain enough pure hormone for clinical trials.[24] During inflight operations the astronaut installed the sample injection syringes and the sample collection tray. The electrophoresis separation was controlled by a special computer, and the astronaut photographed sample streams that were visible through the transparent chamber walls. After each run the sample tray was removed by a crew member and stowed in a temperature-controlled locker.

Protein separations were conducted in microgravity to confirm computer models of the resolution and throughput that could be achieved using the MDAC CFES. A barbital carrier buffer was used (pH 8.3, conductivity = 250 μmho/cm). The conductivity of the sample stream was 970 μmho/cm, mostly contributed by the 25% protein solution. In ground experiments a sample containing 0.1% rat serum albumin and 0.1% ovalbumin (0.2% total protein) was electrophoresed at E_t = 110 V min cm^{-1}. A four-tube peak-to-peak separation was routinely achieved, with a five-tube overlap between the two proteins. When the protein concentration was increased more than 0.2%, the higher sample density resulted in decreased resolution. On shuttle flight STS-4 a sample containing 12.5% rat serum albumin and 12.5% ovalbumin (25% protein w/v) was separated, with a four-tube peak separation similar to the ground controls using a 0.2% protein solution. The chamber thickness and diameter of the sample stream in the flight chamber were twice that of the laboratory chamber, so a fourfold increase in sample volume flow rate was possible. The increased sample flow and sample concentration resulted in 400 × more throughput in microgravity with essentially the same resolution as the ground controls.[24] On STS-6 the throughput was increased to 556 × with a fourfold increase in resolution.

1. Particle Separations (STS-7)

The CFES was used on STS-7 to separate latex particles on the basis of charge and different sizes. The polystyrene latex spheres had diameters of 0.56 μm (red), 0.30 μm (white), and 0.80 μm (blue). These were separated to test for resolution (in the absence of sedimentation), fraction bandwidth, and the effects of conductance discontinuities between the particles and the carrier buffer. Mobilities of the particles were 3.5 + 0.2 SD (red), 2.4 + 0.2 SD (white), and 1.6 + 0.1 SD (blue) μm cm V^{-1} s^{-1}. The buffer was 2.25 mM sodium propionate (pH 5.0, conductivity = 155 μmho/cm). The resultant separations were approximately the same in microgravity and on the ground. Particle sedimentation, if present, did not affect migration significantly. The fractions were measured by electron microscopy and turbidimetry counts, and the bandwidths were found to be two times the standard deviation of the particle mobilities. When the sample conductivity was increased threefold to 455 μmho/cm, significant band broadening was observed. The results confirmed that field distortions could be caused from a mismatch between the conductivity of the sample stream and the carrier buffer. These electrohydrodynamic effects can either pull the sample stream toward the chamber wall or compress it toward the center of the

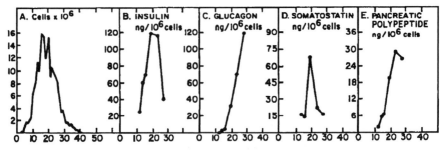

CFES FRACTION NO.

Figure 15-6 Continuous flow electrophoresis system (CFES) separation of pancreatic cells on STS-8 and hormone secretions of the cell fractions. The flight sample concentration was ~1 × 10⁸ cells; hormone levels are expressed in nanograms per 10⁶ cells after 4 days of culture. See text for details. (From Hymer, W. C., Barlow, G. H., Blaisdell, S. J., et al., *Cell Biophys.*, 10, 61, 1987. With permission.)

chamber, depending on the flow of current carrying ions and temperature gradients.[25] This phenomenon is discussed further by Todd in this Chapter 4.[26]

2. Cell Separations (STS-8)

On the STS-8 flight the collection system was reduced from 197 to 50 outlet tubes to accommodate fraction volumes of 11 to 12 ml of living cell suspensions. Three types of cells were separated in two runs each: dog pancreatic islet cells, human kidney cells, and rat pituitary cells. In each experiment approximately 10^8 cells contained in a special injection syringe were loaded in the CFES unit at 4°C and the separations were carried out at $E_t = 150$ V min cm^{-1}. Fractions of separated cells were collected and stored at 4°C for 5 days until return to Earth.

a. Pancreatic Cells

Four major types of endocrine cells are found in the pancreatic islets. Beta cells produce insulin, alpha cells produce glucagon, delta cells produce somatostatin, and F cells produce pancreatic polypeptides. The objective of the microgravity separations was to separate enough pure beta cells for transplantation experiments to demonstrate the commercial feasibility of using space electrophoresis to provide a cure for diabetes ($\sim 10^9$ cells are required for human transplantation). In ground control experiments beta cells were separated from alpha cells by 6 to 15 tubes using $E_t = 150$ V min cm^{-1} (data not shown).[28] In microgravity some 40 cell fractions were collected for each CFE experiment using a sample concentration of 10^8 cells per milliliter and an $E_t = 150$ V min cm^{-1}. The cell distribution is shown in Figure 15-6A. After landing, each cell fraction was cultured for 4 days and hormone levels were determined by alcohol extraction followed by radioimmunoassay (RIA).

Intracellular insulin content (Figure 15-6B), expressed as nanograms of insulin per 10^6 cells, demonstrated the most beta-cell enrichment between fractions 17 and 23.[28] The peak glucagon content was found at fraction 27 (Figure 15-6C). The cells with the most somatostatin content were located in fraction 19 (Figure 15-6D), while most of the polypeptides were found between fractions 23 and 27 (Figure 15-6E). It should be noted that the larger alpha cells migrated faster than the beta cells.[28]

b. Human Kidney Cells

Separations of HEK cells on STS-8 were performed at two different cell concentrations (run 3 = 2.5×10^6 cells per milliliter and run 4 = 8×10^7 cells per milliliter). Postflight

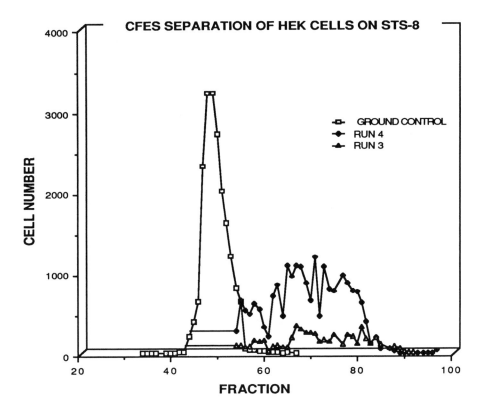

Figure 15-7 Continuous flow electrophoresis system (CFES) separation of human embryonic kidney (HEK) cells on Earth and on STS-8. The sample input concentrations were: ground control = 7×10^6 cells per milliliter, STS-8 run 3 = 2.5×10^6 cells per milliliter, and run 4 = 8×10^6 cells per milliliter. Fraction number is relative to outlet position with no electric field. (Adapted from Reference 29.)

each cell fraction was recovered from the CFES collection tray, grown to 95% confluence on growth medium, and then maintained on serum-free urokinase production medium for 28 days.[28] Medium samples were removed every 4 days and subjected to urokinase-type plasminogen activator (uPA) and tissue-type plasminogen activator (tPA) analysis. Although >45 fractions of HEK cells were collected, low cell numbers and viability in the most mobile cells resulted in only 36 cultured subpopulations. Figure 15-7 shows the cell distributions (as a function of EPM) for the ground control and both flight experiments. Five ground-based CFE experiments showed consistent resolution and band spread of 30 to 35 cell fractions obtained at sample input concentrations up to 2.7×10^6 cells per milliliter.[29] The mean mobility for the flight cells was approximately 30% greater than the ground controls. The heterogeneity and band spread of the EPM distribution were also significantly increased in microgravity.[30] Although run 3 and run 4 had different cell sample concentrations, no reduction in EPM mobility or band spread was noted at the higher cell concentration. The ground control EPM distribution was significantly reduced at the cell concentration of 7.0×10^6 cells per milliliter.

Microscopic classification of cells cultured from the flight fractions showed that they differentiated into four morphological types: 69% small epithelioid, 15% large epithelioid, 9% "domed" (single cells with an upward bulge), and 5% "fenestrated" (one or two large holes through the cytoplasm). There was a correlation with the EPM in that cells cultured from low mobility fractions were rich in domed cells, while those from high mobility fractions were rich in small epithelioid cells.[31]

uPA and tPA are proteolytic enzymes used to dissolve blood clots. The former is produced commercially by purification from urine or from large-scale cultures of HEK cells, while tPA is produced from genetically engineered cells. After ground-based CFE separations of HEK cells, PA assays showed that all cell fractions produced some uPA; however, 5 to 7 (out of 30) fractions produced two to three times more uPA than the rest. Most cell fractions also produced tPA.[32] Gel electrophoresis revealed that the cells with lower mobility produced the 54-kDa form of uPA (HMW) and tPA, while the cell fractions with the highest mobility produced only HMW-uPA, and a 150-kDA form of uPA.[33]

After the CFES experiments on STS-8, up to eight replicate cultures of HEK cells were established from each of the 35 cell fractions that were collected. Fibrinolytic and chromogenic substrate assays on culture medium from the cell fractions showed that all cells produced significant levels of uPA and tPA.[29,33] Figure 15-8 shows the distribution of uPA (upper panel) and tPA (lower panel) produced by the separated fractions (shown in Figure 15-7) collected from run 4 and maintained in stationary culture for 4 days after flight. There were three to five cell fractions that produced two to four times more uPA than the remainder of the fractions. There were five to six cell fractions that also produced high levels of tPA; however, the fractions that produced high levels of uPA generally were not the same as those that produced high levels of tPA.[28,33] The elevated production of uPA in each of the cell fractions continued for at least 12 days.[34] Gel electrophoresis and fibrin overlays on gel slices combined with anti-tPA and anti-uPA antibody reactions showed that the conditioned medium contained only the 54-kDa uPA and tPA. The peak uPA activity in relative fractions 54, 60, 82, and 87 coincided with peak secretion levels of tPA, whereas the peak PA levels in fractions 68 and 75 were almost entirely from uPA (see Figure 15-8, top). Additional fibrinolytic assays, including preincubation with plasmin, showed clearly that there were significantly different levels of urokinase proenzyme present in the medium from the STS-8 cell fractions.[30]

c. Pituitary Cells

Somatotropic cells from the anterior pituitary produce several hormones of commercial interest: growth hormone (GH), prolactin (PRL), and gonadotrophins (follicle-stimulating hormone [FSH] and luteinizing hormone [LH]). In these experiments RIAs were used to determine the intracellular hormone content. Following ground control CFE experiments on rat pituitary the cell viability was 80%. Cells were injected at a concentration of 10^7/ml, but cell recoveries were often only 21%. Most of the GH cells were recovered in fractions 30 to 55 out of a total of 60 fractions (see Figure 15-9A).[29] The low number of viable cells made it necessary to repool some adjacent fractions in order to obtain enough cells for the hormone bioassays. The STS-8 experiments separated the pituitary cells into a 50-tube band spread (see Figure 15-9F) with a significantly different cell distribution (EPM) than the ground control. Postflight cell viability was 53%, and a low cell recovery (20%) was attributed to a problem with the cells clumping in the sample injection syringe. Therefore, adjacent fractions had to be repooled into five fraction groups to conduct the RIA hormone assays. Figure 15-9B and G compare the amount of intracellular GH found in the repooled fractions. It is clear that the lower mobility cells returned from STS-8 contained more than twice the GH as the analogous fraction pools from the ground control experiments. Figure 15-9C and H also show that fractions with GH cells tended to have the highest mobility. In ground controls the cells containing the most PRL tended to be in the lower mobility group, while most of the PRL cells were

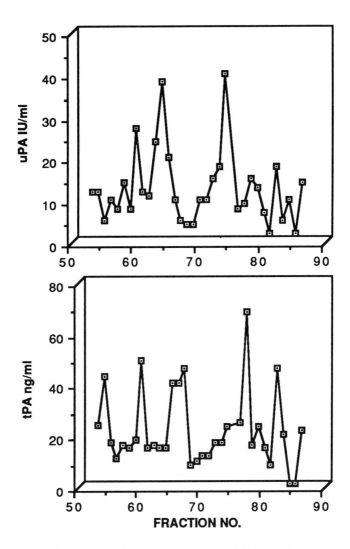

Figure 15-8 Urokinase-type plasminogen activator (uPA) levels (upper panel) and tissue-type plasminogen activator (tPA) levels (lower panel) produced from CFES run 4 cell fractions returned from STS-8. Each activity level is the mean of three to eight replicate cultures per fraction maintained in culture for 4 days. Urokinase activity, expressed in international units per milliliter, was measured by an artificial substrate method, S-2444 (American Diagnostica, New York); tPA activity, expressed in nano-grams per milliliter, was measured by enzyme-linked immunosorbent assay (ELISA; American Diagnostica). (Adapted from References 29 and 33.)

found in the center of the band spread. Intracellular PRL levels were two to four times greater than in the less mobile fractions. The STS-8 experiments confirmed that the more dense GH cells were also the most mobile of the pituitary cell types. This was not clear from previous ground studies, where sedimentation was suspected of increasing the effective residence time.[6] Also, the STS-8 results confirm that the GH cells have a higher EPM than do the PRL cells.[28]

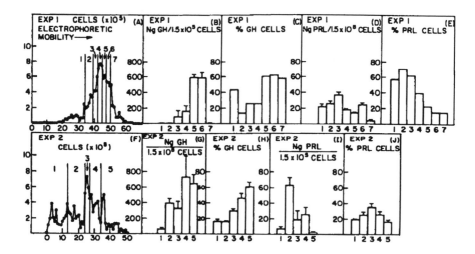

Figure 15-9 Separation of rat pituitary cells on STS-8 and hormone secretions of pooled cell fractions. The sample concentrations were ~3 × 10⁷ cells per milliliter. Cell fractions were pooled to a final concentration of 1.5 × 10⁶ for the ground control (EXP 1) and 2.2 × 10⁶ cells per milliliter in the microgravity experiment (EXP 2). Intracellular hormone levels are expressed as nanograms per 1.5 × 10⁵ cells per milliliter. See text for details. GH = growth hormone; PRL = prolactin. (From Hymer, W. C., Barlow, G. H., Blaisdell, S. J., et al., *Cell Biophysics,* 10, 61, 1987. With permission.)

B. FREE-FLOW ELECTROPHORESIS UNIT EXPERIMENTS (SPACELAB-J)

The Japanese free-flow electrophoresis unit (FFEU) has been developed for experiments on the Spacelab-J mission. This device has a separation chamber 4 mm thick, 6 cm wide, and 10 cm long. It operates at a maximum field strength with a carrier flow rate of 25 ml/min. It has a sample cassette system which allows astronaut change-out of the samples; however, each cassette contains only 0.6 ml of sample. A total of 60 fractions can be collected and monitored by an optical detector system with a resolution of 0.1 mm which monitors the outlet at either 280 or 600 nm wavelength. The optical scanning data are provided to the investigators who select the fraction streams to be collected and stored for postflight analysis.

On Spacelab-J two types of biologicals were separated electrophoretically. Experiment L-3 tested the effects of protein concentration, sample flow rate, and carrier flow rate on the ability to resolve a mixture of cytochrome C, conalbumin, bovine serum albumin, and a trypsin inhibitor at concentrations of 25 and 50 mg/ml.[35] The results were compared with those from a similar ground-based FFEU using a chamber only 0.8 mm thick. Experiment L-8 attempted to separate three strains of *Salmonella typhimurium* that have different cell membrane structures which is reflected in different cell surface charges. A 10 m*M* triethanolamine buffer, pH 7.5, was used with a carrier rate of 6 ml/min and a sample flow rate of 5 ml/min. The field strengths used were either 33.3 or 50 V/cm. Inflight problems with optical data transmission prevented confirmation that the separations were totally successful; however, initial microscopic counts of the recovered fractions indicated that three peaks of total cells were found among the collected fractions. Postflight distributions of individual *S. typhimurium* strains are being determined by outgrowth in differential media.[36]

V. SUMMARY OF SPACEFLIGHT EXPERIMENTS

A. ADVANTAGES
1. Proteins and Macromolecules

Microgravity electrophoresis experiments using stable protein samples have confirmed some of the major restrictions to preparative-scale electrophoresis of soluble biologicals on Earth.[37] Protein separations in space also have illuminated the role of electroosmosis, which causes parabolic flow profiles in both FZE and CFE systems, and have confirmed that major increases in throughput and resolution are possible in the absence of gravity. CFE experiments on STS-4 and STS-6 demonstrated a 718-fold increase in sample throughput of using a chamber that was just twice as thick as the ground unit.[24,28] Enough erythropoietin was isolated directly from conditioned culture medium on STS-61B to begin preclinical studies.[24] Since proteins are much more tolerant of large changes in temperature, pH, and osmotic pressure and they retain their functional characteristics through several freeze-thaw cycles, they have served as good test materials to characterize the electrohydrodynamic effects over a large range of field strengths, carrier flow rates, and sample injection rates.

2. Cells and Particles
a. Particle Separations

Microgravity electrophoresis of stable particles with known size and EPM has been very important in understanding the relative effects of droplet sedimentation, electroosmosis, countercurrent streaming potential, and Poiseuille flow in closed FZE and CFE systems. Test particles also have demonstrated the practical application of higher electric field strengths and larger chambers[28,29] and have confirmed the theoretical advantages of separating live cells in the absence of convection and sedimentation.[38]

b. Cell Concentration Effects

On Earth when the sample concentration exceeds a certain limit in CFE the band spread and resolution are significantly reduced. The mobility distribution of kidney cells is reduced by 50% at sample concentrations greater than 2.7×10^6 cells per milliliter.[29] Figure 15-7 shows clearly that in microgravity the cell concentration effects are absent up to at least 8×10^7 cells per milliliter. The mean mobility of the cells separated in space was some 30% greater and the EPM distribution was 40% greater in bandwidth (thus yielding greater resolution) at a sample concentration of 8×10^7 cell per milliliter as compared to the ground separation using 7.0×10^6 cells per milliliter. Pituitary cells appear to have the same problem at concentrations exceeding 2×10^7 cells per milliliter.[28] Figure 15-10 shows the effect of high sample concentrations on the EPM distribution of cultured SKHEP liver cells separated by CFE in 1 g. A twofold increase in sample concentration reduced the mean mobility by almost 20%, while a fourfold increase in sample concentration reduced the mean mobility by 60% and the EPM range by approximately 45%.[39] Several possible mechanisms have been considered: (1) differences in local ionic concentrations between the sample and carrier buffers (which alter the local electrical field), (2) effects of depleted Ca^{2+} and Mg^{2+} ions on cell clumping (which effectively produces droplet sedimentation); and (3) cell congestion in a highly concentrated sample stream wherein the innermost cells are either screened from the electric field or physically cannot migrate through the other cells toward the anode.[6] Microgravity CFE separations of kidney, pituitary, and pancreas cells confirmed that ranges of operating conditions could be extended in the absence of sedimentation and thermal convection.

c. Cell Function and Electrophoretic Mobility

In microgravity the greater number of fractions, increased resolution, and higher cell throughput make it possible to collect enough cells per fraction to characterize more

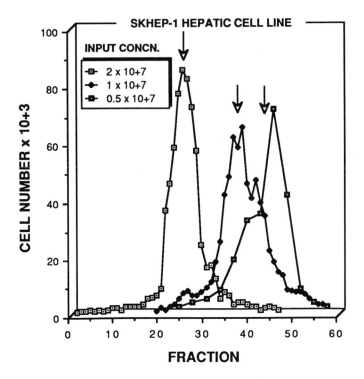

Figure 15-10 Graph of SKHEP showing effect of increasing cell concentration on electrophoretic mobility (EPM) distributions in upward-flowing continuous flow electrophoresis operating in unit gravity. A fourfold increase in sample concentration resulted in a 60% decrease in mean mobility and a 45% reduction of the distribution band spread (EPM). (Redrawn from Reference 39.)

closely the relationship between specific secretory functions and cell surface charge. Flight experiments have confirmed the potential usefulness of employing CFE to separate different cells with specific secretory functions. STS-8 experiments showed that four to six electrophoretic subpopulations of kidney cells make uPA and tPA molecules with different molecular weights[29,40] and that two types of pituitary cells produce two types of growth hormone.[28,41,42]

Ground experiments have shown that the EPM of cells is generally independent of cell cycle and size;[43] however, microscopic studies of cells after microgravity CFE showed a unique morphologic distribution of four cell types of kidney cells that appear to have different physiological functions,[3] and large pancreatic alpha cells migrated farther than the smaller beta cells in the same electric field.[28]

Cell electrophoresis at high concentrations demonstrates that EPM is independent of zone sedimentation.[4,41] However, highly concentrated cells (in a very low conductivity buffer) will themselves carry a significant amount of current when exposed to a strong electric field. The result is increased current flow through the sample zone relative to the carrier buffer, and the effects of this phenomenon can be characterized in microgravity only with very high concentrations and field strengths.[44]

B. DISADVANTAGES

A number of technical problems must be overcome to begin preparative-scale electrophoretic separation of cells in microgravity. The major operational difficulties include chamber cooling (in the absence of convection), gas formation at the electrodes, and sample preservation. Preparative electrophoresis requires relatively high electric fields, and joule heating is a significant problem at potential differences on the order of 100 V in the conductivity ranges (0.01 to 0.2 g ions per liter) of the buffers that are compatible with live cells. Unless optimal cooling is provided, increased temperatures can easily lead to denaturation of proteins or death of the cells. Living cells seldom can tolerate temperatures greater than 41°C for more than 10 min, and even that exposure causes "heat shock" stress which alters the normal functions of the cells for 24 to 48 h.[45] The effects of heat shock or cell stress on EPM has not been well studied. Joule heating also can cause second-order convection (Marangoli flow) and electric field effects.

FFE operations require continuous removal of the hydrogen and oxygen produced at the electrodes. This requires adequate buffer circulation flow though the electrode chambers that is separate from the carrier buffer. The volumes of gas can be large; for example, a CFE chamber 5 mm thick × 15 cm wide × 100 cm long, operating at a voltage gradient of 20 V/cm with a 15 A current flow though a buffer conductivity of 0.015 ohm^{-1} cm^{-1}, would generate 115 ml of hydrogen and 50 ml of oxygen per minute.[46] Gases can be separated from the fluid centrifugally or by phase separation between hydrophobic and hydrophilic membranes, but the hydrogen must be converted catalytically.

Sample storage conditions and operational time lines during spaceflight present some difficulty for electrophoresis experiments and preparative-scale production. Live cells typically must be stored during prelaunch and launch phases of spaceflight, which often means 48 to 72 h before the first sample of cells can be processed. After collection at the end of the run, the fractions usually must be stored for 4 to 10 days before return to Earth. Freezing the samples prior to launch and after electrophoresis reduces cell viability and requires cryopreservatives in the sample buffer. Although dimethyl sulfoxide (DMSO) and other cryopreservatives have been shown not to affect the cell surface charge,[18] freezing and thawing the cells several times often can reduce cell viability and could eliminate certain subpopulations of cells which have been actually separated. Storage of live cells at high sample concentrations has a time limitation, in that most sample systems have relatively small volumes and, therefore, the cells can become deprived of nutrients and begin to suffer from exposure to metabolic waste products if they are stored too long. After CFE the cells are suspended in the fraction volume; however, storage of anchorage-dependent cells for several days often reduces viability and alters their physiology. Fortunately, cell attachment experiments on STS-8 have showed that mammalian cells can readily attach to microcarrier beads and grow while floating in microgravity conditions.[47] Thus, microcarriers can be added to help preserve cell homeostasis in the collected fractions. Also, maintenance cell culture systems and suspension bioreactors have been designed for microgravity[48,49] and could be used to maintain target cells before and after electrophoresis.

VI. FUTURE OF ELECTROPHORESIS IN SPACE

The U.S. held the lead in space electrophoresis for 20 years; however, in 1988 the McDonnell-Douglas Electrophoresis Operations in Space (EOS) program was terminated and all flight experiments ceased using the MDAC CFES for separation of living cells and proteins. More basic research into physical phenomena and theoretical aspects of process improvements was emphasized by NASA since the agency did not have a flight system to replace the CFES until the recent development of the United States Commercial Electrophoresis Program in Space (USCEPS).

A. NEW MICROGRAVITY ELECTROPHORESIS SYSTEMS
The ability of CFE systems to be operated in a continuous, automated mode makes them good candidates for microgravity research and development experiments. Several new

Table 15-2 **Comparison of electrophoresis flight system designs**

System Name	Thickness (mm)	Width (cm)	Length (cm)	Field (V/cm)	No. of Fractions	Ref.
MA-011	6.4 (diam.)		15.2	13.5	28	15
MA-014	3.8	2.8	18.0	60	128[a]	22
AAFE[b]	5.0	5.0	10.0	50	50[a]	3
TEM06-13	0.5	7.0	20.0	143	7[b]	12
CFES	3.0	16.0	120.0	40	197	28
RAMSES	3.0	6.0	30.0	50	40[b]	51
USCEPS[c]	3.0[d]	16.0	120.0	40	99[b]	50
FFEU	4.0	6.0	10.0	100	60[b]	35
DLE[c]	2.0	3.0	10.0	94	10[b]	9

Note: AAFE = ; CFES = continuous flow electrophoresis system; FFEU = free-flow electrophoresis unit; DLE = deflected-lamina electrophoresis.

[a] Detector array.

[b] Fraction collector and optical detector.

[c] Not yet flown.

[d] Varied on different models.

systems have been developed by different countries for microgravity experiments to establish the design requirements for pilot and full-scale production systems on the space station. Table 15-2 compares the design specifications and basic capabilities of various electrophoresis systems developed for space flight.

1. Systems Under Development
a. USCEPS (U.S.)
The Pennsylvania State University Center For Cell Research (CCR) is developing a new CFE system under the NASA-sponsored USCEPS to optimize continuous flow electro-phoresis operations in microgravity. The USCEPS hardware will be similar to that operated by McDonnell Douglas on the space shuttle from 1982 to 1985, but with significant improvements which allow sterilization of the hardware and its buffer mate-rials prior to launch, increased throughput potential, and increased purity potential.[50]

The initial USCEPS design will accommodate six separate 25-ml samples. Samples can be cell culture medium concentrates and blood plasma products, or suspended particles such as fixed or live cells. USCEPS will have a separation chamber 3.0 mm thick, 16 cm wide, and 120 cm long and will be tested at 20 V/cm on Spacehab 4 and 5. Separated fractions will be collected in 99 fraction containers. USCEPS is available to both commercial partners as well as government (including NASA) and academic scien-tists for both scientific research and developmental pilot operations. Future designs will have a detector array and different chamber thicknesses depending on requirements to accommodate larger samples aimed toward a production capability that will be tested on Spacehab 6 and on the space station.

b. TEM06-13 CFE (German)
The TEM06-13 system developed and tested on the Texus rocket also had been adapted to fit into the German Biotek racks on the Spacelab D-2 mission. It did not fly as part of D-2, but is planned for CFE experiments on future Spacelab missions.[23]

c. RAMSES (French)
A free-flow electrophoresis system called Recherches Appliquees sur les Methodes de Separation Electrophoretique Spatiales (RAMSES) has been developed by the French

Centre National d'Etudes Spatiales (CNES) in Paris. This system will be tested first on the International Microgravity Laboratory IML-2 mission on STS-65 in June 1994. This device has interchangeable chambers and an illuminated detector system to enable different types of experiments to study free-zone electrophoresis and applied electrohydrodynamic relationships under a variety of operational conditions. The ground-based system has a chamber 1.5 mm thick, 6 cm wide, and 30 cm long, while the flight system has a 3.0-mm-thick chamber. It will have a field strength capacity of 50 V/cm and also will be used to study isoelectric focusing. Future plans include capabilities to study membrane processes and chromatography in microgravity.[51]

2. Other Novel Microgravity Systems
a. Deflected-Lamina Electrophoresis
The deflected-lamina electrophoresis (DLE) system uses a unique approach for maximizing the operational latitudes afforded by microgravity.[50] In most CFE systems the sample stream is introduced as a thin column near the cathode and migrates laterally across the width of the separation chamber toward the anode. In DFE the sample is introduced as a broad band or sheet of fluid across the width of the chamber. When the electric potential is applied across the thickness (using electrodes near the front and rear face) of the chamber, the trajectory sheet of sample (lamina) is deflected toward the anode (face) of the chamber. A typical chamber is 2 mm thick, 2 cm wide, and 10 cm long. This approach allows much larger throughputs since the sample is injected as a broad band (only 0.1 mm thick) instead of a thin filament. Theoretically, it has reduced energy requirements, smaller temperature gradients, reduced electroosmosis effects, and a resolution adequate for separating mixtures of erythrocytes or kidney cells.[9] Operating at 94 V/cm and using a buffer conductivity of 6.5×10^{-4} ohm^{-1} cm^{-1}, the DFE should consume only 56.8 W and generate 9.3 ml of H_2 and 4.7 ml of O_2 per minute. The original flight design also included an aspect ratio inverter (ARI) at the outlet to improve collection of the separated fractions. Although a flight DLE system has not been completed, the ARI design was used in the MDAC CFES system that flew on the space shuttle.[28]

b. Moving Wall Electrophoresis
Another new approach to increasing resolution and throughput is by use of a thick chamber and outer chamber walls that move at the same velocity as the carrier buffer. Combined with coatings to eliminate electroosmosis, the moving wall design mechanically eliminates flows that disturb the migration velocities of the separands.[52] The use of a thick chamber, however, presents problems with cooling and convection in 1 *g;* therefore, this device must be operated in microgravity to get optimum performance.

B. CANDIDATES FOR MICROGRAVITY ELECTROPHORESIS
Future candidate biologicals for space electrophoresis include mixtures of proteins, particles or fixed cells designed to test operating efficiencies, living cells to be returned for ground studies of the unique functions of target subpopulations, potential commercial macromolecules, and cells that could be used for human transplantation.

The main emphasis for selecting candidate cells is based on ground studies that demonstrate unique subpopulations which cannot be adequately resolved using Earth-based electrophoresis systems or other electrokinetic methods. Target cells include human kidney and pituitary cells that have been flown previously (see Table 15-3), cancer cell lines (wherein abnormal cell functions are reflected in unique cell surface charges), and cells that have intracellular granules or other structures that increase their density. Pituitary cells containing growth hormone granules and prolactin and suspended growth hormone granules will be separated using the FFEU on the IML-2 mission in 1994.[50] Hybridomas that produce unique antibodies also are candidates, especially those new hybridomas that are formed in space by electrofusion methods.[53,54]

Table 15-3 **Free-flow electrophoresis experiments in microgravity**

Mission	System	Cell Type or Particle	Protein	Objective
Apollo-14 (1971)	Static fluid		DNA, hemoglobin	Test resolution
Apollo-16 (1972)	Static fluid	Latex particles		Test resolution
ASTP (1976)	MA-011	RBCs		Test resolution
		Kidney cells		Cell function
		Lymphocytes		Isotachophoresis
	MA-014	Lymphocytes + RBCs		Test resolution
		Bone marrow		Test resolution
		Splenocytes		Test resolution
STS-3 (1981)	MA-011	RBCs		Test resolution
		Kidney cells		Cell function
Salyut-7 (1982)	Tavriya		Serum albumin	Isoelectric focusing
		Bone marrow		Test resolution
STS-4 (1982)	CFES		Albumins	Test resolution
			Erythropoietin	and throughput
STS-6 (1983)	CFES		Hemoglobin	Test resolution
			Erythropoietin	and throughput
			Polysaccharide	
STS-7 (1983)	CFES		Erythropoietin	Test resolution
		Latex particles		and throughput
STS-8 (1983)	CFES	Pituitary		Test resolution
		Kidney		and cell function
		Pancreatic		
STS-41D (1984)	CFES		Erythropoietin	Clinical trials
STS-51D (1985)	CFES		Erythropoietin	Purity tests
STS-61B (1985)	CFES		Erythropoietin	Clinical trials
MIR (1987)	Svetlana		Vaccine protein	Vaccine standard
MIR (1988)	Svetoblok		Test proteins	Gel electrophoresis
Texus rocket (1988)	TEM06-13	Rat, guinea pig, rabbit, RBCs		Test resolution
Spacelab-J (1992)	FFEU		Cytochrome C	(L-3) Separate
			Albumins	Mixed proteins
			Trypsin inhibitor	
		Salmonella sp.		(L-8) Resolution
IML-2 (1994)	FFEU	Pituitary cells		GH and PRL cells
		Intracellular granules		GH granules
	RAMSES	Latex particles		Test electrohydrodynamics
Spacehab 4, 5, and 6 (1995–1996)	USCEPS	Pituitary		GH and PRL cells
			Culture medium	Purify products

Note: ASTP = Apollo-Soyuz Test Project; CFES = continuous flow electrophoresis system; FFEU = free-flow electrophoresis unit; GH = growth hormone; PRL = prolactin.

Table 15-4   **1 *g* vs. microgravity electrophoresis of three commercial products**

Product	Annual Doses ($\times 10^3$)	Value ($M)	Chambers Required 1 *g*	Chambers Required 0 *g*
AHF	250	25	3,878	20
EPO	1,300	1,700	23,042	115
Beta cells (pancreas)	630	100	27	1

From NASA Contract NAS 8-31353, Final Report, McDonnell-Douglas Astronautics Company, St. Louis, MO, 1977.

1. Commercial Biologicals

Commercial materials which are candidates for space electrophoresis are those having high commercial value per unit mass or those which cannot be obtained on Earth in adequate purity and quantity. The number of candidates is limited to those which can justify high production costs, including an estimated transportation cost of $3,000/lb (U.S.). It has been estimated that if the annual market volume of a target biological is at least $10 million (U.S.) and if it can be purified by electrophoresis from a reasonable starting volume, then the costs of space processing could be justified.[24] Table 15-4 shows a comparison of the number of CFE chambers required on Earth vs. in microgravity to purify commercial quantities of three potential therapeutic products by electrophoresis. The required purity and throughput dictate the large number of chambers that usually prohibits commercial-scale electrophoresis of products on Earth.

In a 1985 report of the Center for Space Policy the total space processing market in the year 2000 was estimated at $2.6 to $17.9 billion (U.S.),[54] and the projected Japanese space manufacturing market on the space station could be 200 billion yen (over $1.3 billion [U.S.]).[24] Table 15-5 lists a number of candidate pharmaceutical products which could be produced by space electrophoresis.[24,46]

C. POTENTIAL USES ON SPACE STATION

Microgravity experiments have already contributed greatly to the understanding of the gravity-dependent phenomena that occur in various free-fluid electrokinetic separation systems. As a result, a generalized mathematical description of electrokinetic separation processes has been formulated for static fluids in a column.[56] A major attraction for pilot-scale electrophoresis is the potential dual use in tandem operations that are envisioned for the space station.[54,57] High-resolution cell electrophoresis also can be used to isolate target cells or hybridomas made by space electrofusion. The purified cells could then be used to seed a space bioreactor system, first to grow large numbers of cells and then to maintain the cells under conditions that are conducive to production of secreted hormones, antibodies, or other products. Finally, conditioned culture medium could be harvested from the bioreactor and processed in a high-throughput electrophoresis system to purify the target product in one final step. Figure 15-11 is a flow schematic for such a facility which illustrates the major process steps and microgravity systems that are being developed for exploratory experiments as part of the biotechnology facility on the space station Freedom. In this scenario the potential advantages of cell electrophoresis, cell culture in a low shear environment,[47] and protein purification by electrophoresis can be added together by tandem operations in microgravity. Microgravity research has already demonstrated potential advantages for each of the major processes. High-resolution cell electrophoresis was performed on the STS-8 mission. Electrofusion experiments were demonstrated on the German Texus sounding rocket in 1986.[53] One type of perfusion bioreactor is under

Table 15-5 **Candidate biologicals for commercial electrophoresis in space[4,24]**

Product	Medical Treatment	Estimated No. Patients ($\times 10^3$)	U.S. Market ($M)
Proteins:			
Erythropoietin	Anemia	1,600	244
Urokinase	Thrombosis	1,000	85
Antihemophilic factor VIII and IX	Hemophilia	20	50
Epidermal growth factor	Burns	150	TBD
Interferon-γ	Antitumor/virus	10,000	TBD
Granulocyte stimulating factor	Wound healing	2,000+	TBD
Transfer factor	Multiple sclerosis	550	TBD
Cells:			
Beta cells (insulin)	Diabetes	600	600
Pituitary cells	Dwarfism	850	TBD

development,[49] and another bioreactor was flown on STS-54 to support lymphocyte mobility experiments.[58] Protein hormones have been purified directly from concentrated culture medium by CFE on six shuttle missions.[28] Other competing microgravity methods utilized include recirculating isoelectric focusing (RIEF) experiments conducted on STS-11[59] and aqueous two-phase partitioning studies[60] conducted on STS-26, which also may be used to purify protein products from cultured cells in the space station. Of the five potential subsystems, three have been flown in space and the other two are being developed for space research.

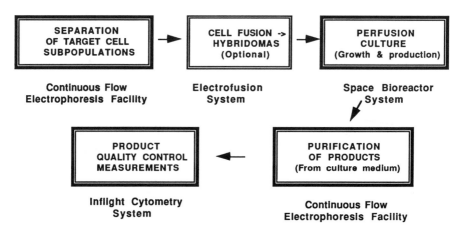

Figure 15-11 Flow chart of process steps for a typical cell bioprocessing and protein product purification in a proposed commercial-scale space station operation. Continuous flow electrophoresis, electrofusion, and a bioreactor culture system have all been tested in microgravity. A flow cytometer prototype has already been developed for use on the space station.

A major consideration for commercial-scale electrophoresis on the space station are the facility support requirements. Requirements for a continuous flow electrophoresis facility (CFEF) on the space station have been developed by MDAC in NASA-sponsored studies.[61] The CFEF will weigh 600 kg, will occupy a volume of 1.5 m³, and will require 1.5 kW of electrical power (1 kW processing + 500 W servicing) and 151 l of ultrapure water (60% must be pyrogen-free) per run. Inflight support requirements include laboratory systems for sterile handling of cells, incubators, refrigeration, contamination control, and sterilization methods. The CFEF would be automated, but inflight operations would still require 5 to 10 h of crew time per day for commercial-scale operations. Resupply of consumables (80 kg) to and transportation of products and biohazardous waste (60 kg) from the space station would depend on shuttle flights every 90 days. Studies indicate that space bioprocessing of a 30% market share for two to four high-value pharmaceuticals could make a profitable commercial enterprise.[24,55]

VII. SUMMARY AND CONCLUSIONS

Three free-zone electrophoresis systems and five continuous flow electrophoresis systems have been flown on 15 manned space flights and several sound rocket flights since 1976. Two new CFE systems are now being prepared for Spacelab and Spacehab flights in 1994 and 1995. Electrophoresis experiments in microgravity have validated the theoretical advantages of eliminating convection and sedimentation to improve both resolution and throughput. Microgravity separations of protein mixtures and test particles have enabled researchers to improve their computer models of the electrohydrodynamic phenomena that influence the efficiency of both free zone and continuous flow electrophoresis. The role of electroosmosis has been characterized, and the development of wall coatings to reduce the chamber wall zeta potential have been shown to effectively eliminate this problem. Microgravity CFE experiments have demonstrated that discontinuities of the buffer ion concentrations between the sample band and the carrier buffer can produce electrohydrodynamic distortions of the sample bands. Microgravity CFE also has helped us to understand sample band spreading and occasional retrograde (cathodal) migration which occurs at high sample concentrations. In commercial CFE experiments erythropoietin was separated from concentrated culture medium at concentrations impossible in 1 g, resulting in 718 times greater throughput and over four times greater resolution than Earth-based controls. Adequate quantities of erythropoietin for preclinical trials were purified by CFE on STS-51D and STS-61B.

Free zone electrophoresis of fixed cells, test particles, and live cells has been very useful in characterizing the phenomenon of droplet sedimentation and has illuminated the possible role of streaming potentials resulting from the movement of charged cells within the carrier buffer. Isotachophoresis of cells in microgravity also has been very useful for understanding the practical applications of highly concentrated samples. Microgravity CFE has shown that live cells can be separated with higher resolution, using concentrations several times greater than that which causes narrowing of the band spread in 1 g, and it appears that the practical upper limit of cell sample concentration has not been reached. CFE experiments have demonstrated that several different subpopulations of human and animal cells produce different types of secreted enzymes and hormones. The higher resolution and throughput makes electrophoretic purification of human cells for therapeutic transplantation conceptually feasible.

The U.S., France and Japan all have CFE experiments manifested on Spacelab flights next year. The Space Station Biotechnology Facility will accommodate several types of electrokinetic separation systems and a space bioreactor for culturing cells. Microgravity electrophoresis is a valuable research tool and potential commercial applications are very promising, but pilot-scale systems have yet to be tested in flight.

REFERENCES

1. **Hannig, K. and Wirth, H.,** Free-flow electrophoresis in space, *Prog. Astronaut. Aeronaut.,* 52, 411, 1977.

2. **Vanderhoff, J. W. and van Oss, C. M.,** Electrophoretic separation of biological cells in microgravity, in *Electrophoretic Separation Methods,* Righetti, P. G., van Oss, C. M., and Vanderhoff, J. W., Eds., Elsevier/North-Holland, Amsterdam, 1977, 257.

3. **McCreight, L. R.,** Electrophoresis for biological production, in *Bioprocessing in Space,* Morrison, D. R., Ed., NASA TMX-58191, National Technical Information Service, Springfield, VA, 1976, 143.

4. **Lanham, J. W.,** Feasibility Study of Commercial Space Manufacturing, NASA Contract NAS 8-31353, Final Report, McDonnell-Douglas Astronautics Company, St. Louis, MO, 1977.

5. **Todd, P.,** Microgravity cell electrophoresis experiments on the space shuttle: a 1984 overview, in *Cell Electrophoresis,* Schütt, W. and Klinkmann, H., Eds., Walter de Gruyter, Berlin, 1985, 3.

6. **Todd, P. and Hjerten, S.,** Free-zone electrophoresis of animal cells. I. Experiments on cell-cell interactions, in *Cell Electrophoresis,* Schütt, W. and Klinkmann, H., Eds., Walter de Gruyter, Berlin, 1985, 23.

7. **Morrison, D. R. and Todd, P.,** NASA's program in microgravity biotechnology — an overview, in *Microgravity Science and Applications,* National Academy of Sciences, Washington, D.C., 1986, 197.

8. **Todd, P.,** Separation physics, in *Progress in Astronautics and Aeronautics: Low Gravity Fluid Dynamics and Transport Phenomena,* Vol. 130, Koster, J. N. and Sani, R. L., Eds., American Institute of Aeronautics and Astronautics, Washington, D.C., 1990, 539.

9. **Strickler, A. and Sacks, T.,** Continuous free-film electrophoresis: the crescent phenomenon, *Prep. Biochem.,* 3, 269, 1973.

10. **Patterson, W. J.,** Development of polymeric coatings for control of electro-osmotic flow, in ASTP MA-011 Electrophoresis Technology Experiment, NASA TMX-73311, National Technical Information Service, Washington, D.C., 1976.

11. **Strickler, A.,** Deflected-lamina electrophoresis (DLE): an approach to high performance separation of biological materials in space, Paper 77-233, 15th AIAA Aerospace Sciences Meet., Los Angeles, 1977.

12. **Mang, V.,** Technical description of the electrophoresis facility for sounding rockets in the Texus program, *Appl. Microgravity Technol.,* 2, 52, 1989.

13. **McKannan, E. C., Krupnick, A. C., Griffin, R. N., and McCreight, L. R.,** Electrophoresis Separation in Space — Apollo 14, NASA TMX-64611, National Technical Information Service, Washington, D.C., 1971.

14. **Snyder, R. S.,** Electrophoresis Demonstration on Apollo-16, NASA TMX-64724, National Technical Information Service, Washington, D.C., 1972.

15. **Allen, R. E., Barlow, G. H., Bier, M., Bigazzi, P. E., Knox, R. J., Micale, F. J., Seaman, G. V. F., Vanderhoff, J. W., van Oss, C. J., Patterson, W. J., Scott, F. E., Rhodes, P. H., Nerren, B. H., and Harwell, R. J.,** Electrophoresis technology experiment MA-011, in Apollo-Soyuz Test Project Summary Science Report, Vol. 1, NASA SP-412, National Technical Information Service, Washington, D.C., 1977, 307.

16. **Allen, R. E., Rhodes, P. H., Snyder, R. S., Barlow, G. H., Bier, M., Bigazzi, P. E., van Oss, C. J., Knox, R. J., Seaman, G. V. F., Micale, F. J., and Vanderhoff, J. W.,** Column electrophoresis on the Apollo-Soyuz Test Project, *Sep. Purif. Methods,* 6, 1, 1977.

17. **Bier, M., Hinckley, J. O. N., and Smolka, A. J. K.,** Potential use of isotachophoresis in space, in *Protides of the Biological Fluids: 22nd Colloquium,* Peeters, H., Ed., Pergamon Press, New York, 1975, 673.
18. **Morrison, D. R. and Lewis, M. L.,** Electrophoresis tests on STS-3 and ground control experiments: a basis for future biological sample selections, Paper 82-152, in Proc. 33rd Int. Astronautics Federation, Paris, 1982.
19. **Sarnoff, B. E., Kunze, M. E., and Todd, P.,** Electrophoretic purification of cells in space: Evaluation of results from STS-3, *Adv. Astronaut. Sci.,* 53, 139, 1983.
20. **Snyder, R. S., Rhodes, P. H., Herren, B. J., Miller, T. Y., Seaman, G. V. F., Todd, P., Kunze, M. E., and Sarnoff, B. E.,** Analysis of free-zone electrophoresis of fixed erythrocytes performed in microgravity, *Electrophoresis,* 6, 3, 1985.
21. **Mitichkin, O. V., Vavirovsky, L. A., Azhitsky, G. Yu., Serebrov, A. A., Savitskaya, S. E., and Fokin, V. E.,** Biotechnological experiment Tavriya carried out aboard Salyut-7, in *Scientific Readings on Cosmonautics and Aviation,* Gagarin, U., Ed., Nauka Press, Moscow, 1984.
22. **Hannig, K., Wirth, H., and Schoen, E.,** Electrophoresis experiment MA-014, in Apollo-Soyuz Test Project Summary Science Report, Vol. 1, NASA SP-412, National Technical Information Service, Washington, D.C., 1977, 335.
23. **Hannig, K. and Bauer, J.,** Free flow electrophoresis in space shuttle program (Biotex), *Adv. Space Res.,* 9, 91, 1989.
24. **Clifford, D. W.,** Commercial prospects for bioprocessing in space, in *Commercial Opportunities in Space, Progress in Astronautics and Aeronautics,* Vol. 110, Shahrokhi, F., Chao, C. C., and Harwell, K. E., Eds., American Institute of Aeronautics and Astronautics, Washington, D.C., 1988, 177.
25. **Snyder, R. S., Rhodes, P. H., Miller, T. Y., Micale, F. J., Mann, R. V., and Seaman, G. V. F.,** Polystyrene latex separations by continuous flow electrophoresis on the space shuttle, *Sep. Sci. Technol.,* 21, 157, 1986.
26. **Todd, P.,** The loading and unloading of cells in electrophoretic separation, in *Cell Separations,* Bauer, J., Ed., CRC Press, Boca Raton, FL, 1994, chap. 4.
27. **Lacy, P. E. and Greider, M. H.,** in *Endocrinology: Endocrine Pancreas,* Stein, D. and Freinkel, N., Eds., American Physiological Society, Washington, D.C., 1972, 77.
28. **Hymer, W. C., Barlow, G. H., Blaisdell, S. J., Cleveland, C., Farrington, M. A., Feldmeier, M., Grindeland, R., Hatfield, J. M., Lanham, J. W., Lewis, M. L., Morrison, D. R., Olack, B. J., Richman, D. W., Rose, J., Scharp, D. W., Snyder, R. S., Swanson, C. A., Todd, P., and Wilfinger, W.,** Continuous flow electrophoretic separation of proteins and cells from mammalian tissues, *Cell Biophys.,* 10, 61, 1987.
29. **Morrison, D. R., Barlow, G. H., Cleveland, C., Farrington, M. A., Grindeland, R., Hatfield, J. M., Hymer, W. C., Lanham, J. W., Lewis, M. L., Nachtwey, D. S., Todd, P., and Wilfinger, W.,** Electrophoresis separation of kidney and pituitary cells on STS-8, *Adv. Space Res.,* 4, 67, 1984.
30. **Morrison, D. R., Lewis, M. L., Cleveland, C., Kunze, M. E., Lanham, J. W., Sarnoff, B. E., and Todd, P.,** Properties of electrophoretic fractions of human embryonic kidney cells separated on space shuttle flight STS-8, *Adv. Space Res.,* 4, 77, 1984.
31. **Todd, P., Kunze, M. E., Williams, K., Morrison, D. R., Lewis, M. L., and Barlow, G. H.,** Morphology of human embryonic kidney cells in culture after spaceflight, *Physiologist,* 28 (Suppl.), S-182, 1985.
32. **Lewis, M. L., Barlow, G. H., Morrison, D. R., Nachtwey, D. S., and Fessler, D. L.,** Plasminogen activator production by human kidney cells separated by continuous flow electrophoresis, in *Progress in Fibrinolysis,* Vol. 6, Churchill Livingstone, Edinburgh, 1983, 143.

33. **Lewis, M. L., Morrison, D. R., Barlow, G. H., Todd, P., Cogoli, A., and Tschopp, T.,** Cell electrophoresis and preliminary cell attachment investigations on STS-8, in Space Life Science Symposium: Three Decades of Life Sciences Space Research, NASA Publication, National Technical Information Service, Washington, D.C., 1987, 186.

34. **Stewart, R. M., Todd, P., Cole, K. D., and Morrison, D. R.,** Further analyses of human kidney cell populations separated on the space shuttle, *Adv. Space Res.,* 12, 223, 1992.

35. **Kuroda, M., Tagawa, K., Tanaka, T., and Uozumi, M.,** Separation of biogenic materials by electrophoresis under zero gravity, in Proc. 7th SL-J Post Flight IWG Six Month Science Report, NASDA Tsukuba Space Center, Ibaraki-Ken, Japan, April 26, 1993.

36. **Yamaguchi, T., Nakajima, H., and Akiba, T.,** Separation of animal cellular organella by means of free-flow electrophoresis, in Proc. 7th SL-J Post Flight IWG Six Month Science Report, NASDA Tsukuba Space Center, Ibaraki-Ken, Japan, April 26, 1993.

37. **Rhodes, P. H.,** High resolution continuous flow electrophoresis in the reduced gravity environment, in *Electrophoresis '81,* Allen, R. C. and Arnaud, P., Eds., Walter de Gruyter, Berlin, 1981, 919.

38. **Snyder, R. S. and Rhodes, P. H.,** Electrophoresis experiments in space, in *Frontiers in Bioprocessing,* Sikdar, S. K., Bier, M., and Todd, P., Eds., CRC Press, Boca Raton, FL, 1989, 245.

39. **Morrison, D. R., Cohly, H. P., Rodkey, L. S., Barlow, G. H., and Hymer, W. C.,** Comparison of bioseparation methods for microgravity experiments, Paper 88-0073, in Proc. AIAA 26th Aerospace Sciences Meet., Reno, NV, Jan. 10, 1988.

40. **Morrison, D. R. and Todd, P.,** An overview of NASA's program and experiments in microgravity biotechnology, in *Microgravity Science and Applications,* National Academy of Sciences, Washington, D.C., 1986, 168.

41. **Todd, P., Plank, L. D., Kunze, M. E., Lewis, M. L., Morrison, D. R., and Barlow, G. H.,** Electrophoretic separation and analysis of living cells from solid tissues by several methods, *J. Chromatogr.,* 364, 11, 1986.

42. **Hymer, W. C., Grindeland, R., Lanham, J. W., and Morrison, D. R.,** Continuous flow electrophoresis (CFE): applications to growth hormone research and development, in *Pharm Tech Conference '85 Proc.,* Aster Publishing, Springfield, OR, 1985, 13.

43. **Todd, P., Kurdyla, J., Sarnoff, B. E., and Elsasser, W.,** Analytical cell electrophoresis as a tool in preparative electrophoresis, in *Frontiers in Bioprocessing,* Sikdar, S. K., Bier, M., and Todd, P., Eds., CRC Press, Boca Raton, FL, 1989, 223.

44. **Todd, P., Morrison, D. R., Barlow, G. H., Lewis, M. L., Lanham, J. W., Cleveland, C., Williams, K., Kunze, M. E., and Goolsby, C. L.,** Kidney cell electrophoresis in space flight: rationale, methods results and flow cytometry applications, in Microgravity Science and Applications Flight Programs, Vol. 1, NASA TM-4069, National Technical Information Service, Washington, D.C., 1, 1988, 89.

45. **Morrison, D. R.,** Cellular changes in microgravity and the design of space radiation experiments, *Adv. Space Res.,* 14(18), 1994.

46. **Bonting, S. J.,** Bioprocessing in space, in Proc. 2nd Eur. Symp. Life Sciences Research in Space, Porz Wahn, Germany, 1984, 75.

47. **Tschopp, A., Cogoli, A., Lewis, M. L., and Morrison, D. R.,** Bioprocessing in space: human cells attach to beads in microgravity, *J. Biotechnol.,* 1, 281, 1984.

48. **Morrison, D. R.,** Bioprocessing in space: an overview, in *The World Biotech Report 1984,* Vol. 2, Online Conference Inc., New York, 1984, 557.

49. **Morrison, D. R.,** Cell separations in microgravity and development of a space bioreactor, in Microgravity Science and Applications Program Tasks, Pentecost, E., Ed., NASA TM-87568, National Technical Information Service, Washington, D.C., 1985, 146.

50. **Hymer, W. C. and Lanham, J. W.,** Pennsylvania State University, personal communication, 1993.

51. **Bozouklian, H., Sanchez, V., Clifton, M., Marsal, O., and Esterle, F.,** Electrokinetic bioprocessing under microgravity in France as illustrated by space bioseparation: a programme initiated in France and in cooperation with Belgium and Spain, *Adv. Space Res.,* 9, 105, 1989.

52. **Rhodes, P. H. and Snyder, R. S.,** Preparative electrophoresis for space, in Microgravity Science and Applications Flight Programs, January–March 1987, Selected Papers, NASA TM-4069, National Technical Information Service, Washington, D.C., 1988, 27.

53. **Zimmerman, U., Buechner, K. H., and Schnettler, R.,** Electrofusion of Yeast Protoplasts, Department of Biotechnology, Julius-Maximilians University, Würzburg, Germany, 1987.

54. **Morrison, D. R. and Hofmann, G. A.,** Cell separation and electrofusion in space, in *Space Commercialization: Platforms and Processing, Progress in Astronautics and Aeronautics,* Shahrokhi, F., Hazelrigg, G., and Bayuzick, R., Eds., American Institute of Aeronautics and Astronautics, Washington, D.C., 1990, 214.

55. **Anon.,** Commercial Space industry in the Year 2000: A Market Forecast, The Center for Space Policy, Cambridge, MA, June 1985.

56. **Bier, M., Palusinski, O. A., Mosher, R. A., and Saville, D. A.,** Electrophoresis: mathematical modeling and computer simulation, *Science,* 219, 1983, 1281.

57. **Morrison, D. R., Johnson, T. S., and Gratzner, H. G.,** Commercial monoclonal antibody production on the Space Station Freedom, in *Proc. Space Station Commercial Users Workshop,* Denver, CO, Oct. 26–28, 1988, 337.

58. **Pellis, N. R.,** Immunology Experiments in Microgravity, Testimony to the United States House of Representatives, Committee on Science, Space and Technology, Subcommittee on Space, Washington, D.C., March 9, 1993.

59. **Bier, M.,** Hormone purification by isoelectric focusing in space, in Microgravity Science and Applications Flight Programs, January–March 1987, Selected Papers, Vol. 1, NASA TM-4069, National Technical Information Service, Washington, D.C., 1988, 133.

60. **Brooks, D. E. and Bamberger, S.,** Studies on aqueous two phase polymer systems useful for partitioning of biological materials, in *Materials Processing in the Reduced Gravity of Space,* Rhindone, E. E., Ed., Elsevier/North-Holland, Amsterdam, 1982, 233.

61. **Clifford, D., Lanham, J. W., Purdy, D., and Richman, D.,** Electrophoresis Technology Development Mission Study, Final Report EOS-DWC-86-1, McDonnell Douglas Astronautics Corp., St. Louis, MO, 1988.

Chapter 16

Cell Electrophoresis in Microgravity

Ulrike Friedrich, Günter Ruyters, and Johann Bauer

CONTENTS

I. INTRODUCTION

The negative surface charge density of cells is an important biological parameter. Distinct densities of negative surface charges are typical for morphologically defined cell populations and in some instances even for cell populations of one lineage which are in different states of activation or differentiation. However, their biological significance is still not completely understood.[1-6] Hence, research is going on with the aims to analyze this cell parameter and to use this surface characteristic for preparative separation of cells.

A very common method for studying the negative surface charge density of cells is to measure the speed of cell movement within an electrical field. Considerable efforts were made during 30 years of research to develop sophisticated methods and implements for determination of the electrophoretic mobility of cells. Free flow electrophoresis is one of these methods.

II. THEORETICAL CONSIDERATIONS

Despite the efforts, even modern free flow electrophoresis machines are severely limited with regard to their throughput and their resolution, so it is difficult to perform more profound studies on the negative surface charge density of cells. The limitations are caused by several physical parameters; two of them are cell sedimentation and thermoconvection. Sedimentation is due to the fact that cells have higher specific densities than the buffers in which they are suspended. Thermoconvection is due to the fact that the buffer is colder near the cooling elements and therefore has a higher specific density than it has farther from the cooling elements, where it is warmer.[7]

Microgravity *(µg)* was expected to solve this problem because density differences within the buffer curtain and between cells and buffer do not play a role under microgravity. Hence, it was suggested that more concentrated cell suspensions could be injected into a separation chamber and that a considerable temperature gradient within the chamber would be tolerated at µg. The latter means that thicker buffer curtains and higher voltages can be used during cell electrophoresis at µg. Thicker buffer curtains and higher cell concentrations would allow the improvement of the throughput. This assumption has been proved by numerous spaceflight and accompanying control experiments. The efforts are summarized in Chapter 15 of this book.[8]

Another consideration is that the residence time of the cells within the electric field can be elongated by reducing the speed of the buffer flow if thermoconvection and

0-8493-8918-6/94/$0.00+$.50
© 1994 by CRC Press, Inc.

sedimentation are eliminated at μg. In this way the resolution achieved by an electrophoresis machine should be improved. During long residence times the paths covered by the cells within an electric field are longer and the points of the chamber bottom line, at which cells with varying negative surface charge densities arrive after separation, are more distant from each other. If the Gaussian distributions of the separated cells are kept narrow, better separation results can be expected.

An improvement of the resolution of electrophoresis machines is very desirable because the range of cell electrophoretic mobility (EPM) values occurring in nature seems to be very narrow.[6] Most mammalian cells have EPM values ranging from 0.7 and 1.4×10^{-4} cm²/Vs. All cell populations investigated so far show distribution curves which have a height that is 5% of the maximal height at 0.05×10^{-4} cm²/Vs below and above the mean value. This means that cell populations with mean EPM differences of 0.05×10^{-4} cm²/Vs may be distinguished analytically. However, satisfactory preparative cell electrophoretic separations are only possible if the cells have EPM differences of at least 0.15×10^{-4} cm²/Vs.

III. TEXUS ROCKET FLIGHT EXPERIMENTS

In order to investigate whether the resolution is really improved in μg, the physical parameters which are important during a cell electrophoresis experiment were determined and quantitatively calculated.[7,9] Based on these calculations and on the experiences from a spaceflight cell electrophoresis experiment, which was performed during the Apollo-Soyuz flight in 1975,[10] a very precisely working machine was developed and manufactured through collaboration between the Max Planck Institute in Martinsried, the space company MBB-ERNO in Bremen, and Hirschmann in Unterhaching (Figure 16-1). It provides optimal separation results in ground laboratories and also can be used in spaceflights.[11] The development and manufacturing of the instrument as well as most of the scientific preparation were performed under contract to the Project Management Division for Space Science and Technology of the German Aerospace Establishment DLR representing the German Federal Ministry of Science and Technology. The electrophoresis machine was then tested during two successful flights in the German Sounding Rocket Program Texus. With these rockets it is possible to achieve 6 min of microgravity with a quality of about 10^{-5} normal earth gravity (g_o).

The aim of the electrophoresis experiments during the Texus flights was to study whether the elimination of cell sedimentation and thermoconvection would show possibilities of improving the resolution of the machine. Mixtures of rabbit, rat, and guinea pig erythrocytes were used for the experiments because guinea pig and rat erythrocytes have very similar electrophoretic mobilities. Their mean values can be distinguished in good laboratory experiments, but a satisfying separation of these cells is not possible because their distribution curves overlap. The rabbit erythrocytes were added to the mixture in order to calibrate the system.

The first cell electrophoresis experiment on a Texus rocket was performed in 1988 (Texus 18/20). It revealed that the machine worked very well under microgravity and that no instrumental problems could be detected due to acceleration forces or vibrations during launch. The separation profiles recorded during the 6 min of μg phase were as good as they were under optimal laboratory conditions.[12]

During another rocket flight in 1989 (Texus 24) it was investigated whether the detrimental effects of cell sedimentation during long residence times and of thermoconvection due to considerable temperature gradients within the chamber could be eliminated at μg. The buffer speed was reduced to 1.6 mm/s and the chamber thickness was enhanced to 0.8 mm; this had the consequence that the cells remained within the

Figure 16-1 Photograph of a free-flow electrophoresis module which was used during the Texus rocket flights. The power supply and all data-regulating implements are located under the base plate. Separation chamber, pumps, and media containers are assembled above this plate. The width of the total module is 44 cm; its height is 127 cm.

separation chamber for 125 s and that a temperature gradient of 5 to 6°C arose across the 0.8-mm-thick chamber. If such conditions are applied during ground electrophoresis experiments, considerable disturbations of known separation profiles are observed. When they were applied during the Texus flight at μg, however, a separation profile was recorded which was as good as separation profiles performed in ground laboratories at optimal separation conditions. The experiment clearly demonstrates that physical parameters which disturb a separation profile at g_o can be eliminated at μg if the buffer speed is reduced to 1.6 mm/s and the chamber thickness is increased to 0.8 mm.[13]

IV. SUMMARY AND CONCLUSIONS

For more than 10 years, space cell electrophoresis experiments were performed in several countries — e.g., in the U.S., Russia, Japan, and France. The main scientific aim of the

Texus flight experiments described above was to determine to what extent resolution and throughput of electrophoresis machines can be promoted in microgravity. Another intention was to develop a cell electrophoresis device for a tool which optimally supplements the equipment of a future space laboratory. Between 1980 and 1987 German scientists together with persons responsible for the German space program elaborated plans to establish a biotechnological laboratory for space conditions ("Bioprocessing in Space"). For that purpose it was necessary to make sure that each of the different tools and methods such as cell cultivation, sample extraction, electrofusion of cells, or cell electrophoresis works well in microgravity. It also seemed desirable that each method or tool could be handled by astronauts. Therefore, a respective project was initiated for a Spacelab mission together with NASA. Unfortunately, due to financial and mission-related restrictions it was not possible to implement the free-flow electrophoresis facility in the German D-2 mission together with all other biological equipment. However, it had indeed been shown by the unmanned Texus rocket flights that such a very complicated instrument works under microgravity and that the theoretical considerations concerning the enhancement of electrophoretic separation effectivity under microgravity conditions could be confirmed.

At the moment and in the near future a successive and intensive use of the different methods appears impossible because of the shortage of future manned missions. Nevertheless, this free-flow electrophoresis machine will be available for use in space, even in cooperation with other space-faring countries, if questions of high scientific interest appear solvable by this method under space conditions.

REFERENCES

1. **Golovanov, M. V.,** Electrophoresis of cells at physiologic ionic strength, in *Cell Electrophoresis,* Bauer, J., Ed., CRC Press, Boca Raton, FL, 1994, chap. 9.
2. **Slivinsky, G. G.,** Simultaneous two-parameter measurements of electrophoretic features of subpopulations of cells and their different sedimentation characteristics, in *Cell Electrophoresis,* Bauer, J., Ed., CRC Press, Boca Raton, FL, 1994, chap. 10.
3. **van Oss, C. J.,** Cell interactions and surface hydrophilicity —influence of Lewis acid-base and electrostatic forces, in *Cell Electrophoresis,* Bauer, J., Ed., CRC Press, Boca Raton, FL, 1994, chap. 11.
4. **Vargas, F. F.,** The negative surface charge density of endothelial cells and its possible role in cell migration, in *Cell Electrophoresis,* Bauer, J., Ed., CRC Press, Boca Raton, FL, 1994, chap. 12.
5. **Schütt, W., Hashimoto, N., and Shimizu, M.,** Application of cell electrophoresis for clinical diagnosis, in *Cell Electrophoresis,* Bauer, J., Ed., CRC Press, Boca Raton, FL, 1994, chap. 13.
6. **Bauer, J.,** The negative surface charge density of cells and their actual state of differentiation or activation, in *Cell Electrophoresis,* Bauer, J., Ed., CRC Press, Boca Raton, FL, 1994, chap. 14.
7. **Boese, F. G.,** Contributions to a mathematical theory of free flow electrophoresis, in *Cell Electrophoresis,* Bauer, J., Ed., CRC Press, Boca Raton, FL, 1994, chap. 1.
8. **Morrison, D. R.,** Cell electrophoresis in space. Past and future, in *Cell Electrophoresis,* Bauer, J., Ed., CRC Press, Boca Raton, FL, 1994, chap. 15.
9. **Hannig, K., Wirth, H., Meyer, B. H., and Zeiller, K.,** Free flow electrophoresis. I. Theoretical and experimental investigation of the influence of mechanical and electrokinetic variables on the efficiency of the method, *Hoppe Seyler's Z. Physiol. Chem.,* 356, 1209, 1975.

10. **Hannig, K. and Wirth, H.,** Free flow electrophoresis in space, *Prog. Astronaut. Aeronaut.,* 52, 411, 1977.
11. **Mang, V.,** Technical description of an electrophoresis facility for sounding rockets in the Texus program, *Appl. Microgravity Technol.,* 2, 52, 1989.
12. **Hannig, K. and Bauer, J.,** Free flow electrophoresis in a space shuttle program (Biotex), *Adv. Space Res.,* 9, 91, 1989.
13. **Hannig, K., Kowalski, M., Klöck, G., Zimmermann, U., and Mang, V.,** Free flow electrophoresis under microgravity: evidence for enhanced resolution of cell separation, *Electrophoresis,* 11, 600, 1990.

INDEX

321